A Most Incomprehensible Thing: Notes Towards a Very Gentle Introduction to the Mathematics of Relativity

Copyright © Peter Collier 2012

First published 2012

Second edition 2014

Publisher by Incomprehensible Books

Peter Collier asserts the moral right to be identified as the author of this work

ISBN 978-0-9573894-5-8

A catalogue record for this book is available from the British Library.

(v84IS)

Email: incomprehensiblething@gmail.com

A Most Incomprehensible Thing

Notes Towards a Very Gentle Introduction to the Mathematics of Relativity

SECOND EDITION

Peter Collier

Incomprehensible Books

For Anne Redfarn, xxx

Contents

Preface

This book is written for the general reader who wishes to gain a basic but working understanding of the mathematics of Einstein's theory of relativity, one of the cornerstones of modern physics.

I must have been eleven or twelve years old when I treated myself to a paperback layperson's introduction to the theory. I enjoyed maths and science at school, vaguely knew that relativity was a difficult but important theory, and was curious to find out more. Being a popular guide, there was very little mathematics in the text, but there were a sprinkling of simple equations from the special theory of relativity (as we'll see, Einstein proposed two theories: special relativity and general relativity). These formulations described relativistic phenomena such as time dilation and length contraction – nice, straightforward equations that even I could understand. I remember wondering what all the fuss was about if those equations were all there were to relativity. The complexity of Einstein's theory had obviously been exaggerated. I read and enjoyed the book and mentally ticked off the theory as something else I had mastered, like long division or factorisation.

I was hopelessly wrong, of course. I'd made the mistake of confusing popular science with the real thing, and believing I'd grasped the theory of relativity when in reality I'd barely scratched its surface. It took me a long time to realise that those equations, though perfectly valid, were but a tiny part of a much larger, more complex and wider ranging description of the physical world.

Although my formal education in maths and physics ended some time ago, I've continued to enjoy dabbling around in the subjects. During the winter of 2010–11, with time on my hands, I came across the excellent series of YouTube general relativity lectures by Professor Leonard Susskind [30] of Stanford University. Much of what the professor said went over my head, but after watching the series, I found myself intrigued with the language of spacetime, gravitation, metrics, tensors and black holes. For the second time in my life I resolved to try to teach myself relativity. This time around I was more aware of the size and shape of the challenge. Plus, of course, I now had access to the resources of the internet (and more pocket money for textbooks!). My goal was to move beyond the popular expositions of relativity and get to grips with the underlying mathematics, the beating heart of the theory (to paraphrase Euclid, there is no royal road to relativity – you have to do the maths). And so my adventure began. For the next twelve months I was obsessed, with almost every spare moment, at home and at work, spent poring over books and websites. It was hard work but great fun, and fortunately my partner was supremely patient with my new infatuation.

Thus also began my quest for my 'ideal' relativity textbook. There are of course various maths-lite popular guides, such as the one I'd read as a boy. Plus there are technically demanding undergraduate and higher-level textbooks. The popularisations may be entertaining, but by excluding the maths they can only give a cursory understanding of the subject. The mathematically rigorous texts, on the other hand, are unreadable for the non-specialist. I was looking for something in-between, a Goldilocks volume pitched just at my level, neither too easy nor too difficult. I didn't want to take a degree in physics, but I did want to get to grips with the essential mathematics of relativity.

I never found my ideal volume but instead had to make use of many different sources, winkling out bits of useful information here and there, struggling to fit the different pieces of the jigsaw into a coherent whole. Along the way, it dawned on me that if I were looking for a self-contained, introductory, mathematical text on relativity, others might be as well; after all, it is one of the most important theories in physics. The book I had in mind would assume little prior mathematical knowledge (even less than my own patchy sixth-form/high school maths if it were to be suitable for the general reader). It would therefore need to begin with a crash course in foundation maths. To give the kind of meaningful mathematical understanding I sought, it would try wherever possible to give the relevant derivations in full, even at the risk of stating what appears to be the blindingly obvious to the more mathematically savvy reader. And it would contain numerous fully worked problems, because in my experience, seeing how the mathematics is used in practice is the best way to understand it. Oh, and it would be written in a user-friendly style with lots of helpful diagrams and pictures. And – given the high cost of many textbooks – it would be inexpensive verging on the downright cheap.

What you see in front of you is my attempt to write such a book, the accessible teach-yourself study aid I would have liked to have got my hands on when I first seriously started to learn Einstein's theory – an introduction to the mathematics of relativity for the enthusiastic layperson. By 'layperson' I mean someone with a minimal mathematical background, though obviously there are no penalties if yours exceeds that – just skip what you know. 'Enthusiastic' suggests this may not be an easy journey, but one that demands some degree of commitment and effort from the reader. Physicists may believe that the language of Nature is mathematics, but in the case of general relativity she might have made that language a touch easier to learn for the average *Homo sapiens*. Even at our basic level, if we really want to understand what's going on in spacetime, we need to tackle the delights of things such as tensors, geodesics and, of course, the Einstein field equations.

There's no escaping the fact that it's not easy learning a technically demanding subject such as relativity on your own. Away from college or university, the self-studier ploughs a lonely furrow, with no structured coursework, lectures or interaction with tutors and fellow students. However, the pursuit of education and understanding for its own sake is an admirable goal and deserves every encouragement. We should all have the chance to appreciate – in the words of Matthew Arnold – 'the best that has been said and thought in the world.' This book tries to offer a helping hand to such intrepid seekers of truth. In a world overflowing with trivia, irrationality and nonsense, relativity is the genuine article, a challenging but fundamentally important scientific theory.

I would suggest that for most readers the best way to approach this book is as an ultra-relaxed marathon rather than a sprint. We cover a lot of material, from elementary mathematics to tensor calculus, so give yourself plenty of time to absorb and thoroughly understand what's being presented. You're bound to get stuck sooner or later. Indeed, you will probably grind to a halt more than once – I know I did. Try not to be discouraged. Einstein was a genius, yet it took even him ten years to progress from special to general relativity. Personally, I found that if I hit a brick wall, instead of struggling on, it helped to put the book I was studying to one side and take a break. Time for reflection, working on something else, maybe a good night's sleep were often all that was necessary for comprehension to dawn. If you are still baffled, you could seek enlightenment via one of the excellent online physics/maths forums (three are listed in the bibliography – [20], [25] and [26]). Of course, the ideal is to finish the book, but if you don't, at least try to have fun getting through as much as you can. Thankfully, you don't need to be a genius to appreciate the wonders of relativity or to ponder the strange, mysterious world that the theory so accurately describes.

- I've made every effort to ensure that there are no errors in this book. However, mistakes can happen. A current list of errata can be found at http://amostincomprehensiblething.

`wordpress.com`. If you find an error that does not appear on that list, please e-mail it to incomprehensiblething@gmail.com.

- Feedback, comments? Email the author at the same address.

- This book was written using LyX, an excellent maths-friendly (and much more) open source document preparation system based on LaTeX.

Note on the second edition   I have been most gratified by the interest and enthusiasm shown towards my 'incomprehensible' book since it was first published four years ago. Readers of all ages and from a wide variety of backgrounds have clearly relished the opportunity of trying to understand Einstein's wonderful theory of relativity.

Despite the most careful preparation, it was inevitable that a work of this scope and complexity would contain some minor errors and typos. Thanks to all those sharp-eyed readers who have taken the trouble to bring these to my attention.

This revised second edition includes corrections and a scattering of amendments, a new chapter on gravitational waves and a new appendix on the Riemann curvature tensor and of some of its symmetries. I only briefly mentioned gravitational waves when I first wrote this book. Space was limited, and at the time there was only indirect evidence for their existence. In February 2016, LIGO physicists announced they had observed gravitational waves from the merger of two black holes. Following that historic discovery, I felt duty-bound to provide more coverage than my original few lines – hence the new short chapter. The Riemann tensor appendix I've included because (a) I think it's rather neat, (b) this object is central to the mathematics of curvature as it relates to general relativity, and (c) the derivations provide a further useful demonstration of covariant differentiation and index manipulation in action.

Peter Collier

March 2016

Introduction

The most incomprehensible thing about the world is that it is at all comprehensible.

ALBERT EINSTEIN

Figure 0.1: Albert Einstein (1879–1955) – photographed in 1921.

During the first few years of the twentieth century, Albert Einstein (Figure 0.1) revolutionised our understanding of the physical world. In 1905 he proposed his special theory of relativity, which fatally undermined long-standing scientific and common sense assumptions about the nature of space and time. Simultaneity, for example: the new theory meant that two events happening at the same time for one observer might well occur at *different* times for another. Henceforth, space and time could no longer be regarded as separate and absolute quantities. Instead they merged – the theory and all available evidence demanded they merge – into a new single entity called spacetime. Furthermore, matter and energy were also joined, in the shape of one of the most famous equations in physics: $E = mc^2$.

This was radical stuff, but much more was to come. Special relativity deals with the motion of objects and of light in the absence of gravity. For over two hundred years Newtonian gravitation had proved itself a theory of astonishing accuracy. And it still is, allowing the precise calculation of the motion of a falling cup or an orbiting planet. But though of immense practical use, Newton's theory isn't compatible with special relativity (Newtonian gravity is instantaneous; special relativity imposes a natural speed limit – the speed of light). It took Einstein a further ten years to reconcile gravity and special relativity. That synthesis, his general theory of relativity, was published in 1916. General relativity explains gravity as an effect of the bending of spacetime in the vicinity of a massive object. The general theory describes phenomena as diverse as non-Newtonian deviations in planetary orbits, gravitational time dilation, and gravitational bending and redshift of light. It has been used to predict the existence and properties of black holes and is at the heart of modern cosmology – the study of the history and structure of the universe. Necessarily, given the appropriate conditions, general relativity is smart enough to reduce to both special relativity and Newtonian theory.

Special and general relativity can be summarised as follows:

- The speed of light in a vacuum has the same value for all uniformly moving observers.

- Mass and energy curve spacetime.

- Mathematically, spacetime can be represented by a curved space that is locally flat.

- An equation called a metric describes the curvature of such a space. The metric will vary from region to region depending how the space curves.

- Light and free particles follow paths, called geodesics, through spacetime determined by how the spacetime is curved.

- In the appropriate circumstances general relativity should approximate both to special relativity and to Newtonian gravitation.

- The laws of physics must take the same form in all coordinate systems.

Or, more succinctly, the famous quotation from physicist John Archibald Wheeler (1911–2008) states that

> 'Matter tells space how to curve.
> Space tells matter how to move.'

The astrophysicist Kim Griest [12] asserted the fundamental importance of general relativity when he wrote:

> 'Most physicists don't study general relativity [GR] because it only differs from Newton's gravity theory and from special relativity in a few cases. But GR is Nature's choice – whenever GR differs from Newton, GR has been shown to be right. It is how Nature actually works, and requires a radical rethinking of physical reality.'

In short, relativity is a triumph of human reason, and as such deserves the widest possible audience. However, to really understand even the basics of the theory we have to tackle some quite challenging mathematics, and that is what we are going to attempt to do.

Chapter 1 sets the ball rolling by introducing the necessary foundation mathematics needed to progress through the rest of the book, from the basic definition of a function, through calculus and simple vectors, to our first metric tensor. For those with little mathematical background this chapter will be a baptism of fire. Emerging from the flames, we advance invigorated into Chapter 2. I don't see how it's possible to get to grips with relativity without understanding at least some of the physics it supplanted. To that end, this chapter comprises a brisk discussion of Newtonian mechanics, with more time spent, deservedly I trust, on Newton's wonderful theory of gravitation, including a neat little detour on how to plot the orbit of a hypothetical planet around the Sun. In Chapter 3 we move on to special relativity and the strange world of Minkowski spacetime, including the counter-intuitive phenomena of time-dilation and length contraction. After developing our spacetime intuition with the geometrical assistance of spacetime diagrams we progress to a more algebraic approach using the Lorentz transformations. We end this chapter by looking at how special relativity reformulates the laws of mechanics.

The next three chapters develop the mathematics that underpins general relativity. First, a brief introduction to the concept of the manifold and the all important metric tensor $g_{\mu\nu}$. Next we look at vectors, one-forms and tensors in order to ease us into the mathematics of curvature, including connection coefficients, parallel transport of vectors, geodesics and the Riemann curvature tensor. Chapter 7 pulls these various strands together to take us to the star of the show: the Einstein field equations. On the way we meet the equivalence principle (Einstein's 'happiest thought'), geodesics in spacetime and the energy-momentum tensor. In Chapter 8 we see how, given appropriate non-relativistic conditions, the equations of general relativity approximate to the ultra-successful formulations of Newtonian mechanics. The next chapter introduces the Schwarzschild solution, the first and most important exact solution to the Einstein field equations. This solution provides a good approximation to the gravitational field of slowly rotating bodies such as the Sun and Earth. We derive the Schwarzschild solution and use it to discuss the four classical tests of general relativity. The Schwarzschild solution can be used to predict and describe the simplest type of black hole, which we do in Chapter 10. Here we investigate the weird nature of time and distance as experienced by an observer unfortunate enough to fall into a black hole, compared to how that same journey appears to a distant observer.

Chapter 11 gives a brief introduction to relativistic cosmology. We start with four key observed properties of the universe, including the cosmological principle – the assumption that on a very large scale the universe looks the same to all observers, wherever they are. The Robertson-Walker metric and the Friedmann equations together establish the theoretical framework that then allows us to discuss several simple cosmological models and gain insight into the history and evolution of our own universe.

The final chapter is a short add-on following the first direct observation of gravitational waves in September 2015. We outline the derivation of the basic gravitational wave equation using linearized theory, briefly discuss the historic 2015 LIGO detection and give a little background to this emerging, potentially revolutionary field.

- Note, in this book, you are going to (pun alert) see a lot of c, the symbol that denotes the speed of light in a vacuum and which (approximately) equals 300,000,000 metres per second. Be aware that in relativity life is often made simpler by defining c as being equal to 1. We sort of do this ourselves when we start using spacetime diagrams in special relativity. This is perfectly legitimate, all we are doing is playing around with units of measurement. Some authors then take this practice a step further and go on to omit c from their equations. For example, the Lorentz factor (3.4.1), which in this book is given by

$$\gamma = \frac{1}{\sqrt{1 - (v/c)^2}},$$

would become (assuming $c = 1$)

$$\gamma = \frac{1}{\sqrt{1 - v^2}}.$$

To the learner, it's not obvious that these two equations mean the same thing (which they do). There's scope for confusion here, which is the reason I include c in all the relevant equations in this book.

- Also note that whenever we refer to 'light' we aren't of course only referring to the narrow range of the electromagnetic spectrum visible to the human eye, but to all electromagnetic radiation, such as gamma rays, X-rays, visible light, radio waves, etc.

1 Foundation mathematics

Do not worry about your difficulties in mathematics. I can assure you mine are still greater.

ALBERT EINSTEIN

1.1 Introduction

In order to make sense of what follows, we need to introduce some essential maths. There's quite a lot of it, and those with a limited mathematical background may find this chapter somewhat of a challenge. Furthermore, lack of space permits only a brief discussion of a wide range of topics. So although I've tried to include as much information as seems necessary, you may need to forage elsewhere for additional insights and support (see, for example, [21] and [24] in the bibliography). But take heart. With enthusiasm and perseverance we can now begin to familiarise ourselves with a broad sweep of fundamental mathematical ideas. By the end of this chapter we'll have met functions, exponents, coordinate systems, calculus, vectors, matrices and more, including our very first metric tensor, the central object of study in relativity.

1.2 Definitions and symbols

I'll discuss most of the mathematics and physics used in this book as we go along. However, rather than clutter up the text with explanations of basic terms and symbols, I thought it best to gather them together here, in a somewhat indigestible dollop, at the start of the first chapter. This is a reference section – you aren't expected to learn all this stuff straight off. Some we'll define and discuss in greater detail later, so don't worry if they make no sense at the moment. A few are for background interest only. Others you really need to know from the outset: you won't get far using even the most elementary mathematics if you don't know what a decimal number is or the meaning of the equals $=$ or divided by / symbols.

Terms

- Constant: a number, term or expression that doesn't change. If $y = x + 7$ then 7 is a constant. The area of a circle equals πr^2 where r is the radius (a variable), and π (pi) is a constant.

- Cube: the cube of a number is the number multiplied by itself twice, eg 3 cubed $= 3^3 = 3 \times 3 \times 3 = 27$.

- Decimal: a number containing a decimal point, eg 1.3, 20.07, 0.9134 .

- Denominator: see fraction.

- Factor: a number or expression that divides exactly into another number or expression. $1, 2$ and 5 are factors of 10; x and $(2x + 3)$ are factors of $2x^2 + 3x$ because $x(2x + 3) = 2x^2 + 3x$.

- Fraction: a number such as $1/2$, $2/3$, $17/2$. Can also be written with a horizontal line, eg $\frac{1}{3}$, $\frac{-23}{2}$, $\frac{1098}{901}$. The number above the line (called the numerator) is divided by the number below the line (called the denominator).

- Infinity: symbol ∞. Something larger than any real number.

- Integer: a negative or positive whole number (including zero) such as -34, -5, 0, 1, 17, 1021.

- Irrational number: a real number that is not rational, eg π, Euler's number e.

- Negative number: a number less than zero, eg $-19, -2.5, -0.04$. The product of (ie the result of multiplying) a positive number and a negative number is negative, eg $3 \times$ -7 = -21. The product of two negative numbers is positive, eg -3 $\times$ -7 = 21.

- Numerator: see fraction.

- Positive number: a number greater than zero, eg $1, 56, 1007$.

- Product: the result of multiplying together two or more numbers or things, eg the product of 4 and 5 is 20.

- Quotient: the result of dividing one number by another. The quotient of 14 and 7 is 2 as $14/7 = 2$.

- Rational number: a number that can be made by dividing one integer by another, eg $1/2$, $5/7$, $12/108$. Integers, 12 for example, are also rational because $12/1 = 12$.

- Real number: the normal numbers we use, including decimals, fractions, integers, negative numbers, positive numbers, irrational numbers, etc. They are called real because they aren't complex numbers (strange things that involve the square root of -1). Apart from a single mention of the 'most beautiful theorem in mathematics' we don't use complex numbers in this book.

- Square: the square of a number is the number multiplied by itself, eg 5 squared $= 5^2 = 25$.

- Square root: the square root of a number is a number that when multiplied by itself gives the original number. The square root of 36 is ± 6. The square root symbol is $\sqrt{}$, so $\sqrt{25} = \pm 5$.

- Variable: a quantity that may change, eg the circumference of a circle equals $2\pi r$ where π (pi) is a constant and r, the radius, is a variable.

Symbols

$=$ is equal to, eg $6 = 6$.

$\approx$ is approximately equal to, eg $\pi \approx 3.14$.

$\sim$ is very approximately equal to, eg $2 \sim 5$.

$\neq$ is not equal to, eg $12 - 3 \neq 7$.

$+$ plus, eg $13 + 56 = 69$.

$-$ minus, eg $13 - 3 = 10$.

$\times$ times or multiplied by, eg $2 \times 13 = 26$. Another way of saying the same thing is that 26 is the product of 2 and 13.

$\div$ or $/$ divided by, eg $7 \div 2 = 3.5$ or $7/2 = 3.5$.

$\pm$ plus or minus, eg $5 \pm 2 = 7$ or 3.

$<$ is less than, eg $5 < 10$.

$\ll$ is much less than, eg $0.12 \ll 1,000,000$.

$>$ is greater than, eg $6.3 > 5.9$.

$\gg$ is much greater than, eg $30,000 \gg 0.008$.

$\leq$ is less than or equal to.

$\geq$ is greater than or equal to.

$\propto$ is proportional to, eg if $y = 7x$ then $y \propto x$ (in other words, if x is doubled then so is y).

$\sqrt{}$ the square root of, eg $\sqrt{25} = \pm 5$.

$!$ factorial, eg $6! = 6 \times 5 \times 4 \times 3 \times 2 \times 1 = 720$.

$\Rightarrow$ implies that, eg $x = 5 \Rightarrow 3x = 15$.

∞ infinity – think of a big number. Infinity is bigger than that.

$||$ absolute value, eg $|-7| = |7| = 7$.

$\mathbf{B}$ a two or three dimensional vector.

$\hat{e}_x, \hat{e}_y, \hat{e}_z$ Cartesian unit basis vectors.

$\hat{\mathbf{r}}$ a unit vector in the radial direction.

$\vec{V}$ a four-vector.

$\tilde{V}$ a one-form.

Δx a small increment of x. (If we square Δx, we write this as Δx^2.)

$\frac{dy}{dx}, \frac{df}{dx}, f'(x), y'$ derivative, eg if $y = x^2$ then $\frac{dy}{dx} = 2x$.

$\frac{\partial y}{\partial x}, \frac{\partial f}{\partial y}, f_x, \partial_x f$ partial derivative, eg if $z = x^2 + xy^2$ then $\frac{\partial z}{\partial x} = 2x + y^2$.

$\int$ integral, eg $\int 2x\,dx = x^2 + C$.

Quantity	Name	Symbol	In terms of other SI units
length	metre	m	
mass	kilogram	kg	
time	second	s	
angle	radian	rad	
frequency	Hertz	Hz	$1/s$ or s^{-1}
force	newton	N	$kg\,m/s^2$ or $kg\,m\,s^{-2}$
energy, work	joule	J	$N\,m$ or $kg\,m^2s^{-2}$
area	square metre		m^2
volume	cubic metre		m^3
speed, velocity	metre per second		m/s or $m\,s^{-1}$
acceleration	metre per second squared		m/s^2 or $m\,s^{-2}$
Celsius temperature	degree Celsius	°C	
thermodynamic temperature	kelvin	K	
density	kilogram per cubic metre		kg/m^3 or $kg\,m^{-3}$

Table 1.1: SI base units and derived units used in this book.

1.3 Units and constants

The standard units used in physics and other sciences are the **Système Internationale** or **SI units**. These are the units used throughout this book (see Table 1.1). Occasionally, hopefully to add clarity, I've also used non-SI units such as hours or miles, which are still in common use in the UK and US. The convention is that rates of something or other tend to be written in exponential (see Section 1.8.1) form. So, velocity, for example, in metres per second is written as $m\,s^{-1}$ not m per s or m/s. Similarly, acceleration, in metres per second per second is written as $m\,s^{-2}$ not m/s^2.

Two temperature scales are used in physics: kelvin and Celsius (which used to be known as the centigrade scale). 1 K has the same magnitude as 1°C, so an increase in temperature of 1 K is the same as an increase of 1°C. Absolute zero, the lowest temperature theoretically attainable, occurs at 0 K on the Kelvin scale and at −273.15°C on the Celsius scale. So, water freezes at 0°C = 273.15 K, and a pleasantly warm summer's day in the UK registering 20°C would equal 293.15 K.

Units can be prefixed to produce a multiple of the original unit (Table 1.2). For example: a kilometre = 1000 m = 1 km; a nanosecond equals a billionth of a second = 0.000000001 s = 1 ns. Multiples of the kilogram are named as if the gram were the base unit, even though it isn't.

Multiple	Prefix	Symbol
10^9	giga	G
10^6	mega	M
10^3	kilo	k
10^2	hecto	h
10^1	deca	da
10^{-1}	deci	d
10^{-2}	centi	c
10^{-3}	milli	m
10^{-6}	micro	μ
10^{-9}	nano	n

Table 1.2: Standard SI prefixes.

The physical constants used in this book are given in Table 1.3.

Name of constant	Symbol	SI value
gravitational constant	G	$6.673 \times 10^{-11} \, \text{N m}^2 \, \text{kg}^{-2}$
speed of light in a vacuum	c	$2.998 \times 10^8 \, \text{m s}^{-1}$
Planck's constant	h	$6.63 \times 10^{-34} \, \text{J s}$
mass of the Sun	$M_\odot$	$1.99 \times 10^{30} \, \text{kg}$
radius of the Sun	$R_\odot$	$6.96 \times 10^8 \, \text{m}$
mass of the Earth	$M_\oplus$	$5.97 \times 10^{24} \, \text{kg}$
radius of the Earth	$R_\oplus$	$6.37 \times 10^6 \, \text{m}$
light-year	ly	$9.461 \times 10^{15} \, \text{m}$
parsec	pc	$3.086 \times 10^{16} \, \text{m}$

Table 1.3: Constants.

1.4 Expressions, equations and functions

In mathematics we come across collections of symbols, numbers, etc called expressions, equations and functions.

- **Expressions** are mathematical phrases containing numbers, **variables** (t, x, y, z, etc) and **operators** ($+$, $-$, $\div$, $\times$, etc). For example $3z + 7$, $2x^2 - y$, $23xy^2 - 13$ are all expressions.

- **Equations** are statements that say two mathematical expressions are equal to each other. For example, we can combine two of the above expressions to obtain the equation $3z + 7 = 2x^2 - y$. Expressions by themselves are meaningless. Equations are of fundamental importance because they tell us the relationship between different quantities.

- **Functions** are equations of the form $y = $ (various expressions), where when we evaluate the right-hand side of the equation we end up with exactly one value of y. If we get two or more values of y then the equation is not a function.

The equation $y = 3x + 7$, for example, is a function because any value of x will only give one value of y. The equation $y^2 = x$ is not a function because there can be two solutions of y for one value of x. For example, if $x = 9$ y can equal 3 or -3 because $(3)^2 = 9$ and $(-3)^2 = 9$.

For example, the function $y = 3x$ tells us that we need to multiply x by three to find the corresponding value of y. So if we let x equal the whole numbers from 0 to 5, the values of y are as follows:

x	0	1	2	3	4	5
y	0	3	6	9	12	15

Another simple function is $y = x^2$. Here are the values of y when we let x equal the whole numbers from -3 to +3. Don't forget that a negative number multiplied by a negative number gives a positive number, so $-3 \times -3 = 9$.

x	-3	-2	-1	0	1	2	3
y	9	4	1	0	1	4	9

Here's a function containing five variables (where we are only considering non-negative values of $\sqrt{s}$):

$$q = 3 + t\frac{u}{v} - 5\sqrt{s}.$$

If we were given the values of t, u, v, s we could then solve for q. For example, if $t = 6, u = 2$, $v = 0.3, s = 9$ then

$$q = 3 + \left(\frac{6 \times 2}{0.3}\right) - 5\sqrt{9} = 3 + (20 \times 2) - (5 \times 3) = 28.$$

A function is often denoted by the letter f, so you might read in a textbook that the 'function f' does this, that or the other. You will also often see functions written as $y = y(x)$ or $y = f(x)$ or $y = g(x)$ etc All three mean the same thing: that y is a function of x. In these examples, because the value of y only depends on x, we say y is a **function of one variable**.

Alternatively, if y is a function of several variables, we can write $y = y(a, b, c)$ or $y = f(a, b, c)$ meaning y is a function of a, b, c. In these examples, y is a function of three variables. A few more examples:

- If $u = x^2 + 5x$, we say $u = u(x)$ or $u = f(x)$, ie u is a function of x.

- If $q = \sin(t)$ (trigonometric functions are discussed in Section 1.7), we say $q = q(t)$ or $q = f(t)$, ie q is a function of t.

- If $w = 4r + p^2 - b$, we say $w = w(r, p, b)$ or $w = f(r, p, b)$, ie w is a function of r, p, b.

1.5 Getting a feel for functions

Now we can recognise when a thing is a function, we need to acquire at least a basic sense of its underlying structure. Whenever we meet a function for the first time, we should ask ourselves a few simple questions along the lines of: how does the function change if the variables change; does it ever equal zero; does it give a meaningful answer for all variable values? By attempting to answer these questions, we'll develop a deeper feel of what the function's about. Here's a list of some of the more obvious properties of a few elementary functions.

Function 1. $y = 2x$.

- y becomes bigger as x becomes bigger.

- y is proportional to x, ie $y \propto x$ (for example, if we double x, we double y).

- When $x = 0$, $y = 0$.

- There's a value of y for every value of x.

Function 2. $y = x^2$.

- y becomes bigger as x becomes positively bigger.

- Because y is a function of x^2, y also becomes larger as x **negatively** increases (remember, a minus times a minus gives a plus). For example, when $x = -1$ then $y = 1$, when $x = -5$ then $y = 25$, when $x = -10$ then $y = 100$.

- Because y is a function of x^2 y grows much faster than x and is not proportional to x.

- When $x = 0$ then $y = 0$.

- There's a value of y for every value of x.

Function 3. $y = \frac{1}{x}$.

- y becomes smaller as x becomes bigger.

- y never equals zero.

- When $x = 0$, the function becomes impossible to calculate. Dividing by zero is **undefined**, not possible in other words.

- Therefore, there's a value of y for every value of x, except when $x = 0$.

Now let's take a real example from special relativity. Equation (3.4.1)

$$\gamma = \frac{1}{\sqrt{1 - (v/c)^2}}$$

is the Lorentz factor, denoted by the Greek letter gamma γ, and is used extensively as part of a set of equations known as the Lorentz transformations. c is the speed of light and is a constant; v is the relative velocity between two inertial frames of reference and is always less than c, ie $v < c$. So we can now say that γ is a function of v or, more succinctly, $\gamma = f(v)$. Just by examining this function, and without knowing anything about what it actually means, we can observe that:

- If $v = 0$ then $(v/c)^2 = 0$, $\sqrt{1 - (v/c)^2} = 1$, and therefore $\gamma = 1$.

- As v increases then γ will also increase. However, for 'ordinary' velocities (a car travelling at 70 miles per hour, for example) the value of $(v/c)^2$ will be very small (because the speed of light is a lot faster than 70 miles per hour) and γ will only be a tiny bit more than 1. Only as v approaches c will γ start to significantly increase.

- γ cannot equal zero because the numerator (the top bit of the fraction) is a non-zero constant, ie equals 1.

1.5.1 Graphs of simple functions

A more thorough way of understanding a function is to graph or plot it. At the time of writing there are available several 2D/3D online (GraphFunc Online [11], for example) and free downloadable function plotters (Winplot for Windows [32], for example) that, as well as being fun to play with, will improve your insight into how functions behave.

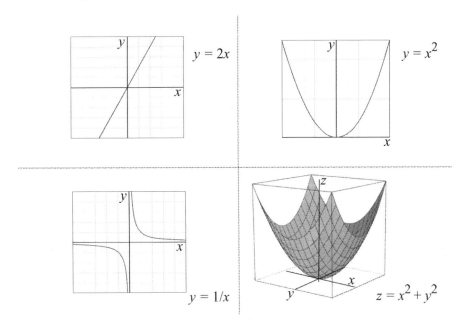

Figure 1.1: Graphs of four simple functions.

Figure 1.1 shows the graphs of four simple functions. The first three are curves, plotted using the functions of one variable that we looked at in the previous section: $y = 2x$, $y = x^2$ and $y = 1/x$.

The graphs nicely illustrate the properties we listed. For example, the first two functions are both valid at $x = 0$, meaning the functions have values at this point. This isn't the case for the function $y = 1/x$, where the curve doesn't pass through $x = 0$. Instead, the curve approaches the y axis (ie the line where $x = 0$) but never actually meets it. There's also a function of two variables – the fourth example $z = x^2 + y^2$. Functions of two variables typically don't produce curves but surfaces in three-dimensional space, in this case a sort of 'bag-shaped' surface.

The first function $y = 2x$ gives a straight line. The **gradient** or **slope** of this line is found by dividing its change in height by its change in horizontal distance. We don't need to take a ruler to the graph to work this out. The general equation of *any* straight line is given by

$$y = ax + b, \tag{1.5.1}$$

where a is the gradient and b is a constant (zero in the case of $y = 2x$). So the gradient of $y = 2x$ is 2. Similarly, the functions $y = 0.7x - 12$, $y = -4x + 3$ and $y = 37x$ are all equations of straight lines and have gradients of 0.7, -4 and 37 respectively.

We need to wait until we look at differential calculus to see how to find the changing gradient of a curve, such as $y = x^2$.

1.6 Simplifying and rearranging equations and expressions

Simplifying and rearranging equations can often make them easier to solve. For example, say we are given the equation

$$y = \frac{2x}{t} + 17, \tag{1.6.1}$$

which gives y in terms of x and t. That's fine if we know x and t and want to solve for y, but what if we know the values of x and y and want to solve for t. In that case, we need to rearrange the equation to give t in terms of y and x:

$$t = \frac{2x}{y - 17}. \tag{1.6.2}$$

Alternatively, if we knew the values of y and t and wanted to solve for x, we could rearrange to give x in terms of y and t:

$$x = \frac{t(y - 17)}{2}. \tag{1.6.3}$$

A simple example: average speed equals distance travelled divided by time taken. If we call speed s, distance d and time t, we can write

$$s = \frac{d}{t}. \tag{1.6.4}$$

If it takes us 2 hours to drive the 120 miles from London to Birmingham, for example, we could calculate our average speed as

$$s = \frac{120}{2} = 60 \, \text{mph}.$$

If we only knew the speed and time, we could rearrange (1.6.4) to find the distance. To do this we need to end up with d by itself on the left-hand side of the equation.

The rule is:

- What we do to one side of the equation we must do to the other side of the equation.

Step by step, we multiply *both* sides of equation (1.6.4) by t to give

$$s \times t = \frac{d \times t}{t}.$$

On the right-hand side, we have a fraction with t on the top and bottom, so we can cancel the t's to give

$$s \times t = d,$$

which we can rearrange to give

$$d = st.$$

Now we can find the distance d if we already know the speed s and the time t. For example, if we drive at 70mph for 2 hours, we travel a distance of

$$d = 70 \times 2 = 140 \, \text{miles}.$$

Returning to equation (1.6.1)

$$y = \frac{2x}{t} + 17,$$

which we want to rearrange so t is on the left-hand side. We first want to have the term containing t (ie the $2x/t$ term) by itself on one side of the equation, so we subtract 17 from both sides to give

$$y - 17 = \frac{2x}{t}.$$

Next we want to bring t to the top (at present we have $2x$ divided by t on the right-hand side) so we multiply both sides by t to give

$$(y - 17)\, t = \frac{2xt}{t} = 2x,$$

and divide both sides by $(y - 17)$ to give

$$\frac{(y - 17)\, t}{(y - 17)} = \frac{2x}{(y - 17)}.$$

The $(y - 17)$ terms cancel on the left-hand side to give what we want, the equation in terms of t

$$t = \frac{2x}{y - 17}.$$

Another two important rules for simplifying equations are known as

- multiplying out the brackets, and
- collecting like terms.

Problem 1.1.

Simplify and if possible solve the equation for b

$$2\,(a + 7b) + 3b = \frac{4\,(a + 8)}{2}.$$

First, multiply out the brackets

$$2a + 14b + 3b = \frac{4a + 32}{2}.$$

Next, multiply both sides by 2 to get rid of the 2 on the bottom of the right-hand side fraction

$$4a + 28b + 6b = 4a + 32.$$

Collect the terms

$$28b + 6b = 4a + 32 - 4a$$

$$34b = 32$$

$$b = \frac{32}{34} = 0.9412.$$

If we need to multiply two brackets of the form

$$(a + b)(c + d)$$

we use the

- **FOIL** (First, Outside, Inside, Last) method, ie

Multiply the first terms $(= ac)$, the outside terms $(= ad)$, the inside terms $(= bc)$, the last terms $(= bd)$ to give

$$(a + b)(c + d) = ac + ad + bc + bd.$$

Problem 1.2. Multiply the following (a) $(3 + 4)(9 - 2)$, (b) $(x - 3)(5 + x^2)$, (c) $(z + 10)(z - 10)$

(a)
$$(3 + 4)(9 - 2) = 27 - 6 + 36 - 8 = 49.$$

(b)
$$(x - 3)(5 + x^2) = 5x + x^3 - 15 - 3x^2.$$

(c)
$$(z + 10)(z - 10) = z^2 - 10z + 10z - 100 = z^2 - 100.$$

Problem 1.3. Solve

$$49 = \frac{(3x + 8)^2}{x^2}$$

It might be tempting to rush in and multiply out the brackets, but such a step would be unnecessarily complicated. Pause and note that it's possible to take easy square roots of both sides (remember, what we do to one side of the equation we must do to the other)

$$\sqrt{49} = \frac{\sqrt{(3x + 8)^2}}{\sqrt{x^2}}$$

$$\pm 7 = \frac{3x + 8}{x}.$$

Taking the left-hand side as $+7$, multiply both sides by x

$$7x = 3x + 8$$

$$4x = 8$$

$$x = \frac{8}{4} = 2.$$

If we take $\sqrt{49} = -7$, we obtain another solution, ie $x = -4/5$.

1.7 Trigonometric functions

1.7.1 Basic trigonometric functions

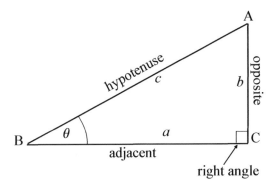

Figure 1.2: A right-angled triangle.

Trigonometric functions are functions of angles. Figure 1.2 shows a right-angled triangle (ie a triangle with one angle of 90°) with another angle denoted by the Greek letter theta θ. The sides of the a right-angled triangle are called the **adjacent side** (next to θ), the **opposite side** (opposite to θ) and the **hypotenuse** (opposite the right-angle).

For any right-angled triangle, the ratios of the various sides are constant for any particular value of θ. The basic trigonometric functions are:

$$\sin \theta = \frac{\text{opposite}}{\text{hypotenuse}} = \frac{b}{c},$$

$$\cos \theta = \frac{\text{adjacent}}{\text{hypotenuse}} = \frac{a}{c},$$

$$\tan \theta = \frac{\text{opposite}}{\text{adjacent}} = \frac{b}{a},$$

these ratios being the same for any shape or size of right-angled triangle. So, for example, if we measured a right-angled triangle to have sides $a = 4$, $b = 3$ and $c = 5$, the above trigonometric functions would equal $\sin\theta = 3/5 = 0.6$, $\cos\theta = 4/5 = 0.8$ and $\tan\theta = 3/4 = 0.75$. Conversely, if we know the angle θ plus the length of one of the sides, we can calculate the length of the remaining side. So, if we know the opposite side has a length of 8 and we know that $\theta = 25°$, we can look up $\sin 25°$ on a calculator (it equals 0.4226) and find the length of the hypotenuse thus:

$$\sin\theta = \frac{\text{opposite}}{\text{hypotenuse}} = \frac{b}{c},$$

$$\sin 25° = 0.4226 = \frac{8}{c}$$

$$c = \frac{8}{0.4226} = 18.930.$$

The trigonometric functions can also be defined for angles that are greater than $90°$. Things then become a bit more complicated because the sign of the function changes as shown in Figure 1.3.

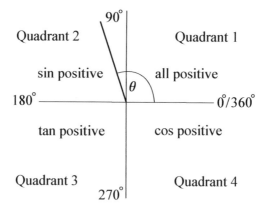

Figure 1.3: The four quadrants.

For example, $\theta = 140°$ is in Quadrant 2, and therefore $\sin\theta = 0.643$ is positive, but $\cos\theta = -0.766$ and $\tan\theta = -0.839$ are both negative.

1.7.2 Radians

Degrees are OK for simple calculations, but for more advanced work we need to measure angles in **radians**, the standard unit of angular measure. An angle measures 1 radian (see Figure 1.4) when the arc length AB equals the circle's radius r. In general, an angle in radians equals the ratio between the length of an arc and its radius. As the circumference of a circle equals $2\pi r$, an angle of $360°$ equals $2\pi r/r$ equals 2π radians.

Radians allow us to use real numbers in the trigonometric functions, rather than degrees, which are an arbitrary angular measurement. The use of real numbers in trigonometric functions is essential in more advanced mathematics, calculus for example, hence the need to use radians. Figure 1.5 shows the graphs of the three basic trigonometric functions $y = \sin x$, $y = \cos x$ and $y = \tan x$ with the x axes in radians.

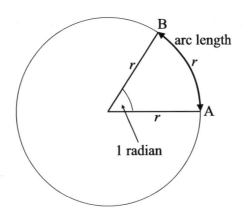

Figure 1.4: An angle of 1 radian.

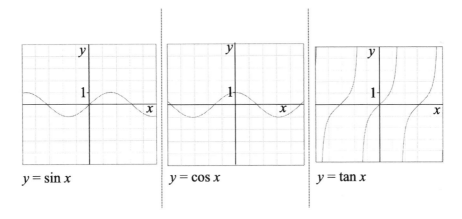

Figure 1.5: Graphs of basic trigonometric functions.

1.7.3 Trigonometric identities

An **identity** is a relationship between two expressions that is always true. For example, $x^2 - y^2 = (x + y)(x - y)$ is true for all values of x and y. There are many trigonometric identities, probably the most frequently used being the **Pythagorean trigonometric identity**

$$\cos^2 \theta + \sin^2 \theta = 1, \tag{1.7.1}$$

that expresses Pythagoras' theorem (1.14.1) in terms of trigonometric functions.

1.8 Powers and logarithms

1.8.1 Powers

If we multiply 3 by itself 4 times we get

$$3 \times 3 \times 3 \times 3 = 81.$$

A more concise way of writing this is to say

$$3^4 = 81,$$

where 3 is the **base** and the superscript 4 is called the **power** or **exponent**. The usual way to describe this is to say we are 'raising 3 to the power of 4', or simply '3 to the 4'. Another example would be

$$10^6 = 10 \times 10 \times 10 \times 10 \times 10 \times 10 = 1{,}000{,}000,$$

which we would describe as '10 to the power of 6', or '10 to the 6'. In general terms, this process of raising something to the power of something is called **exponentiation**, which we can write as x^p where x is the base and p the power or exponent.

These are the basic rules for manipulating powers:

1. $x^0 = 1$ where $x \neq 0$, eg $12^0 = 1$, $0.098^0 = 1$.

2. $x^1 = x$, eg $45^1 = 45$, $5.6^1 = 5.6$.

3. $x^p \times x^q = x^{p+q}$, eg $2^3 \times 2^5 = 8 \times 32 = 256 = 2^{3+5} = 2^8$.

4. $\frac{x^p}{x^q}$ (where $x \neq 0$) $= x^{p-q}$, eg $\frac{3^8}{3^2} = \frac{6561}{9} = 729 = 3^{8-2} = 3^6$.

5. $(x^p)^q = x^{pq}$, eg $\left(5^2\right)^3 = (25)^3 = 15625 = 5^{2 \times 3} = 5^6$.

6. $(xy)^p = x^p y^p$, eg $(3 \times 7)^2 = 21^2 = 441 = 3^2 \times 7^2 = 9 \times 49$.

7. $x^{-p} = \frac{1}{x^p}$, eg $8^{-2} = \frac{1}{8^2} = \frac{1}{64} = 0.015625$.

8. $x^{1/q} = \sqrt[q]{x}$ (where q is a positive integer). For example, $27^{1/3} = \sqrt[3]{27} = 3$, because $3 \times 3 \times 3 = 27$.

9. $x^{p/q} = \sqrt[q]{x^p}$ (where p and q are positive integers). For example, $4^{5/2} = \sqrt[2]{4^5} = \sqrt{1024} = 32$.

1.8.2 Logarithms

A **logarithm** (log or logs for short) goes in the opposite direction to a power by asking the question: what power produced this number? So, if $2^x = 32$, we are asking, 'what is the logarithm of 32 to base 2?' We know the answer: it's 5, because $2^5 = 32$, so we say the logarithm of 32 to base 2 is 5.

In general terms, if $x^p = a$, then we say the logarithm of a to base x equals p, or

$$\log_x (a) = p.$$

For example, $10^4 = 10{,}000$, so we say the logarithm of 10,000 to base 10 equals 4, or

$$\log_{10} (10{,}000) = 4.$$

We can take logarithms of any positive number, not just whole ones. So, as $10^{3.4321} = 2704.581$ we say the logarithm of 2704.581 to base 10 equals 3.4321, or

$$\log_{10}(2704.581) = 3.4321.$$

Logarithms to base 10 are called **common logarithms**. Older readers may remember, many years ago – after the dinosaurs, but before calculators and computers were widely available – doing numerical calculations laboriously by hand using tables of common logarithms and anti-logarithms.

The properties of logarithms are based on the aforementioned rules for working with powers. Assuming that $a > 0$ and $b > 0$ we can say:

1. $\log_x(ab) = \log_x a + \log_x b$. For example, $\log_{10}(1000 \times 100) = \log_{10} 100{,}000$
   $= 5 = \log_{10} 1000 + \log_{10} 100 = 3 + 2$.

2. $\log_x(1/a) = -\log_x a$. For example, $\log_3(1/27) = -\log_3 27 = -3$.

3. $\log_x(a/b) = \log_x a - \log_x b$. For example, $\log_2(128/8) = \log_2 16 = 4 = \log_2 128 - \log_2 8 = 7 - 3$.

4. $\log_x(a^y) = y\log_x a$. For example, $\log_5(25)^3 = \log_5 15{,}625 = 6 = 3 \times \log_5 25 = 3 \times 2$.

1.8.3 The constant e and natural logarithms

There is a fundamentally important mathematical constant e that is approximately equal to 2.71828. Like that other famous constant π (pi), e is an **irrational number**, meaning it cannot be written in the form a/b (ie as a fraction) where a and b are both whole numbers. e is sometimes known as the **exponential constant** or **Euler's number** after the great Swiss mathematician Leonhard Euler (1707–1783).

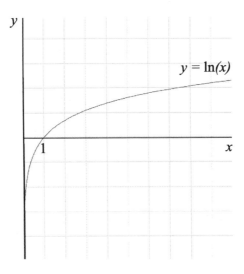

Figure 1.6: $y = \ln x$.

Logarithms to base e are called **natural logarithms**, which are written as $\ln(x)$ or $\log_e(x)$ or (if it's obvious that base e is being used) $\log(x)$. In other words, natural logarithms are defined by

$e^{\ln(x)} = x$ or, equivalently, $\ln(e^x) = x$. Natural logarithms are of great importance and pop up all over the place in mathematics and the sciences. We'll be meeting them later in this book. Figure 1.6 shows the graph of $y = \ln(x)$.

Although e has many interesting properties and applications, one straightforward way to understand it is in terms of compound interest. Say we invest £1.00 in an exceedingly generous bank that pay 100% interest per annum. If the bank calculated and credited the interest at the end of one year, our investment would then be worth $1 + 1 = £2.00$. But what if the bank credits the interest more frequently than once a year? If interest is calculated and added every six months, at the end of that period the balance would equal $1 + \frac{1}{2} = £1.50$ and at the end of one year the total amount would be $\left(1 + \frac{1}{2}\right)\left(1 + \frac{1}{2}\right) = £2.25$. Calculated three times a year, the final balance would be $\left(1 + \frac{1}{3}\right)\left(1 + \frac{1}{3}\right)\left(1 + \frac{1}{3}\right) = £2.370$. In general, if interest is calculated n times a year, the balance x after one year is given by

$$x = \left(1 + \frac{1}{n}\right)^n.$$

Table 1.4 shows the value of x after one year for different values of n.

n	$\left(1 + \frac{1}{n}\right)^n.$
1	2
2	2.25
3	2.37037
4	2.44141
5	2.48832
10	2.59374
100	2.70481
1000	2.71692
100,000	2.71827
1,000,000	2.71828
10,000,000	2.71828

Table 1.4: Value of $\left(1 + \frac{1}{n}\right)^n$ for increasing values of n.

We can see that as n increases, the value of the function $x = \left(1 + \frac{1}{n}\right)^n$ appears to settle down to a number approximately equal to 2.71828. It can be shown that as n becomes infinitely large, it does indeed equal the constant e. The mathematically succinct way of saying this introduces the important idea of a **limit** and we say

$$e = \lim_{n \to \infty} \left(1 + \frac{1}{n}\right)^n, \tag{1.8.1}$$

where $\lim_{n \to \infty}$ means the limit of what follows (ie $\left(1 + \frac{1}{n}\right)^n$) as n approaches infinity (symbol ∞). In other words, e approaches the value of $\left(1 + \frac{1}{n}\right)^n$ as n approaches infinity.

A brief digression. Not relevant to the subject matter of this book, but well worth mentioning in passing whilst we are on the subject of Euler's number is **Euler's identity**

$$e^{i\pi} + 1 = 0,$$

which connects the five most important constants of mathematics and has been called the 'most beautiful theorem in mathematics.' All the numbers we use in this book are **real numbers**, which can be thought of as points on a line. Real numbers are fine for most practical purposes, but they do not allow solutions to equations resembling $x^2 = -1$. Complex numbers – which are written in the form $a + bi$ (where a and b are real) and can be thought of as point on a plane, the complex plane – are the solution to this problem. Real numbers are themselves a subset of complex numbers, where i, the imaginary unit, is defined by $i = \sqrt{-1}$.

1.8.4 The exponential function

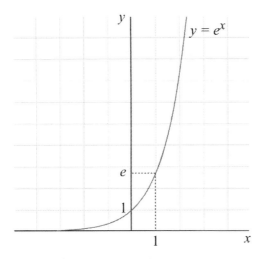

Figure 1.7: $y = e^x$.

The exponential function $f(x) = e^x$, often written as $\exp x$, (see Figure 1.7) arises whenever a quantity grows or decays at a rate proportional to its size: radioactive decay, population growth and continuous interest, for example. The exponential function is defined, using the concept of a limit, as

$$e^x = \lim_{n \to \infty} \left(1 + \frac{x}{n}\right)^n. \tag{1.8.2}$$

1.9 Coordinate systems

Coordinate systems are used to uniquely define the position of a point or other geometric or physical object. An everyday use of a coordinate system is the grid reference used to locate places on a map. So far we've used basic x, y, z coordinates to plot the graphs of various functions: x, y in two dimensions, x, y, z in three.

To be useful, a coordinate system must define a point's position using a unique set of numbers. There are an infinite number of possible ways of doing this, meaning there are an infinite number of possible coordinate systems. That said, some coordinate systems are definitely more useful than others and for the next few chapters, until we encounter the coordinate free-for-all that is differential geometry, we'll be using one of three common coordinate systems: Cartesian, plane polar and spherical. Although the choice of coordinate system is arbitrary, we can make our lives significantly easier by using the simplest possible system for a given situation. For example, we could use three-dimensional x, y, z coordinates to plot the location of points on the Earth's surface, but this method would be horribly complicated compared to the conventional use of longitude and latitude coordinates.

1.9.1 Cartesian coordinates

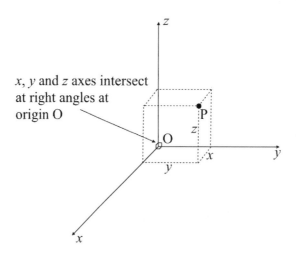

Figure 1.8: Three-dimensional Cartesian coordinates.

The x, y and x, y, z coordinates we've used so far, where the **axes** (singular – **axis**, as in 'the x axis') intersect at right angles, are known as the **Cartesian** or **rectangular coordinate system**, with the point O where the axes intersect called the **origin**. Figure 1.8 shows a three-dimensional Cartesian coordinate system.

A **plane** is a flat two-dimensional surface, so Cartesian coordinates in two dimensions are often known as the **Cartesian plane**.

We define the position of a point, P for example, in terms of its x, y and z **coordinates**. So if P's position is at $x = 2$, $y = 3$, $z = 4$ we say $(2, 3, 4)$ are the coordinates of P. The coordinates of various points on the Cartesian plane are shown in Figure 1.9.

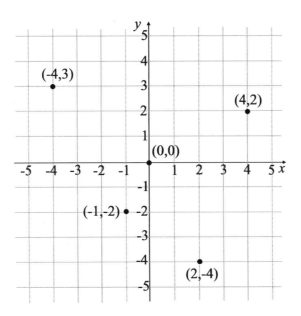

Figure 1.9: Points on the Cartesian plane.

1.9.2 Plane polar coordinates

Although Cartesian coordinates are a nice, simple coordinate system, they aren't so suitable when circular, cylindrical, or spherical symmetry is present. In those circumstances, in two dimensions, **plane polar coordinates** are a better choice, where the position of a point on the plane is given in terms of the distance r from a fixed point and an angle θ from a fixed direction (see Figure 1.10). So, if point P was a distance $r = 6$ from the origin, and the angle $\theta = 120°$, the coordinates of P would be $(6, 120°)$.

It's easy enough to convert from plane polar to Cartesian coordinates by superimposing the axes (see Figure 1.11). Using the simple trigonometric functions, we can now see that

$$x = r\cos\theta \tag{1.9.1}$$

and

$$y = r\sin\theta$$

and it is straightforward to go the other way and convert from Cartesian to plane polar coordinates:

$$r = \sqrt{x^2 + y^2},$$

$$\tan\theta = \frac{y}{x}.$$

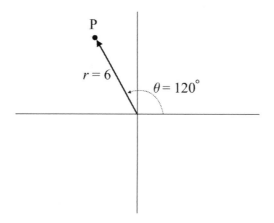

Figure 1.10: Plane polar coordinates.

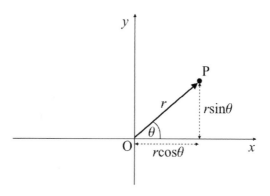

Figure 1.11: Converting from polar to Cartesian coordinates.

1.9.3 Spherical coordinates

In three dimensions, the position of a point P in space can be defined using **spherical coordinates** (r, θ, ϕ) as shown in Figure 1.12.

To go from spherical coordinates to Cartesian coordinates we see that

$$OB = r \sin \theta,$$

and therefore

$$x = OD = r \sin \theta \cos \phi,$$

$$y = OC = r \sin \theta \sin \phi,$$

$$z = PB = r \cos \theta.$$

In summary:

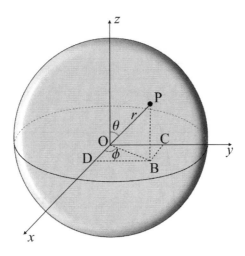

Figure 1.12: Spherical coordinates.

$$x = r \sin \theta \cos \phi, \qquad\qquad (1.9.2)$$
$$y = r \sin \theta \sin \phi,$$
$$z = r \cos \theta.$$

1.10 Calculus

If I'm driving at a constant *velocity*, say 60 km per hour, the *distance* I travel is changing in a constant manner, ie every hour I cover 60 km. We can say the rate of change of the distance I travel with respect to time is constant. However, if my car is moving with constant *acceleration*, its *velocity* is changing in a constant manner, but the distance I travel is not changing constantly – because I'm accelerating. We can say the rate of change of the distance I travel, with respect to time, is *not* constant.

Differential calculus is concerned with the precise mathematical description of such rates of change. In other words, how quickly a variable (distance, velocity, x, y, z, etc) is changing with respect to another variable (time, x, y, z, etc). If you see things like dy, dx, dt or $\partial x, \partial y, \partial t$, etc, then you are dealing with differential calculus.

The other main branch of the subject is called **integral calculus** (using the symbol $\int$), which is more or less the opposite of differential calculus.

For example, say we have a function that tells us the acceleration of an object after a certain time. If we can integrate that function we obtain a new function that tells us the velocity of the object after a certain time. If we integrated the second function, we'd get a third function that tells us the distance covered after a certain time. But we can also do that sequence of calculations in reverse, by starting with the function that tells us distance covered after a certain time. We can differentiate that function to find the object's velocity and differentiate again to find its acceleration.

In order to differentiate or integrate a function (assuming that this is possible – sometimes it isn't) we use various rules of calculus, which vary in their complexity. Most of the ones needed in this book are detailed below. There is also the wonderful invention of online calculus calculators (the WolframAlpha Calculus and Analysis Calculator [33], for example) that hugely simplify the process of differentiation, integration and other calculations, bearing in mind the apposite warning: 'garbage in, garbage out'.

1.10.1 Differential calculus

1.10.1.1 The derivative

Figure 1.13 shows the curve of the function $y = x^3 - 2x^2$. The slope of this curve changes continuously and has been indicated at four points A, B, C, D by tangents (the straight lines) of the curve passing through those points. Those tangents show the rate of change of the function at that particular point. We use the Greek letter Delta Δ to denote a small bit or increment of a variable, so the gradient or slope of the tangent passing through A equals $\Delta y/\Delta x$ (just as the gradient of a hill can be found by dividing the vertical height gain by the horizontal distance).

Differential calculus allows us to find the exact slope of a curve, in other words a function's rate of change, at any particular point. We are assuming that the function is differentiable in the first place and that the gradient doesn't go shooting off to infinity. Figure 1.14 shows the curve of the function $y = x^2$. The gradient at A is, roughly, given by $\Delta y/\Delta x$. The closer we move point B along the curve towards A, the more accurately $\Delta y/\Delta x$ equals the gradient of the curve at A.

The slope of AB is equal to

$$\frac{BC}{AC} = \frac{\Delta y}{\Delta x} = \frac{f(x + \Delta x) - f(x)}{\Delta x},$$

where $f(x + \Delta x)$ is the value of y at B, and $f(x)$ is the value of y at A. The quotient

$$\frac{f(x + \Delta x) - f(x)}{\Delta x} \qquad (1.10.1)$$

is known as the **difference quotient**, or the **Newton quotient**, and its limit defines the derivative of a function. The limit refers to the value of the difference quotient as we make Δx smaller and smaller and smaller as it approaches, but never quite equals, zero. In other words, we shrink the triangle ABC until it gives us the exact gradient of the tangent to the curve at A.

This limit is known as the **derivative** and is the heart of differential calculus. In this example, where $y = x^2$, the derivative could be denoted by

$$\frac{dy}{dx},$$

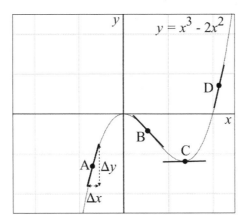

Figure 1.13: Varying gradients of the curve $y = x^3 - 2x^2$.

which gets across the idea of a ratio of two infinitesimal quantities, dy and dx.

There are several alternative ways of denoting the derivative, including: $\frac{df}{dx}$ for a general function f, and also by using prime notation (the little mark $'$ is called the prime symbol) where we would write $f'(x)$ or y', where $y = f(x)$. Using the limit concept we define the derivative of a function as

$$\left(\frac{dy}{dx} = \frac{df}{dx} = f'(x) = y' \right) = \lim_{\Delta x \to 0} \frac{\Delta y}{\Delta x} = \lim_{\Delta x \to 0} \frac{f(x + \Delta x) - f(x)}{\Delta x}, \qquad (1.10.2)$$

where $\lim_{\Delta x \to 0}$ means the limit of $\frac{f(x + \Delta x) - f(x)}{\Delta x}$ as Δx approaches zero.

Problem 1.4. If $y = x^2$ use the difference quotient to show that the derivative $\frac{dy}{dx} = 2x$.

We can simplify the difference quotient (1.10.1)

$$\frac{f(x + \Delta x) - f(x)}{\Delta x}$$

by substituting h for Δx

$$\frac{f(x + h) - f(x)}{h},$$

$$\frac{dy}{dx} = \lim_{\Delta x \to 0} \frac{f(x + h) - f(x)}{h},$$

which, as $y = x^2$, we can write as

$$\frac{dy}{dx} = \lim_{\Delta x \to 0} \frac{(x + h)^2 - x^2}{h}$$

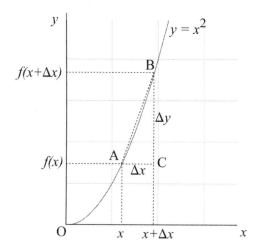

Figure 1.14: Defining the derivative with the difference quotient.

$$\frac{dy}{dx} = \lim_{\Delta x \to 0} \frac{x^2 + 2hx + h^2 - x^2}{h} = \frac{2hx + h^2}{h}$$

$$\frac{dy}{dx} = \lim_{\Delta x \to 0} 2x + h,$$

which, as $h \to 0$, becomes

$$\frac{dy}{dx} = 2x.$$

This example is a special case of the rule that if $f(x) = x^k$ then $f'(x) = kx^{k-1}$, ie for $y = x^2$, $k = 2$, so $f'(x) = kx^{k-1} = 2x^{(2-1)} = 2x$. Some common derivatives are shown in Table 1.5

Problem 1.5. From Figure 1.13, find the gradient of the curve at point D if $x = 2.5$ at D.

The function is $y = x^3 - 2x^2$. Differentiating this function (using the third rule from Table 1.5) gives

$$\frac{dy}{dx} = 3x^2 - 4x.$$

Substituting $x = 2.5$

$$\frac{dy}{dx} = 3(2.5)^2 - (4 \times 2.5) = 18.75 - 10 = 8.75.$$

The gradient of the curve at D equals 8.75.

Problem 1.6. Differentiate (a) $y = x^5 + x - 12$, (b) $k = \sin t + 1/t^2$, (c) $s = e^x$.

$f(x)$	$f'(x)$
$k(\text{constant})$	0
x	1
x^k	kx^{k-1}
$\sin x$	$\cos x$
$\cos x$	$-\sin x$
e^{kx}	ke^{kx}
$\ln x$	$1/x$
$a^x,\ (a>0)$	$a^x \ln a$

Table 1.5: Common derivatives.

(a) Using the first, second and third rules from Table 1.5

$$\frac{dy}{dx} = 5x^4 + 1.$$

(b) Rewriting the function

$$k = \sin t + 1/t^2 = \sin t + t^{-2}.$$

Using the third and fourth rules from Table 1.5

$$\frac{dk}{dt} = \cos t - 2t^{-3} = \cos t - \frac{2}{t^3}.$$

(c) Using the sixth rule from Table 1.5

$$\frac{ds}{dx} = e^x.$$

The derivative (rate of change) of the exponential function is the exponential function itself. In fact, e^x is the *only* function which is its own derivative.

1.10.1.2 The product rule

We use this rule when differentiating a product of two or more functions. For two functions the product rule is

$$\frac{d(uv)}{dx} = v\frac{du}{dx} + u\frac{dv}{dx}. \tag{1.10.3}$$

For three functions the product rule is

$$\frac{d(uvw)}{dx} = vw\frac{du}{dx} + uw\frac{dv}{dx} + uv\frac{dw}{dx}.$$

Problem 1.7. Differentiate the function $y = x\left(x^2 + 1\right).$

45

let $y = uv, u = x, v = x^2 + 1$,

therefore $\frac{du}{dx} = 1$ and $\frac{dv}{dx} = 2x$,

and $\frac{dy}{dx} = \frac{d(uv)}{dx} = \left((x^2 + 1) \times 1\right) + (x \times 2x) = x^2 + 1 + 2x^2 = 3x^2 + 1$.

This example is easy enough to check by noting that $y = x\left(x^2 + 1\right) = x^3 + x$,

and therefore $\frac{dy}{dx} = 3x^2 + 1$, which is the same answer.

1.10.1.3 The chain rule

This rule is used if a function is a function of another function. So, if y is a function of u and u is a function of x, then the derivative of y with respect to x is equal to the derivative of y with respect to u multiplied by the derivative of u with respect to x, ie

$$\frac{dy}{dx} = \frac{dy}{du} \times \frac{du}{dx}. \tag{1.10.4}$$

Problem 1.8. Differentiate the function $y = \left(1 + x^2\right)^{100}$ (a) using the chain rule, and (b) using an online calculus calculator.

(a) Let $u = 1 + x^2$, then $\frac{du}{dx} = 2x$.

Then $y = u^{100}$ and $\frac{dy}{du} = 100u^{99}$.

Using (1.10.4)

$$\frac{dy}{dx} = \frac{dy}{du} \times \frac{du}{dx}$$

$$= 100u^{99} \times 2x = 200x\left(1 + x^2\right)^{99}.$$

(b) Using the WolframAlpha Calculus and Analysis Calculator [33], type 'derivative of (1+x^2)^100' (omit the quotes) into the input box to obtain

$$\frac{d\left(\left(1 + x^2\right)^{100}\right)}{dx} = 200x\left(1 + x^2\right)^{99}.$$

1.10.1.4 Partial derivatives

We've seen how to differentiate a function of one variable to find its rate of change, ie the gradient of the function's curve at any particular point. Now we need to look at how to differentiate functions of more than one variable. We mentioned earlier that functions of two variables typically don't produce curves but surfaces in three-dimensional space. One such function is shown in Figure 1.15. Any one point on the surface of $z = x^2 + y^2$ will have an infinite number of curves passing through it, each with a different gradient. How to make sense of all those different rates of change? The answer is **partial derivatives**.

Taking partial derivatives involves using the same rules as for ordinary differentiation, but allowing one variable to change whilst keeping all the other variables fixed, ie treating those variables as

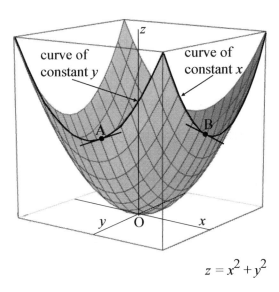

$$z = x^2 + y^2$$

Figure 1.15: Curves of constant y and x for $z = x^2 + y^2$.

constants. The notation is a little different to ordinary differentiation because we use the symbol ∂, derived from the Greek letter delta, to indicate partial derivatives.

We can formally define the partial derivative by first recalling that we defined the ordinary derivative (1.10.2) as

$$\frac{dy}{dx} = f'(x) = \lim_{\Delta x \to 0} \frac{f(x + \Delta x) - f(x)}{\Delta x},$$

where Δx denotes a small increment of x. For a function of two variables $f(x, y)$, we first take partial derivatives with respect to x by keeping y constant and say

$$\frac{\partial f}{\partial x} = \lim_{\Delta x \to 0} \frac{f(x + \Delta x, y) - f(x, y)}{\Delta x}. \tag{1.10.5}$$

We next take partial derivatives with respect to y by keeping x constant and say

$$\frac{\partial f}{\partial y} = \lim_{\Delta y \to 0} \frac{f(x, y + \Delta y) - f(x, y)}{\Delta y}. \tag{1.10.6}$$

This is actually a lot more straightforward than it looks. In order to take partial derivatives of the function $z = x^2 + y^2$, for example, we first treat y as a constant and differentiate with respect to x to obtain

$$\frac{\partial z}{\partial x} = 2x. \tag{1.10.7}$$

Note that as we've treated y as a constant and the derivative of a constant is zero, y doesn't appear in this particular partial derivative. Next we treat x as a constant and take partial derivatives with respect to y to obtain

$$\frac{\partial z}{\partial y} = 2y. \tag{1.10.8}$$

Figure 1.15 helps us visualise what the partial derivatives are telling us. Equation (1.10.7) gives us the slope (ie z's rate of change) at any point on a surface curve of constant y (such as the x axis), point A for example. Equation (1.10.8) gives us the rate of change at any point on a surface curve of constant x (such as the y axis), point B for example. So, if the value of x at A were (picking a number out of thin air) 12, we could find the rate of change of z with respect to x at A by plugging 12 into (1.10.7), ie

$$\frac{\partial z}{\partial x} = 2x = 2 \times 12 = 24.$$

The procedure is the same for functions of more than two variables, which are of course much harder to visualise. For example, taking partial derivatives of the function $z = wx^2 + 3wxy + x^2 + 5y + y^3$ we obtain

$$\frac{\partial z}{\partial w} = x^2 + 3xy,$$

$$\frac{\partial z}{\partial x} = 2wx + 3wy + 2x,$$

$$\frac{\partial z}{\partial y} = 3wx + 5 + 3y^2.$$

1.10.1.5 Total differential and total derivative

The partial derivatives of a function $f(x, y)$ give us the function's rate of change along the x and y axes. We used the example of $z = x^2 + y^2$, so the partial derivatives tell us how z changes if we keep x constant or if we keep y constant. But what happens if we allow both x and y to change? How does this affect z? We describe how a function changes in an **arbitrary** direction using something called the **total differential**.

If we have a function $f(x, y)$ and change it ever so slightly by changing x by Δx and y by Δy, we can see the change in $f(x, y)$ is given by

$$\Delta f = f(x + \Delta x, y + \Delta y) - f(x, y).$$

We now add and subtract the quantity $f(x, y + \Delta y)$ to the right-hand side:

$$\Delta f = f(x + \Delta x, y + \Delta y) - f(x, y + \Delta y) + f(x, y + \Delta y) - f(x, y).$$

Then multiply and divide by Δx and Δy:

$$\Delta f = \left[\frac{f(x + \Delta x, y + \Delta y) - f(x, y + \Delta y)}{\Delta x} \right] \Delta x + \left[\frac{f(x, y + \Delta y) - f(x, y)}{\Delta y} \right] \Delta y,$$

which, on closer examination, is the same as saying

$$\Delta f = \left[\frac{f(x + \Delta x, y) - f(x, y)}{\Delta x} \right] \Delta x + \left[\frac{f(x, y + \Delta y) - f(x, y)}{\Delta y} \right] \Delta y. \tag{1.10.9}$$

But the quantities in brackets are very similar to our earlier ((1.10.5) and (1.10.6)) definitions of partial derivatives $\frac{\partial f}{\partial x}$ and $\frac{\partial f}{\partial y}$. If we make the quantities Δx and Δy infinitesimally small, (1.10.9) becomes

$$df = \frac{\partial f}{\partial x} dx + \frac{\partial f}{\partial y} dy,$$

which is the total differential df of the function $f(x, y)$. The total differential tells us the infinitesimal change in a function caused by infinitesimal changes in the function's variables.

For a function of four variables $(f(w, x, y, z))$ the total differential would be

$$df = \frac{\partial f}{\partial w}dw + \frac{\partial f}{\partial x}dx + \frac{\partial f}{\partial y}dy + \frac{\partial f}{\partial z}dz. \qquad (1.10.10)$$

Problem 1.9. Find the total differential of $z = 5x^2 + 6xy + 3y^2$.

To go through all the steps, we start by taking partial derivatives:

$$\frac{\partial z}{\partial x} = 10x + 6y$$

$$\frac{\partial z}{\partial y} = 6x + 6y$$

and put them together to get the total differential

$$dz = \frac{\partial z}{\partial x}dx + \frac{\partial z}{\partial y}dy$$

$$dz = (10x + 6y)\,dx + (6x + 6y)\,dy.$$

We can refine the total differential in the case where a function's variables are themselves functions of a variable. For example, in our earlier function $f(w, x, y, z)$ the total differential (1.10.10) was

$$df = \frac{\partial f}{\partial w}dw + \frac{\partial f}{\partial x}dx + \frac{\partial f}{\partial y}dy + \frac{\partial f}{\partial z}dz.$$

But, say all the variables depended on time t. We could then, effectively, divide by dt to obtain the new relationship

$$\frac{df}{dt} = \frac{\partial f}{\partial w}\frac{dw}{dt} + \frac{\partial f}{\partial x}\frac{dx}{dt} + \frac{\partial f}{\partial y}\frac{dy}{dt} + \frac{\partial f}{\partial z}\frac{dz}{dt}, \qquad (1.10.11)$$

which is called the **total derivative** of f with respect to t.

1.10.1.6 Second derivatives

If we find the derivative of a derivative of a function f we have found the **second derivative** of f. The second derivative tells us the rate of change of the first derivative, and can be denoted by $\frac{d}{dx}\left(\frac{dy}{dx}\right)$ or $\frac{d^2y}{dx^2}$. For example, the first derivative of the function $y = 5x^3 - 2x + 7$ is

$$\frac{dy}{dx} = 15x^2 - 2.$$

We differentiate again to find the second derivative

$$\frac{d}{dx}\left(\frac{dy}{dx}\right) = \frac{d^2y}{dx^2} = 30x.$$

We can also use the prime notation, so the second derivative of a function $f(x)$ is denoted by $f''(x)$.

With partial derivatives, the notation tells us the order in which the second derivatives were taken. So, for a function of two variables $f(x, y)$, if we take partial derivatives first with respect to x and then with respect to y, there's a choice of notation:

$$\frac{\partial}{\partial y}\left(\frac{\partial f}{\partial x}\right) = \frac{\partial^2 f}{\partial y \partial x} = f_{xy} = \partial_{yx} f.$$

All mean the same thing and are known as **second order partial derivatives**.

Problem 1.10. Calculate the second order partial derivatives for $f(x, y) = x^3 + 2xy$.

$$\frac{\partial}{\partial x}\left(\frac{\partial f}{\partial x}\right) = \frac{\partial^2 f}{\partial x \partial x} = f_{xx}$$

$$f_{xx} = \frac{\partial}{\partial x}\left(\frac{\partial\left(x^3 + 2xy\right)}{\partial x}\right) = \frac{\partial\left(3x^2 + 2y\right)}{\partial x} = 6x$$

$$f_{yy} = \frac{\partial}{\partial y}\left(\frac{\partial\left(x^3 + 2xy\right)}{\partial y}\right) = \frac{\partial\left(2x\right)}{\partial y} = 0$$

$$f_{xy} = \frac{\partial}{\partial y}\left(\frac{\partial\left(x^3 + 2xy\right)}{\partial x}\right) = \frac{\partial\left(3x^2 + 2y\right)}{\partial y} = 2$$

$$f_{yx} = \frac{\partial}{\partial x}\left(\frac{\partial\left(x^3 + 2xy\right)}{\partial y}\right) = \frac{\partial\left(2x\right)}{\partial x} = 2.$$

1.10.2 Integral calculus

1.10.2.1 Indefinite integral

The importance of integral calculus from our point of view is that, in the course of this book, we are going to come across equations scattered with various d something or others ($dx, dy, d\phi$, etc), which we want to solve to find x, y, ϕ, etc. We do this by the process of integration, which due to an important result known as the **fundamental theorem of calculus** is the reverse of differentiation. For example, we know how to find the derivative of the function $y = 2x^2 + x - 1$, it's

$$\frac{dy}{dx} = 4x + 1.$$

But what if we want to go in the opposite direction and ask what function or functions could have $4x + 1$ as a derivative or, in other words, we want to solve the equation $dy = (4x + 1)\,dx$. We do this by finding the **indefinite integral** (also known as the **integral** or **antiderivative**) of $4x + 1$.

We already know one answer, the original function $y = 2x^2 + x - 1$. The correct notation for this procedure is

$$\int (4x + 1)\, dx = 2x^2 + x + C,$$

where the symbol $\int$ denotes the integral. Notice that we haven't been able to obtain our exact original function $y = 2x^2 + x - 1$. This is because as the derivative of a constant is zero, there are an infinite number of functions that could have $4x + 1$ as a derivative, eg $y = 2x^2 + x + 231$, $y = 2x^2 + x - 1102$, $y = 2x^2 + x + 12$, etc. We therefore need to include a constant C, known as the **constant of integration** in our answer.

More generally, we can say

$$\int f(x)\, dx = F(x) + C,$$

where $F(x)$ is the indefinite integral or antiderivative of $f(x)$, the quantity being integrated, which is known as the **integrand** (ie the expression after the $\int$ symbol).

Some common integrals (which you can see are the opposite of the derivatives in Table 1.5) are shown in Table 1.6.

Problem 1.11. Calculate the following indefinite integrals:
(a) $\int \left(2x + 5x^2 - 3\right) dx$, (b) $\int \left(\frac{2}{x} + 5\sin x\right) dx$, (c) $\int (3t)^x\, dx$.

(a) Using the first, second and third rules from Table 1.6

$$\int \left(2x + 5x^2 - 3\right) dx = \frac{2x^2}{2} + \frac{5x^3}{3} - 3x + C = x^2 + \frac{5x^3}{3} - 3x + C.$$

(b) Using the fourth and fifth rules from Table 1.6

$$\int \left(\frac{2}{x} + 5\sin x\right) dx = 2\left(\ln |x|\right) - 5\cos x + C.$$

(c) Using the last rule from Table 1.6

$$\int (3t)^x dx = \frac{(3t)^x}{\ln (3t)} + C.$$

Recalling that $\int f(x)\, dx = F(x) + C$, we can check the answers by taking the derivative of $F(x)$, which must take us back to the integrand $f(x)$. For example, where

$$F(x) = x^2 + \frac{5x^3}{3} - 3x + C$$

$$\frac{dF}{dx} = 2x + 5x^2 - 3,$$

which is the original $f(x)$ in example (a).

$f(x)$	$\int f(x)\,dx$
$k(\text{constant})$	$kx + C$
x	$\frac{x^2}{2} + C$
$x^k,\,(k \neq -1)$	$\frac{x^{k+1}}{k+1} + C$
$1/x$	$(\ln|x|) + C$ (†)
$\sin x$	$-\cos x + C$
$\cos x$	$\sin x + C$
$e^{kx},\,(k \neq 0)$	$\frac{e^{kx}}{k} + C$
$\ln x$	$x(\ln(x) - 1) + C$
$a^x,\,(a > 0, a \neq 1)$	$\frac{a^x}{\ln a} + C$

Table 1.6: Common integrals.

(†) $|x|$ means the absolute value of x, ie the positive value of x, so if $x = -3$, then $|x| = 3$, and if $x = 7$, then $|x| = 7$, etc.

1.10.2.2 Definite integrals

An indefinite integral is a function (eg $\int (4x + 1)\,dx = 2x^2 + x + C$), a machine for producing numbers, but is not itself a number. Only when we substitute numbers for the x's and other variables do we turn an indefinite integral into an actual number. When we do this, we change an indefinite integral into a **definite integral**.

The fundamental theorem of calculus relates the indefinite and definite integrals as follows:

$$\int_a^b f(x)\,dx = F(b) - F(a). \tag{1.10.12}$$

The number a is called the **lower limit**, and b is called the **upper limit**. All that (1.10.12) means is that we first find $F(x)$, the indefinite integral of $f(x)$. We next feed a into $F(x)$ to obtain $F(a)$ and similarly feed b into $F(x)$ to obtain $F(b)$. Finally, subtracting $F(a)$ from $F(b)$ provides the required answer. An example should make this clearer.

Problem 1.12. Evaluate the definite integral $\int_0^2 (x^2 + 1)\,dx$.

We first find the indefinite integral

$$F(x) = \frac{x^3}{3} + x + C.$$

Next we feed in $a = 0$ to find $F(a)$

$$F(a) = \frac{0^3}{3} + 0 + C.$$

Do the same with $b = 2$ to find $F(b)$

$$F(b) = \frac{(2)^3}{3} + 2 + C = \frac{8}{3} + 2 + C = \frac{14}{3} + C,$$

and subtract $F(a)$ from $F(b)$ to give the answer

$$F(b) - F(a) = \frac{14}{3} + C - C = \frac{14}{3}.$$

Notice how the constant of integration C cancels out in the final line.

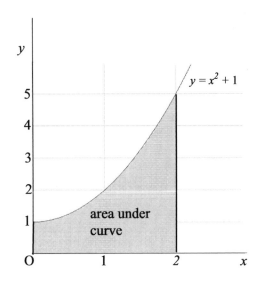

Figure 1.16: Area under the curve $y = x^2 + 1$.

When evaluating a definite integral we often use the convenient square bracket notation

$$\int_a^b f(x)\, dx = [F(x)]_a^b$$

to keep track of the upper and lower limits. Note, there is no mention of the constants of integration as they cancel each other. We could then evaluate Problem 1.12 on one line as

$$\int_0^2 (x^2 + 1)\, dx = \left[\frac{x^3}{3} + x\right]_0^2 = \left(\frac{14}{3}\right) - (0) = \frac{14}{3}.$$

Another way to understand the definite integral $\int_0^2 (x^2 + 1)\, dx$ in Problem 1.12 is that it tells us the area under the curve of $y = x^2 + 1$ between $x = 0$ and $x = 2$, as shown in Figure 1.16, ie the shaded area equals $14/3 = 4.667$.

It's worth pointing out that the definite integral is actually founded on the notion of an infinite sum, hence the symbol $\int$ based on the old-fashioned elongated S (standing for *summation*). Figure 1.17 shows the curve of an arbitrary function $f(x)$. We wish to find the area under the curve between $x = a$ and $x = b$. We can divide the area into n strips of equal width Δx_i. Treating each strip as a rectangle, the area of each strip is equal to the rectangle's height multiplied by its width, ie to $f(x_i) \times \Delta x_i$. The total area under the curve is approximately equal to the sum of the areas of the strips. As we increase n, we increase the number of strips, but decrease the width of each strip. As we make n infinitely large, the width of each strip becomes infinitely small and the area under the curve becomes *exactly* equal to the sum of the areas of the strips. The definite integral can then be *defined* as

$$\int_a^b f(x)\,dx = \lim_{n \to \infty} \sum_{i=1}^{n} f(x_i)\,\Delta x_i,$$

where the Greek letter Sigma $\sum$ denotes the sum of all the different areas $f(x_i)\,\Delta x_i$ from $i = 1$ to $i = n$.

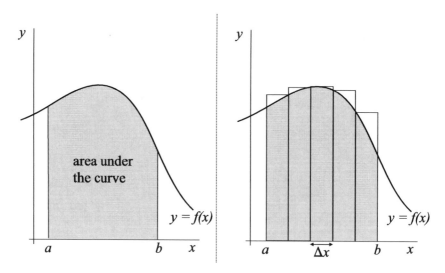

Figure 1.17: Area under the curve.

Problem 1.13. An object with initial velocity u and initial displacement x_0 moves with uniform linear acceleration a. After time t its velocity is $v(t)$ and it has moved a distance $x(t)$. Show that after time $t = T$, (a) the object's velocity is given by $v(T) = u + aT$, and (b) the object's displacement is given by $x(T) = x_0 + uT + \frac{1}{2}aT^2$.

(a) Acceleration equals rate of change of velocity with respect to time, ie

$$a = \frac{dv}{dt}.$$

Using the fundamental theorem of calculus (1.10.12)

$$\int_a^b f(x)\,dx = F(b) - F(a),$$

we can write

$$\int_{t=0}^{t=T} a\,dt = v(T) - v(0) = v(T) - u,$$

and using the square bracket notation

$$[at]_0^T = v(T) - u$$

gives

$$aT = v(T) - u$$

and

$$v(T) = u + aT, \tag{1.10.13}$$

which is often simply written as

$$v = u + at.$$

(b) Velocity equals rate of change of distance with respect to time, ie

$$v(t) = \frac{dx}{dt},$$

which we can rewrite using the fundamental theorem of calculus (1.10.12)

$$\int_{t=0}^{t=T} v(t)\,dt = x(T) - x(0) = x(T) - x_0$$

into which we can substitute 1.10.13 (letting $t = T$) to give

$$\int_{t=0}^{t=T} (u + at)\,dt = x(T) - x_0$$

$$\left[ut + \frac{1}{2}at^2\right]_0^T = x(T) - x_0$$

$$uT + \frac{1}{2}aT^2 = x(T) - x_0$$

giving

$$x(T) = x_0 + uT + \frac{1}{2}aT^2. \tag{1.10.14}$$

1.10.2.3 Equations of uniform motion

Incidentally, (1.10.13) and (1.10.14) are two of a well known set of equations that go by various names including the constant acceleration formulae, the equations of uniform motion, the SUVAT equations (after the variables s, u, v, a, t), or the kinematic equations. They are often written in a slightly more simplified form than what we've used, using total displacement $s = x(T) - x_0$, giving

$$v = u + at, \tag{1.10.15}$$

$$s = ut + \frac{at^2}{2}.$$

Plus there's a couple of others:

$$v^2 - u^2 = 2as$$

and

$$s = \frac{1}{2}(u + v)\,t.$$

1.10.2.4 Integration by substitution

Sometimes an algebraic substitution can transform the given integral into an easier one. This rule is useful if the integrand can be expressed as a multiple of another function u and its derivative $\frac{du}{dx}$, multiplied by a constant multiple of u.

$$\int f(x)\,dx = \int u\left(\frac{du}{dx}\right)\,dx. \qquad (1.10.16)$$

Problem 1.14. Find $\int \frac{1}{(2x+3)^2}\,dx$.

Substitute $u = 2x + 3 \Rightarrow du = 2dx$.

We can now express $\frac{1}{(2x+3)^2}\,dx$ as $\frac{1}{2u^2}\,du$.

The constant multiple is $\frac{1}{2}$, giving

$\frac{1}{2}\int \frac{1}{u^2}\,du$.

The integral of $\frac{1}{u^2}$ is $-\frac{1}{u}$, so

$\frac{1}{2}\int \frac{1}{u^2}\,du = -\frac{1}{2u} + C$.

Substitute back for $u = 2x + 3$ gives

$$\int \frac{1}{(2x+3)^2}\,dx = -\frac{1}{2(2x+3)} + C = -\frac{1}{4x+6} + C.$$

1.10.2.5 Integration by parts

This rule is used when integrating the product of two functions

$$\int v\frac{du}{dx}\,dx = uv - \int u\frac{dv}{dx}\,dx. \qquad (1.10.17)$$

Problem 1.15. Find $\int x\cos x\,dx$ (a) using integration by parts, and (b) using an online calculus calculator.

(a) Let $v = x$ and $\frac{du}{dx} = \cos x$.

Therefore, $\frac{dv}{dx} = 1$ and $u = \sin x$

giving

$$\int x\cos x\,dx = x\sin x - \int \sin x \times 1\,dx = x\sin x + \cos x + C.$$

(b) Using the WolframAlpha Calculus and Analysis Calculator [33], type 'integrate x cos x dx' (omit the quotes) into the input box to obtain

$$\int x\cos x\,dx = x\sin x + \cos x + C.$$

1.10.3 Taylor series

A **Taylor series** is a very useful tool that provides a way of expressing a suitable function as a power series calculated from the function's derivatives. In other words, if we have a difficult function, we can approximate it in a more user-friendly form as a series, with as many terms as we care to include.

Taylor series can be calculated by hand or, more easily, by using a suitable online calculus calculator ([33], for example).

A Taylor series of a suitable function $f(x)$ is centred on a particular point a and is given by

$$f(x) = f(a) + \frac{f'(a)}{1!}(x-a) + \frac{f''(a)}{2!}(x-a)^2 + \frac{f'''(a)}{3!}(x-a)^3 \ldots$$

where the primes $'$ indicate first, second, third, etc derivatives and the $n!$ notation denotes the **factorial** of n, ie the product of all positive integers less than or equal to n. So, for example, $3! = 3 \times 2 \times 1 = 6$.

The most straightforward Taylor series are those centred on zero (ie $a = 0$) and are known as **Maclaurin series**. If you've lost your calculator, for example, you can calculate $\sin x$ to whatever degree of accuracy you like by using the Maclaurin series

$$\sin x = x - \frac{x^3}{3!} + \frac{x^5}{5!} - \frac{x^7}{7!} \ldots$$

The three dots symbol $\ldots$ (called an ellipsis) shows that the pattern continues, in this case to infinity. Another common example is the Maclaurin series for the exponential function e^x

$$e^x = 1 + \frac{x}{1!} + \frac{x^2}{2!} + \frac{x^3}{3!} \ldots$$

Or, final example, the function $f(x) = \sqrt{1+x}$, which can be expressed as

$$\sqrt{1+x} = 1 + \frac{x}{2} - \frac{x^2}{8} + \frac{x^3}{16} - \frac{5x^4}{128} \ldots$$

1.11 Parametric equations

It is often useful to define equations in terms of another variable, called a parameter. For example, the function $y = x^2$ can be expressed in terms of a parameter t by letting $x = t$ and $y = t^2$. We could then find the values of x and y for certain values of t. For example, if $0 \leq t \leq 4$, then x and y would be given as

t	0	1	2	3	4
x	0	1	2	3	4
y	0	1	4	9	16

As another example, the equation of a unit circle $x^2 + y^2 = 1$ can be expressed using the parameter θ as $x = \cos\theta$ and $y = \sin\theta$.

We saw earlier in Section 1.5.1 that if we graph a function of two variables (we used the example of $z = x^2 + y^2$) we obtain a surface in three-dimensional space. If we graph parametric equations of x, y, z using just one parameter (t for example) we get a *curve*, not a surface. You could think of this curve as a length of sinuous wire bending through space, or the path of a moving object. Some online function graphers allow you to input parametric equations and it's instructive to play around with these plotters generating different curves.

Figure 1.18 shows a helix defined by the parametric equations $x = 3\cos(3t)$, $y = 3\sin(3t)$ and $z = t$.

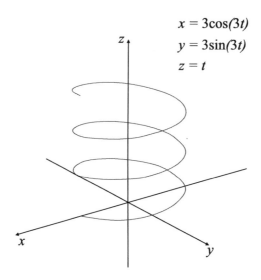

Figure 1.18: Parametric curve.

It's important to understand the concept of parameterised curves as we'll be meeting them later when looking at things such as four-velocity, contravariant vectors and geodesics through spacetime.

Parametric equations are especially useful when describing the motion of a particle or object where we let the parameter t equal time. If we do this, using Cartesian coordinates for example, and can express the position of the object in terms of three functions ($x = f(t), y = f(t), z = f(t)$), we can plug in a value of t and find the object's position at that time.

By differentiating the functions $x = f(t), y = f(t), z = f(t)$ with respect to time we can also find the velocity of the object in the three x, y, z directions.

Problem 1.16. A ball is thrown with a horizontal velocity of $5\,\mathrm{m\,s^{-1}}$ off the top of a $50\,\mathrm{m}$ tall vertical cliff. Assuming that the acceleration due to gravity is $10\,\mathrm{m\,s^{-2}}$ – (a) how long will the ball take to hit the ground? (b) What will be the ball's position after 2 seconds, expressed in Cartesian (x, y) coordinates, with the origin at the base of the cliff? (c) Sketch a curve showing the ball's path through space. (d) Find the ball's velocity in the x and y directions after 3 seconds? Ignore air resistance.

To solve this problem, we need Newton's first law of motion – 'an object will remain at rest or in uniform motion in a straight line unless acted upon by an external force' (Section 2.4.1), which tells us the ball's velocity in the horizontal direction is constant (ignoring air resistance). This might appear counter-intuitive because the ball is also accelerating vertically downwards due to gravity. Nonetheless, if we could separate out the ball's horizontal and vertical motion, the former would be a constant velocity of $5\,\mathrm{m\,s^{-1}}$, the latter would be a constant acceleration of $10\,\mathrm{m\,s^{-2}}$.

We start by using the distance equation of uniform motion (1.10.15) given in Section 1.10.2.3

$$s = ut + \frac{at^2}{2}.$$

Where s equals distance, u equals initial velocity, a equals acceleration and t equals time. We need to change this equation into a parametric equation using Cartesian (x, y) coordinates. To do this, first consider the motion of the ball in the y direction.

At the moment the ball is thrown $t = 0$; the initial velocity in the y direction is zero; the acceleration $a = -10$ (because the acceleration due to gravity is in the opposite direction to increasing y), and the distance $s = 0 = y - 50$ (because when distance $s = 0, y = 50$), so

$$y - 50 = 0t - \frac{10t^2}{2}$$

$$y = 50 - 5t^2. \tag{1.11.1}$$

In the x direction, the initial velocity is $5\,\mathrm{m/s}$, the only acceleration is downwards in the y direction, so

$$x = 5t + \frac{0t^2}{2} = 5t. \tag{1.11.2}$$

(a) The ball will hit the ground when $y = 0$, from (1.11.1)

$$0 = 50 - 5t^2$$

$$t = \sqrt{\frac{50}{5}} = \sqrt{10} = 3.16,$$

ie the ball will hit the ground after $3.16\,\mathrm{s}$.

(b) After $t = 2$ seconds, from (1.11.1)

$$y = 50 - \left(5 \times 2^2\right)$$

$$y = 50 - 20 = 30.$$

And from (1.11.2)

$$x = 5 \times 2 = 10.$$

(c) We can calculate the movement of the ball every $0.5t$ as follows

t	0	0.5	1.0	1.5	2.0	2.5	3.0	3.16
x	0	2.5	5.0	7.5	10.0	12.5	15.0	15.8
y	50	48.7	45.0	38.7	30.0	18.7	5.0	0

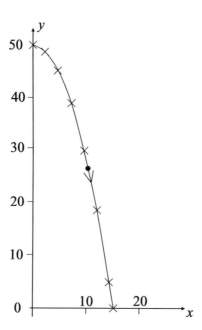

Figure 1.19: Path of the ball over the cliff.

The plot of these points, showing the path of the ball over the cliff, is given in Figure 1.19.

(d) We differentiate 1.11.1 with respect to time to find the ball's velocity in the y direction

$$\frac{dy}{dt} = -10t.$$

Plugging in $t = 3$ we get $\frac{dy}{dt} = -10 \times 3 = -30\,\mathrm{m\,s^{-1}}$.

And we differentiate 1.11.2 with respect to time to find the ball's velocity in the x direction

$$\frac{dx}{dt} = 5,$$

so, in the x direction, the velocity is a constant $\frac{dx}{dt} = 5\,\mathrm{m\,s^{-1}}$.

Note that just as we had a minus sign in front of the acceleration ($a = -10$) we also have a minus sign in front of the velocity $\left(\frac{dy}{dt} = -10t\right)$ in the y direction. The minus sign is there for the same reason in both cases – we have the origin of our Cartesian coordinate system at the base of the cliff, and therefore y is getting smaller as acceleration and velocity increase. For an observer on the ground, the velocity of the ball in the y direction would obviously be increasing as it falls and, at $t = 3$, would be a straightforward $30\,\mathrm{m\,s^{-1}}$.

1.12 Matrices

A **matrix** (plural – matrices) is a means of organising data in columns and rows. Formally, we say that matrices are rectangular arrays of symbols or numbers enclosed in parentheses. An $m \times n$

matrix has m rows and n columns. This, for example, is a 3×4 matrix

$$\begin{pmatrix} 1 & 5 & 2 & 7 \\ 0 & 3 & 6 & 1 \\ 12 & 4 & 21 & -1 \end{pmatrix}.$$

This is a 4×1 matrix

$$\begin{pmatrix} a \\ d \\ c \\ a \end{pmatrix},$$

this is a 3×3 matrix

$$\begin{pmatrix} 1 & 3 & 12 \\ 1 & 5 & 8 \\ 7 & 21 & 0 \end{pmatrix},$$

and this is a 1×3 matrix

$$\begin{pmatrix} \text{apples} & \text{pears} & \text{oranges} \end{pmatrix}.$$

If $m = n$ then the matrix is square. The entries in a matrix are called **elements** and labelled according to their row and column number. So, in the matrix

$$\begin{pmatrix} 1 & 3 & 12 \\ 1 & 5 & 8 \\ 7 & 21 & 0 \end{pmatrix}$$

element $a_{13} = 12$ and $a_{33} = 0$.

1.12.1 Matrix multiplication

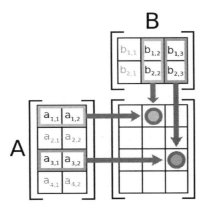

Figure 1.20: Matrix multiplication.

We can multiply two matrices together to give another matrix providing the matrix on the left has as many columns as the matrix on the right has rows.

Matrix multiplication is not usually **commutative**, meaning the order in which the operation is carried out is important. Here's an example of how to multiply two matrices, A and B, where

$$A = \begin{pmatrix} a & b \\ c & d \end{pmatrix}$$

and

$$B = \begin{pmatrix} e & f \\ g & h \end{pmatrix}$$

then

$$AB = \begin{pmatrix} a & b \\ c & d \end{pmatrix} \begin{pmatrix} e & f \\ g & h \end{pmatrix} = \begin{pmatrix} ae + bg & af + bh \\ ce + dg & cf + dh \end{pmatrix}.$$

The rule is to multiply each row in the first matrix by each column in the second matrix. If we multiply the first row by the second column, the result goes in the intersection of the first row and second column of the answer matrix. In general, if we multiply the ith row by the jth column, the answer goes in the intersection of the ith row and the jth column of the answer matrix (see Figure 1.20).

Problem 1.17. Find the product AB of the following matrices:

$$A = \begin{pmatrix} 1 & 2 & 3 \\ 4 & 5 & 6 \end{pmatrix}$$

$$B = \begin{pmatrix} 1 & 2 \\ 3 & 4 \\ 5 & 6 \end{pmatrix}.$$

$$AB = \begin{pmatrix} 1 & 2 & 3 \\ 4 & 5 & 6 \end{pmatrix} \begin{pmatrix} 1 & 2 \\ 3 & 4 \\ 5 & 6 \end{pmatrix}$$

$$= \begin{pmatrix} (1 \times 1) + (2 \times 3) + (3 \times 5) & (1 \times 2) + (2 \times 4) + (3 \times 6) \\ (4 \times 1) + (5 \times 3) + (6 \times 5) & (4 \times 2) + (5 \times 4) + (6 \times 6) \end{pmatrix}$$

$$= \begin{pmatrix} 22 & 28 \\ 49 & 64 \end{pmatrix}.$$

1.12.2 Identity matrix

The identity matrix I_n is an $n \times n$ (ie it's square) matrix where each element of the main diagonal equals 1 and all other elements $= 0$, for example

$$\begin{pmatrix} 1 & 0 & 0 \\ 0 & 1 & 0 \\ 0 & 0 & 1 \end{pmatrix}.$$

The identity matrix can be thought of as being 'equivalent' to 1 because for any $n \times m$ matrix A, $I_n A = A$ and $A I_m = A$.

We can also define the identity matrix using a strange looking function called the **Kronecker delta** δ_{ij}, which has only two values, 1 and 0, such that

$$\delta_{ij} = \left\{ \begin{array}{l} 1 \ \text{if} \ i = j \\ 0 \ \text{if} \ i \neq j \end{array} \right. . \tag{1.12.1}$$

We can then say

$$I_{ij} = \delta_{ij}.$$

So if we have a 3×3 matrix I_{ij}, for example, when $i = j$ $\delta_{ij} = 1$ and $I_{ij} = 1$ (ie $I_{11} = I_{22} = I_{33} = 1$), but when $i \neq j$ $\delta_{ij} = 0$ and $I_{ij} = 0$, which is the same as saying

$$I_3 = \left(\begin{array}{ccc} 1 & 0 & 0 \\ 0 & 1 & 0 \\ 0 & 0 & 1 \end{array} \right).$$

1.12.3 Inverse matrix

To have an inverse, a matrix must be square. The inverse of a matrix A is a matrix X such that

$$AX = I, \tag{1.12.2}$$

where I is the identity matrix.

1.12.4 Symmetric matrices

A symmetric matrix must be square and can informally be defined as a matrix that is symmetrical about the main diagonal. For example, these are both symmetrical matrices:

$$\left(\begin{array}{cc} a & b \\ b & a \end{array} \right)$$

$$\left(\begin{array}{ccc} a & b & c \\ b & d & e \\ c & e & f \end{array} \right).$$

1.12.5 Diagonal matrices

The most important matrices we come across in this book are those used to represent objects called metric tensors (or metrics), functions that define the distance between two points in a particular space. The metric tensors we'll be encountering are all examples of diagonal matrices, where the off-diagonal elements equal zero. For example, the Minkowski metric (3.5.2) used in special relativity can be represented as

$$[\eta_{\mu\nu}] = \left(\begin{array}{cccc} 1 & 0 & 0 & 0 \\ 0 & -1 & 0 & 0 \\ 0 & 0 & -1 & 0 \\ 0 & 0 & 0 & -1 \end{array} \right).$$

Note that the symbol for the metric $\eta_{\mu\nu}$ is enclosed in square brackets [], telling us we are referring to the entire metric tensor matrix and not to any particular metric tensor component such as $\eta_{11} = 1, \eta_{22} = -1, \eta_{13} = 0$, etc.

1.13 Introducing vectors

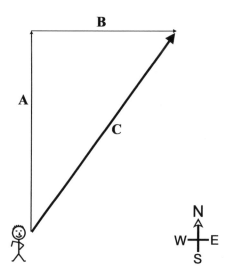

Figure 1.21: Addition of vectors.

The formal definition of a vector is something that is an element of a **vector space**. Basically, if you have a group of things which you can add and subtract, or multiply by an ordinary number and still end up with another thing of the same type, then those things are vectors.

Simple vectors can be used to represent physical quantities with magnitude and direction, such as velocity or force. We can physically draw these vectors as **directed line segments** – a line with an arrow at one end pointing in the direction of the vector's direction. A wind map, for example, uses lots of little arrows, each representing the strength and direction of the wind at various points. Figure 1.21 shows the travels of a little person walking at a constant speed. He first walks 4 km due north, then 3 km east. This motion is represented by vector **A** (which is 4 units long), and vector **B** (which is 3 units long). We add **A** to **B** by joining the head of **A** to the tail of **B** to form vector **C** and we can say

$$\mathbf{A} + \mathbf{B} = \mathbf{C}.$$

If we measured the length of **C** (with a ruler, or we could use Pythagoras' theorem – 1.14.1) we would find it to be 5 units long, so our traveller would have ended up at the same point if he had walked 5 km in the direction shown by vector **C**. The length of a vector is known as its **magnitude** (denoted by the symbol $||$), so we say $|\mathbf{A}| = 4$, $|\mathbf{B}| = 3$ and $|\mathbf{C}| = 5$.

Note the notation: in this book we use a bold typeface to show ordinary two or three-dimensional vectors ($\mathbf{A}$, $\mathbf{F}$, $\mathbf{V}$) and show four-dimensional vectors, which we'll meet later, with a little arrow on top of the letter ($\vec{A}$, $\vec{F}$, $\vec{V}$).

In contrast to **vector quantities**, such as velocity, there are also things called **scalar quantities**, which have magnitude but no direction. Speed is an example of a scalar quantity, referring only to how fast an object is moving. Velocity, as we've seen, is a vector quantity that refers not only to how fast something is moving but in which direction it is moving as well. If we say a car is travelling at 70 mph, we are describing its speed. If we say a car is travelling at 70 mph in a northerly direction, we are describing its velocity. Temperature and mass are other examples of scalar quantities – neither has a meaningful sense of direction.

A vector consists of the product of its components and objects known as **basis vectors**, which define the components' direction. In two dimensions, we need two basis vectors, in three dimension, we need three etc. If you think of a vector as an arrow just hanging there in space, we can describe its position using any coordinate system we like – Cartesian, plane polar, spherical, etc. For now, we'll use nice, simple Cartesian coordinates, which are the simplest to work with because the basis vectors are constant.

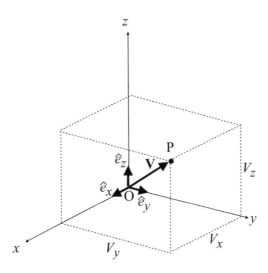

Figure 1.22: Components and basis vectors in Cartesian coordinates.

In Cartesian coordinates (see Figure 1.22), a three-dimensional vector $\mathbf{V}$ consists of the product of its components (V_x, V_y, V_z) and a set of basis vectors $(\hat{e}_x, \hat{e}_y, \hat{e}_z)$ – the little hat means each basis vector is one unit long – pointing along the x, y, z axes respectively (another way of saying this is that the basis vectors are tangent to the coordinate axes), and we can write

$$\mathbf{V} = V_x\hat{e}_x + V_y\hat{e}_y + V_z\hat{e}_z. \tag{1.13.1}$$

Note that the x, y, z subscripts allow us to identify the three different components (V_x, V_y, V_z). With Cartesian vectors, the convention is to use subscript indices for both the components and basis vectors.

The magnitude of $\mathbf{V}$ (the distance OP) is given (using Pythagoras' theorem – 1.14.1) by

$$|\mathbf{V}| = \sqrt{(V_x)^2 + (V_y)^2 + (V_z)^2}. \tag{1.13.2}$$

Problem 1.18. Find the magnitude of vector $\mathbf{V} = 10\hat{e}_x + 3\hat{e}_y + 1\hat{e}_z$.

Using (1.13.2) we say

$$|\mathbf{V}| = \sqrt{(V_x)^2 + (V_y)^2 + (V_z)^2}$$

$$= \sqrt{10^2 + 3^2 + 1^2}$$

$$= \sqrt{100 + 9 + 1}$$

$$|\mathbf{V}| = 10.49.$$

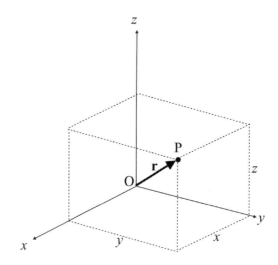

Figure 1.23: Position vector $\mathbf{r}$ of point P.

A particularly useful type of vector is the **position vector** or **radius vector r** from the origin O to the point (x, y, z), which is used to define the position of a point or particle (see Figure 1.23). The position vector has magnitude

$$r = |\mathbf{r}| = \sqrt{x^2 + y^2 + z^2}. \tag{1.13.3}$$

Sometimes we use a unit radial vector $\hat{\mathbf{r}}$, which is simply a radius vector of length 1.

There are several ways of multiplying vectors. The two commonest are the **scalar product** (or **dot product**) and the cross product. We are only interested in the scalar product which, given the Cartesian components of two vectors, is equal to

$$\mathbf{A} \cdot \mathbf{B} = A_x B_x + A_y B_y + A_z B_z \tag{1.13.4}$$

and is a scalar (an ordinary number, with no direction) not a vector.

Problem 1.19. Find the scalar product $\mathbf{A} \cdot \mathbf{B}$ where $\mathbf{A} = 5\hat{e}_x + 3\hat{e}_y - 1\hat{e}_z$ and $\mathbf{B} = -2\hat{e}_x + 7\hat{e}_y - 4\hat{e}_z$.

Using (1.13.4) we say
$$\mathbf{A} \cdot \mathbf{B} = A_x B_x + A_y B_y + A_z B_z$$
$$= (5 \times -2) + (3 \times 7) + (-1 \times -4)$$
$$\mathbf{A} \cdot \mathbf{B} = 15.$$

You will often see the same equation or law written in either vector or scalar form. For example, Newton's second law (2.4.1) can be written as a vector equation
$$\mathbf{F} = m\mathbf{a},$$
where $\mathbf{F}$ is the force vector and $\mathbf{a}$ is the acceleration vector. Because $\mathbf{F}$ and $\mathbf{a}$ are vectors, they can be broken down into their respective components, eg (F_x, F_y, F_z) for $\mathbf{F}$. Sometimes we aren't interested in the direction of a vector, just in its magnitude. We can then write the same law as a scalar equation (2.4.2)
$$F = ma,$$
where F and a are the magnitudes of the vectors $\mathbf{F}$ and $\mathbf{a}$.

1.13.1 Vector fields

Imagine a room filled with moving air. The air might move faster near an open window or above a hot radiator and slower in the corners or behind a bookcase. Say we could measure the velocity of the air at every point in the room and we could express that velocity in terms of the x, y, z Cartesian coordinates of the room. The velocity $\mathbf{V}$ is a vector quantity (it has magnitude and direction) and we can write
$$\mathbf{V} = \mathbf{V}(x, y, z) = (V_x)\hat{e}_x + (V_y)\hat{e}_y + (V_z)\hat{e}_z, \tag{1.13.5}$$
meaning $\mathbf{V}$ is a function of x, y, z. Specifically, the components V_x, V_y, V_z are each functions of x, y, z.

We have described a **vector field**, where every point in the room has a vector associated with it that tells us the speed and direction of the air at that point.

Let's just dream up a function for our vector field, ie
$$\mathbf{V} = (xy)\hat{e}_x + (2y + x + 3z)\hat{e}_y + (2y)\hat{e}_z,$$
so at the position $x = 1, y = 2, z = 3$
$$\mathbf{V} = (1 \times 2)\hat{e}_x + (2 \times 2 + 1 + 3 \times 3)\hat{e}_y + (2 \times 2)\hat{e}_z$$
$$= 2\hat{e}_x + 14\hat{e}_y + 4\hat{e}_z.$$
Similarly, at the position $x = 0, y = 0, z = 0$
$$\mathbf{V} = 0\hat{e}_x + 0\hat{e}_y + 0\hat{e}_z,$$
meaning the velocity of the air at $x = 0, y = 0, z = 0$ equals zero.

An example of a vector field is shown in Figure 1.24. The little arrows indicate the direction and magnitude of vectors at a selection of points.

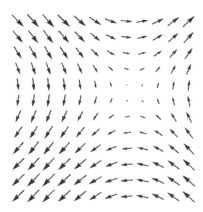

Figure 1.24: An example of a vector field.

1.13.2 Gradient of a scalar field

The nabla symbol ∇ is known as the **differential operator**

$$\nabla = \left(\frac{\partial}{\partial x}, \frac{\partial}{\partial y}, \frac{\partial}{\partial z} \right)$$

and only has meaning when it acts on something (hence the empty slots after the above ∂'s). The differential operator is used in three important operations known as **divergence** (div), **gradient** (grad) and curl. We aren't interested in curl but will be looking at the gradient in this section and divergence in the next. (Nabla is from the Greek word for a Phoenician harp, which nineteenth century mathematicians thought the inverted Delta symbol ∇ resembled.)

Starting with a **scalar field** ϕ, we can find an important vector field known as the gradient or grad, denoted by $\nabla \phi$. For example, assume that the air temperature T varies from point to point inside a room. The temperature may be higher near a radiator for example, or colder near an open window. Say we describe the position of any point in the room using three Cartesian coordinates x, y, z using units of metres (m). Now assume that we've cleverly worked out an equation that gives us the temperature for any point x, y, z, giving us a function $T(x, y, z)$. T is just a number (unlike a vector, it has no direction), is an example of a scalar quantity, and we have described a scalar field. A **scalar function**, not surprisingly, defines a scalar field, though we won't worry too much about the distinction and will tend to use the terms interchangeably. The Greek letter phi ϕ is often used to indicate both a scalar field and a scalar function.

In ordinary Euclidean space, for a point in a scalar field ϕ, the gradient is a vector that points in the direction of greatest increase of ϕ. Using Cartesian coordinates, the gradient is given by

$$\nabla \phi = \frac{\partial \phi}{\partial x} \hat{e}_x + \frac{\partial \phi}{\partial y} \hat{e}_y + \frac{\partial \phi}{\partial z} \hat{e}_z \tag{1.13.6}$$

or, more succinctly, (the Cartesian basis vectors are implied, so we don't bother to write them)

$$\nabla \phi = \left(\frac{\partial \phi}{\partial x}, \frac{\partial \phi}{\partial y}, \frac{\partial \phi}{\partial z} \right). \tag{1.13.7}$$

In other words, the gradient of scalar field is a vector field. The magnitude of the gradient tells us how fast ϕ is changing in that direction.

Problem 1.20. Returning to our room, the temperature T in degrees Celsius ($^\circ$C) is given by the equation $T = x^2 + 3y - zx$. What is the gradient at the point $x = 10, y = 3, z = 5$.

From (1.13.6) (changing ϕ to T)

$$\nabla T = \frac{\partial T}{\partial x}\hat{e}_x + \frac{\partial T}{\partial y}\hat{e}_y + \frac{\partial T}{\partial z}\hat{e}_z.$$

Taking partial derivatives gives $\frac{\partial T}{\partial x} = 2x - z, \frac{\partial T}{\partial y} = 3, \frac{\partial T}{\partial z} = -x$,

so

$$\nabla T = (2x - z)\,\hat{e}_x + 3\hat{e}_y - (x)\,\hat{e}_z.$$

For $x = 10, y = 3, z = 5$

$$\nabla T = (15, 3, -10).$$

The magnitude of the gradient is (using 1.13.2)

$$|\nabla T| = \sqrt{(15)^2 + (3)^2 + (-10)^2} = \sqrt{225 + 9 + 100} = 18.3.$$

The physical interpretation of this answer is that, at the point $x = 10, y = 3, z = 5$, the vector that points in the direction of greatest increase of temperature has components $(15, 3, -10)$ and a magnitude of 18.3. If we are using metre units for our coordinate axes we can therefore say that the temperature at this point is changing at $18.3^\circ\,\text{C}\,\text{m}^{-1}$.

1.13.3 Divergence of a vector field

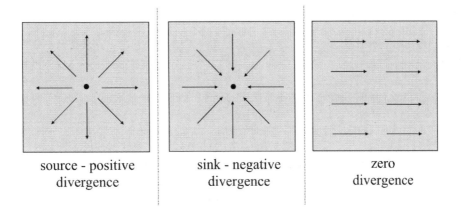

source - positive   sink - negative   zero
divergence   divergence   divergence

Figure 1.25: Divergence of a vector field.

We can imagine a vector field as representing the flow of some fluid. The divergence of that vector field (see Figure 1.25) is a measure of the net flow of fluid at a particular point. If more fluid is entering than leaving, the point is a source (think of a running tap as a source of water) and the divergence is positive. If more fluid is leaving than entering, the point is a sink (think of water flowing down a drain) and the divergence is negative. If the same amount of fluid is entering as leaving there is zero divergence.

For a Cartesian vector field

$$\mathbf{V} = (V_x)\,\hat{e}_x + (V_y)\,\hat{e}_y + (V_z)\,\hat{e}_z,$$

the divergence is given by

$$\nabla \cdot \mathbf{V} = \frac{\partial V_x}{\partial x} + \frac{\partial V_y}{\partial y} + \frac{\partial V_z}{\partial z}. \tag{1.13.8}$$

Problem 1.21. If $\mathbf{V} = \left(x^2 y\right)\hat{e}_x + (2y + x + 3z)\,\hat{e}_y - \left(2y + z^3\right)\hat{e}_z$, find $\nabla \cdot \mathbf{V}$ (or div $\mathbf{V}$) at the point $(1, 3, 2)$.

$$\nabla \cdot \mathbf{V} = \frac{\partial V_x}{\partial x} + \frac{\partial V_y}{\partial y} + \frac{\partial V_z}{\partial z}$$

$$= \frac{\partial\left(x^2 y\right)}{\partial x} + \frac{\partial\left(2y + x + 3z\right)}{\partial y} + \frac{\partial\left(-\left(2y + z^3\right)\right)}{\partial z}$$

$$= 2xy + 2 - 3z^2.$$

For the point $(1, 3, 2)$

$$\nabla \cdot \mathbf{V} = (2 \times 1 \times 3) + (2) - 3 \times (2)^2 = 8 - 12$$

$= -4$, ie the point is a sink.

If $\mathbf{V}$ is the gradient of a scalar field ϕ, ie $\mathbf{V} = \nabla\phi$, then

$$\nabla \cdot \mathbf{V} = \nabla \cdot \nabla\phi = \nabla^2 \phi,$$

where

$$\nabla^2 \phi = \nabla \cdot \nabla\phi = \frac{\partial^2 \phi}{\partial x^2} + \frac{\partial^2 \phi}{\partial y^2} + \frac{\partial^2 \phi}{\partial z^2}, \tag{1.13.9}$$

the $\nabla \cdot \nabla$ and ∇^2 symbols denoting a widely used operator known as the **Laplacian** or **Laplace operator**, which tells us the divergence of the gradient of a scalar field.

1.14 Euclidean geometry

Figure 1.26: Euclid in Raphael's School of Athens.

Euclid (Figure 1.26) was an outstanding Greek mathematician who lived and taught at Alexandria about 300 BC. Euclidean geometry is the familiar geometry we are taught in school when learning about triangles, lines, points and angles, etc. Described at length in his thirteen book treatise *The Elements*, we usually take Euclid's assumptions for granted when doing geometrical calculations in our everyday world.

Some of the key results from Euclidean geometry include:

- The internal angles of a triangle add up to 180°.

- Two straight lines that are initially parallel remain parallel when extended (known as Euclid's fifth postulate or the parallel postulate).

- The area of a circle is given by πr^2, where r is the radius.

- Pythagoras' theorem – the length of the hypotenuse (the longest side – see Figure 1.2) of a right angled triangle is given by

$$a^2 + b^2 = c^2, \tag{1.14.1}$$

  where c is the hypotenuse and a and b are the other two sides.

We can define the type of space where Euclidean geometry works as **Euclidean space**, which may be two-dimensional (a flat tabletop for example), three-dimensional (a working approximation to the 'ordinary' space we live in), or even of higher dimensions. This may seem a statement of the obvious, but as we'll see, Euclidean space is just one of many possible spaces. Curved surfaces, for example, are not Euclidean. On the surface of a sphere, lines of longitude start off parallel on the equator but meet at the poles. Also, the internal angles of a triangle on a sphere can all equal 90°.

In relativity we deal with two types of four-dimensional non-Euclidean space:

- the flat spacetime of special relativity (where Euclid's parallel postulate holds), and

- the curved spacetime of general relativity.

Returning to Euclidean space, Pythagoras' theorem can also be used in three-dimensions to calculate the corner to opposite corner length of a rectangular cuboid (a cardboard box, for example). So, in Figure 1.27 the length of diagonal l is given by

$$l^2 = a^2 + b^2 + c^2 \tag{1.14.2}$$

$$l^2 = 2^2 + 1^2 + 1^2$$

$$l^2 = 6$$

$$l = \sqrt{6} = 2.45.$$

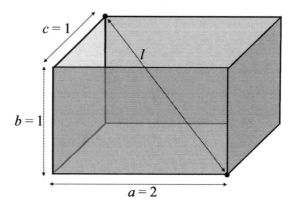

Figure 1.27: A rectangular cuboid, eg a cardboard box.

Using Cartesian coordinates in three dimensions allows us to write (1.14.2) to give the distance l between any two points x_1, y_1, z_1 and x_2, y_2, z_2 as

$$l^2 = (x_2 - x_1)^2 + (y_2 - y_1)^2 + (z_2 - z_1)^2 \tag{1.14.3}$$

or, alternatively, in terms of the intervals $\Delta l, \Delta x, \Delta y, \Delta z$

$$\Delta l^2 = \Delta x^2 + \Delta y^2 + \Delta z^2, \tag{1.14.4}$$

which is the Euclidean distance function. (Note, we are using the convention that $\Delta x^2 = (\Delta x)^2$.)

1.14.1 The Euclidean metric

Equation (1.14.4), by giving the distance between two points in a space, actually defines that space to be Euclidean. If (1.14.4) applies for any interval, then we must be working in a Euclidean space.

This is our introduction to a concept central to the mathematics of curved spaces – the metric.

The metric is a function that defines the distance between two points in a particular space. Once we know the metric of a space, we (theoretically, at least) know everything about the geometry of that space – that's why the metric is of fundamental importance.

Because of the nature of Euclidean space, the distance function (1.14.3) is constant, it doesn't change no matter where you are in that space. That means you can plug in any values of the separations $(x_2 - x_1)$, $(y_2 - y_1)$ and $(z_2 - z_1)$ and end up with the right answer, the distance between the points x_1, y_1, z_1 and x_2, y_2, z_2. That isn't the same for other spaces, where the distance function isn't constant. To get over this problem we express distances in terms of infinitesimally small displacements, dl, dx, dy, etc., which we met when looking at differential calculus. These tiny displacements are called the **coordinate differentials**, and when we use them to describe the distance between two infinitesimally close points in a space we call the distance function a **line element**. The Euclidean line element, in Cartesian coordinates, therefore becomes

$$dl^2 = dx^2 + dy^2 + dz^2, \tag{1.14.5}$$

which is saying exactly the same as (1.14.3) only in coordinate differential form.

Notice that there is $1 \times dx^2$ term, $1 \times dy^2$ term and $1 \times dz^2$ term in (1.14.5). These 1's are the metric coefficients. We can arrange them into a 3×3 matrix (3×3 because we are dealing with three-dimensional space) that is known as the **metric** or **metric tensor**

$$[g_{ij}] = \begin{pmatrix} 1 & 0 & 0 \\ 0 & 1 & 0 \\ 0 & 0 & 1 \end{pmatrix}. \tag{1.14.6}$$

As previously mentioned, we enclose g_{ij} in brackets [] to remind us we are referring specifically to the entire metric matrix and not to any particular metric component.

The matrix in (1.14.6) tells us how to multiply the differentials dx, dy, dz. We can see this by writing the matrix as a table

$$[g_{ij}] = \begin{array}{c|ccc} - & dx & dy & dz \\ \hline dx & 1 & 0 & 0 \\ dy & 0 & 1 & 0 \\ dz & 0 & 0 & 1 \end{array}$$

and seeing that $dx \times dx \times 1 = dx^2, dy \times dy \times 1 = dy^2, dz \times dz \times 1 = dz^2$ and all the other products of dx, dy, dz equal zero.

The indices i, j after the g identify, by reference to rows and columns, the metric coefficients. So $g_{11} = 1, g_{22} = 1, g_{33} = 1, g_{13} = 0$, etc.

We saw in Section 1.9 that we can use different coordinate systems to define points on a plane or in space. Although the metric will be different for each coordinate system, the distance between any two points will obviously not change as all we are doing is using different coordinate systems to describe the same distance.

Problem 1.22. Find the line element and metric for Euclidean two-dimensional space using polar coordinates.

Equations (1.9.1) tell us how the polar angle θ and radius r are defined in terms of Cartesian (x, y) coordinates

$$x = r \cos \theta,$$

$$y = r \sin \theta.$$

Using the rule for differentiating products (1.10.3) plus the total differential equation (1.10.10) we find

$$dx = \cos\theta dr - r\sin\theta d\theta,$$

$$dy = \sin\theta dr + r\cos\theta d\theta.$$

As there is no z coordinate, we use a simplified version of (1.14.5)

$$dl^2 = dx^2 + dy^2$$

$$= (\cos\theta dr - r\sin\theta d\theta)^2 + (\sin\theta dr + r\cos\theta d\theta)^2$$

$$= \cos^2\theta dr^2 + r^2\sin^2\theta d\theta^2 - 2\cos\theta dr \times r\sin\theta d\theta$$

$$+ \sin^2\theta dr^2 + r^2\cos^2\theta d\theta^2 + 2\cos\theta dr \times r\sin\theta d\theta$$

$$= \cos^2\theta dr^2 + r^2\sin^2\theta d\theta^2 + \sin^2\theta dr^2 + r^2\cos^2\theta d\theta^2$$

$$= \left(\cos^2\theta + \sin^2\theta\right) dr^2 + \left(\cos^2\theta + \sin^2\theta\right) r^2 d\theta^2. \tag{1.14.7}$$

We know from the Pythagorean trigonometric identity (1.7.1) that $\cos^2\theta + \sin^2\theta = 1$, so we can rewrite (1.14.7) as the line element for Euclidean two-dimensional space using polar coordinates

$$dl^2 = dr^2 + r^2 d\theta^2 \tag{1.14.8}$$

giving a metric tensor of

$$[g_{ij}] = \begin{pmatrix} 1 & 0 \\ 0 & r^2 \end{pmatrix}.$$

Problem 1.23. Find the line element, metric and inverse metric for Euclidean three-dimensional space using spherical coordinates.

(1.9.2) tells us how the spherical polar angles θ, ϕ and radius r are defined in terms of Cartesian coordinates (x, y, z) by

$$x = r\sin\theta\cos\phi,$$

$$y = r\sin\theta\sin\phi,$$

$$z = r\cos\theta.$$

Using the rule for differentiating products (1.10.3) plus the total differential equation (1.10.10) we find

$$dx = \sin\theta\cos\phi dr + r\cos\theta\cos\phi d\theta - r\sin\theta\sin\phi d\phi \tag{1.14.9}$$

$$dy = \sin\theta\sin\phi dr + r\cos\theta\sin\phi d\theta + r\sin\theta\cos\phi d\phi$$

$$dz = \cos\theta dr - r\sin\theta d\theta.$$

We use (1.14.5)

$$dl^2 = dx^2 + dy^2 + dz^2$$

and square the (1.14.9) terms and again use the Pythagorean trigonometric identity (1.7.1) to eventually find the line element for Euclidean three-dimensional space using spherical coordinates

$$dl^2 = dr^2 + r^2 d\theta^2 + r^2 \sin^2 \theta d\phi^2, \tag{1.14.10}$$

meaning the metric can be written as

$$[g_{ij}] = \begin{pmatrix} 1 & 0 & 0 \\ 0 & r^2 & 0 \\ 0 & 0 & r^2 \sin^2 \theta \end{pmatrix} \tag{1.14.11}$$

and the inverse metric is

$$[g^{ab}] = \begin{pmatrix} 1 & 0 & 0 \\ 0 & \frac{1}{r^2} & 0 \\ 0 & 0 & \frac{1}{r^2 \sin^2 \theta} \end{pmatrix}. \tag{1.14.12}$$

As previously mentioned, all the metric tensors we encounter in this book are in the form of nice, simple diagonal matrices (ie where the off-diagonal elements equal zero). Finding the inverse of such a metric simply involves replacing each of the diagonal matrix elements with its reciprocal, as shown above. Note that from (1.12.2) $AX = I$, multiplying (1.14.11) by (1.14.12) gives the identity matrix I_3

$$I_3 = \begin{pmatrix} 1 & 0 & 0 \\ 0 & 1 & 0 \\ 0 & 0 & 1 \end{pmatrix}.$$

1.15 Index notation

We've seen that when using Cartesian coordinates, a three-dimensional vector $\mathbf{V}$ consists of the product of its components (V_x, V_y, V_z) plus a set of basis vectors $\hat{e}_x, \hat{e}_y, \hat{e}_z$ each of which points along the direction of a coordinate axis, and can be written as (1.13.1)

$$\mathbf{V} = V_x \hat{e}_x + V_y \hat{e}_y + V_z \hat{e}_z.$$

The x, y, z subscripts are indices, allowing us to identify the three different components V_x, V_y, V_z. Similarly, the x, y, z subscripts let us identify the basis vectors $\hat{e}_x, \hat{e}_y, \hat{e}_z$. A more concise way of writing $\mathbf{V}$ would be

$$\mathbf{V} = V_i e_i,$$

where the index i can take values of $1, 2$ or 3. So the components of $\mathbf{V}$ would then be written as $V_x = V_1, V_y = V_2$ and $V_z = V_3$, and the basis vectors as $\hat{e}_x = \hat{e}_1, \hat{e}_y = \hat{e}_2$ and $\hat{e}_z = \hat{e}_3$. In other words,

$$\mathbf{V} = V_i e_i = V_1 \hat{e}_1 + V_2 \hat{e}_2 + V_3 \hat{e}_3 = V_x \hat{e}_x + V_y \hat{e}_y + V_z \hat{e}_z.$$

The expression $V_i e_i$ is an example of **index notation**, and we'll be using an awful lot of it throughout this book. The advantage of index notation is that it allows us to write complicated vector expressions and equations in a concise and convenient form. This will be especially useful when we come to four-dimensional vectors and vector-like objects in special and general relativity.

So far we've been using nice, simple Cartesian vectors with subscript indices for both the components and basis vectors. Shortly, life becomes more complicated as we need to introduce two more general types of vector called **contravariant vectors** and **covariant vectors** (also known as **one-forms**). We'll define these things later. For now, we note that the position of the index allows us to tell them apart: contravariant vector components have an upper index (also called a superscript or upstairs index, eg V^a), and covariant vectors/one-form components have a lower (also called a subscript or downstairs index, eg V_a). To remember the difference between contravariant and covariant, think 'co' for covariant, which rhymes with 'low' – where the index goes.

Just as with Cartesian vectors, both contravariant and covariant vectors consist of components and basis vectors. However, we often aren't explicitly concerned about labelling basis vectors. The indices then only refer to the components of the vector in the particular coordinate system we are using. For example, say we have a vector **A** in spherical polar coordinates (where position is expressed in terms of the three coordinates r, θ, ϕ) that has components $r = 4$, $\theta = 2$, $\phi = 3$. Generically, we could call this vector A^b, where the index b refers to any of the components r, θ, ϕ. We would then refer to the individual components of **A** as $A^r = 4$, $A^\theta = 2$, $A^\phi = 3$. More succinctly, we can also refer to the three components of **A** as A^1, A^2, A^3, where the indices $1, 2, 3$ refer to the coordinates r, θ, ϕ.

We'll shortly find that spacetime has four coordinates (ct, x, y, z). The convention is to use a Greek index $(\alpha, \beta, \mu, \text{etc})$ to refer to spacetime coordinates, the index taking a value of $0, 1, 2, 3$, where 0 refers to the time coordinate, and $1, 2, 3$ refer to the three spatial coordinates.

The **Einstein summation convention** means that when an index occurs twice in the same expression, the expression is implicitly **summed over** all possible values for that index. Returning to Cartesian vectors, for example, we've seen that the scalar product of the vectors **A** and **B** is given by (1.13.4)

$$\mathbf{A} \cdot \mathbf{B} = A_x B_x + A_y B_y + A_z B_z.$$

In index notation we would write

$$A_i B_i = A_1 B_1 + A_2 B_2 + A_3 B_3,$$

$A_i B_i$ being a lot easier to write than $A_x B_x + A_y B_y + A_z B_z$. The index i to which the Einstein summation convention applies is known as a **dummy** or **summation index**. Another example, of a dummy index (α in this case), using a four-dimensional contravariant vector A^α and covariant vector B_α, is

$$A^\alpha B_\alpha = A^0 B_0 + A^1 B_1 + A^2 B_2 + A^3 B_3.$$

Tensors, which we'll look at in more detail later, can have more than one index, eg $X_c{}^{ab}$, $A^x{}_y$, $T^\alpha{}_{\beta\phi}$, etc. Here's an example of two tensors X_{jk} and Y^k being summed over for $k = 1, 2, 3$:

$$X_{jk} Y^k = X_{j1} Y^1 + X_{j2} Y^2 + X_{j3} Y^3.$$

The index k occurs twice in the left-hand term, is summed over and is therefore a dummy index. Notice that the index j only occurs once in a any given term and is not summed over. This type of index is known as a **free index**. Another example of free indices occurs with the Kronecker delta (1.12.1), which has only two values, 1 and 0, such that

$$\delta_{ij} = \begin{cases} 1 & \text{if } i = j \\ 0 & \text{if } i \neq j \end{cases},$$

and which we earlier used to define the 3×3 identity matrix. However, from now on we will be using the Kronecker delta as a tensor, with an upper and lower index. For example, we can write the Kronecker delta as δ_ν^ν. The index ν now appears twice, can be summed over and is a dummy index. The Greek letter ν implies four-dimensional spacetime, and so $\nu = 0, 1, 2$ or 3 and we would write

$$\delta_\nu^\nu = \delta_0^0 + \delta_1^1 + \delta_2^2 + \delta_3^3 = 1 + 1 + 1 + 1 = 4.$$

We earlier met the gradient of a scalar field (1.13.7)

$$\nabla\phi = \left(\frac{\partial\phi}{\partial x}, \frac{\partial\phi}{\partial y}, \frac{\partial\phi}{\partial z} \right).$$

In index notation this can be written as

$$\nabla\phi = \frac{\partial\phi}{\partial x^i} e_i,$$

where the index i can take values of $1, 2$ or 3.

In a nutshell, here are the rules of index notation:

1. The convention is to use a Greek index (α, β, μ, etc) to refer to spacetime coordinates.

2. The index labels are not important. Subject to rules 3 and 4, indices can be renamed without loss of meaning. For example, V^α can be rewritten as V^β or V^γ etc.

3. Free indices only appear as either superscript or subscript, never as both, and they must occur exactly once in every term. So $X^a = Y^{ba} Z_b$ is OK (a is the free index). $X^a = Y^{bc} Z_b$ is wrong, as is $X^a = Y^{ba} Z_b^a$.

4. Dummy indices appear twice in a term, either as subscripts in Cartesian vectors/tensors, or once as superscript and once as subscript in the more general vectors/tensors we tend to use in this book. To avoid confusion dummy indices must never have the same label as free indices.

Problem 1.24. Which of these are valid expressions or equations expressed using the Einstein summation convention? Write out the correct equations assuming three-dimensions, with indices running from 1 to 3.

(a) $A^i B_i$

(b) $A^{ij} B_i$

(c) $A_i = B_{ji} C_j$

(d) $A_i B_j + A_j B_k + A_k B_i = 0$

(e) $A_{ii} = 6$

(f) $A_i + B_{ij} A_j + C_{kik} = 0$

(g) $A_i = B_{ij} C_j D_j$

(a) Valid:

$$A^i B_i = A^1 B_1 + A^2 B_2 + A^3 B_3.$$

(b) Valid:

$$A^{ij}B_i = A^{1j}B_1 + A^{2j}B_2 + A^{3j}B_3.$$

(c) Valid:

$$A_1 = B_{11}C_1 + B_{21}C_2 + B_{31}C_3,$$
$$A_2 = B_{12}C_1 + B_{22}C_2 + B_{32}C_3,$$
$$A_3 = B_{13}C_1 + B_{23}C_2 + B_{33}C_3.$$

(d) Invalid because the terms have different free indices. For example the first term has two free indices i and j, while the second term has free indices j and k.

(e) Valid:

$$A_{11} + A_{22} + A_{33} = 6.$$

(f) Valid:

$$A_1 + B_{11}A_1 + B_{12}A_2 + B_{13}A_3 + C_{111} + C_{212} + C_{313} = 0,$$
$$A_2 + B_{21}A_1 + B_{22}A_2 + B_{23}A_3 + C_{121} + C_{222} + C_{323} = 0,$$
$$A_3 + B_{31}A_1 + B_{32}A_2 + B_{33}A_3 + C_{131} + C_{232} + C_{333} = 0.$$

(g) Invalid because an index cannot appear more than twice in a term.

2 Newtonian mechanics

I seem to have been only like a boy playing on the sea-shore, and diverting myself in now and then finding a smoother pebble or a prettier shell than ordinary, whilst the great ocean of truth lay all undiscovered before me.

ISAAC NEWTON

2.1 Introduction

Figure 2.1: Isaac Newton (1642–1727) – Godfrey Kneller's 1689 portrait of Newton, age 46.

Before Isaac Newton (Figure 2.1) there was little in the way of coherent, evidence-based theory explaining the motion of objects on Earth and in the heavens. Apart from supernatural influences such as sorcery and spirits, the prevailing 'rational' explanation as to why things moved was that of Aristotle and involved the idea of objects moving because they sought their 'proper place' in the cosmos. Newton's genius was to invent a new science of **mechanics** – a precise description of how things move – based on the concept of **forces**, which push or pull objects according to mathematically formulated laws.

What is a force? Before answering that question we need to realise that the natural state of any object not subject to a force is either at rest or in uniform motion (ie moving at a constant speed in a constant direction). This is Galileo's **principle of inertia** and is also a statement of Newton's first law. It's true because – well, because apparently that's just the way the universe is. Of course, if we shove an object on Earth, kick a brick for example, it will hardly move at all. But if we could eliminate the frictional forces between the brick and the ground, and the brick and the surrounding air, it would carry on moving, forever if nothing got in its way.

- A force is something – think push or pull – that when applied to an object at rest or in uniform motion causes that object to accelerate. The object will continue to accelerate for as long as the force is applied. As we will see with Newton's second law, the acceleration is inversely proportional to the object's mass.

Newtonian mechanics is based on Newton's three laws of motion, his law of universal gravitation and certain implicit assumptions about the nature of space and time. After Newton, a huge range of physical phenomena became explicable in terms of these laws. For the first time the motion of cannonballs, pendulums, comets and planets could be accurately explained and predicted. Engineers could calculate the stresses involved in building ever more complex machines and buildings, paving the way for the Industrial Revolution. As the poet Alexander Pope proclaimed:

> 'Nature and Nature's laws lay hid in night,
> God said, let Newton be! and all was light.'

Newton's equations are still perfectly satisfactory for describing most situations in our ordinary, everyday world. The navigational calculations used in NASA's Apollo programme, for example, were all based on the Newtonian theory of gravity.

2.2 Newtonian space and time

Newton assumed that both space and time were absolute and universal. Absolute means space and time are unaffected by physical phenomena and would remain unchanged if no phenomena were occurring. Universal means they are the same for all observers, no matter where they are or how they are moving. Space and time are therefore simply the backdrop or stage on which things happen. Geometrically, Newtonian space is an endless expanse of Euclidean three-dimensional space. Newtonian time can be likened to a uniformly flowing invisible stream.

Effectively, this means space and time are reasonably straightforward concepts when doing Newtonian mechanics. We need to use spatial and temporal measurements in our calculations, but we don't need to worry about the underlying structure of space and time. Make the most of these simplicities because we most certainly cannot ignore space and time when we start discussing relativity.

2.3 Newtonian inertial frames of reference

If you are on board a train that is travelling with a constant velocity and you decide to have a game of table tennis or pool you don't need to make any allowance whatsoever for the motion of the train whilst playing (we are assuming that the rails are smooth, so the train isn't bumping about).

Figure 2.2: Galileo Galilei (1564–1642) – portrait of Galileo by Giusto Sustermans.

The Italian astronomer and physicist Galilei Galileo (Figure 2.2) referred to this phenomenon in his ship thought experiment. This quote is from Galileo's *Dialogue Concerning the Two Chief World Systems* (1632):

'Shut yourself up with some friends in the main cabin below decks on some large ship, and have with you there some flies, butterflies, and other small flying animals. Have a large bowl of water with some fish in it; hang up a bottle that empties drop by drop into a wide vessel beneath it.

With the ship standing still, observe carefully how the little animals fly with equal speed to all sides of the cabin. The fish swim indifferently in all directions; the drops fall into the vessel beneath; and, in throwing something to your friend, you need throw it no more strongly in one direction than another, the distances being equal; jumping with your feet together, you pass equal spaces in every direction.

When you have observed all these things carefully (though doubtless when the ship is standing still everything must happen in this way), have the ship proceed with any speed you like, so long as the motion is uniform and not fluctuating this way and that. You will discover not the least change in all the effects named, nor could you tell from any of them whether the ship was moving or standing still. In jumping, you will pass on the floor the same spaces as before, nor will you make larger jumps toward the stern than toward the prow even though the ship is moving quite rapidly, despite the fact that during the time that you are in the air the floor under you will be going in a direction opposite to your jump. In throwing something to your companion, you will need no more force to get it to him whether he is in the direction of the bow or the stern, with yourself situated opposite.

The droplets will fall as before into the vessel beneath without dropping toward the stern, although while the drops are in the air the ship runs many spans. The fish in their water will swim toward the front of their bowl with no more effort than toward the back, and will go with equal ease to bait placed anywhere around the edges of the bowl. Finally the butterflies and flies will continue their flights indifferently toward every side, nor will it ever happen that they are concentrated toward the stern, as if tired out from keeping up with the course of the ship, from which they will have been separated during long intervals by keeping themselves in the air.'

Both the train and the ship are examples of what are known as **inertial frames** or **inertial frames of reference**. Any inertial frame, Newtonian or relativistic, is one where objects move in straight lines unless acted on by an external force. This definition, as we'll see shortly, is the same as saying objects in inertial frames obey Newton's first law.

The distinguishing feature of a **Newtonian inertial frame** is that it treats gravity as just another force. For Newtonian mechanics, if we ignore friction and air resistance, the Earth's surface (and thus our hypothetical train and ship) are good approximations to inertial frames. Everything on the Earth's surface is acted upon by gravity, but if this force is taken into account what remains is (a good approximation to) an inertial frame. We saw an example of this in Problem (1.16) where we threw a ball off a cliff and calculated its position knowing that, because of Newton's first law (and the implicit assumption we are working in an inertial frame), we could treat its horizontal velocity as constant. In Newtonian celestial mechanics, we can use the 'fixed' stars as a frame of reference, assume that the Solar System is an inertial frame and work out the various movements (ie accelerations) of the planets using Newton's laws of motion and gravitational law.

The Earth and Solar System aren't exactly inertial because they are rotating and are therefore accelerating. In the case of the Earth, we can visualise the effects of the Earth's rotation by imagining we had an infinitely smooth table in our laboratory. Place a ball on the table and hit it lightly so it very slowly moves across the table at a uniform velocity (here we are again assuming that we can magic away friction and air resistance). If we were in a strict inertial frame the ball would move in a straight line across the table. Instead, because of the Earth's rotation, it curves ever so slightly across the surface of the table, an effect easily demonstrated using a device known as a Foucault pendulum. Apparent forces, which are not caused by any physical interaction but are due to an observer using a non-inertial reference frame, are known as **inertial** or, somewhat misleadingly, **fictitious forces** – the force that pushes you to the back of your seat in an accelerating car, or throws you from side to side in a sharply turning vehicle, for example. A uniformly rotating reference frame, such as the Earth, produces two inertial forces, known as the centrifugal and Coriolis force. However, discounting these (relatively small) inertial forces, the Earth approximates to a Newtonian inertial frame. Because inertial forces result from the acceleration of a reference frame, they are (in accordance with Newton's second law (2.4.1), which we meet shortly) proportional to the mass of the object acted upon.

Galileo realised that anyone sealed up in the hold of his imaginary ship would have no way of knowing whether the vessel was stationary or moving with a uniform velocity. Galileo, and later Newton, extended that observation to all inertial reference frames, with what is now known as the **Galilean relativity** principle, which states that if you are in an inertial frame and have no communication with the outside world, there is no experiment in Newtonian mechanics that will tell you whether your frame is stationary or moving with a uniform velocity. More formally, Galilean relativity can be expressed in terms of there being no preferred frame of reference for describing the laws of mechanics.

Two other important properties of inertial frames are:

- Any frame that moves with constant velocity relative to an inertial frame must itself be an inertial frame.

- A frame that is accelerating or rotating (rotation is a form of acceleration as it involves constantly changing velocity) relative to an inertial frame cannot itself be an inertial frame.

We've spent some time looking at inertial frames because they are fundamental to both Newtonian and relativistic physics. For now, we continue with our discussion of Newtonian mechanics.

2.4 Newton's three laws of motion

Newtonian mechanics is built on the foundation of his three laws of motion.

2.4.1 Newton's first law

States that an object will remain at rest or in uniform motion in a straight line unless acted upon by an external force. Because inertia is the property of a body to resist any change in its state of rest or uniform motion, this law is also known as the **law of inertia**. In this sense, the mass of a body is a measure of its inertia and is called the **inertial mass** of the body. We've already met this law in Problem (1.16) where we threw a ball off a cliff and calculated its position by knowing that, due to Newton's first law, its horizontal velocity is constant.

2.4.2 Newton's second law

States that if a net force acts on an object, it will cause an acceleration of that object. The relationship between an object's mass m, its acceleration $\mathbf{a}$ and the applied force $\mathbf{F}$ is given by the vector equation

$$\mathbf{F} = m\mathbf{a} \qquad (2.4.1)$$

or, as a scalar equation,

$$F = ma. \qquad (2.4.2)$$

The first (vector) form of this law is also saying that the force equals the rate of change of **momentum** $\mathbf{p}$ of the object, where momentum equals the mass of the object multiplied by its velocity ($\mathbf{p} = m\mathbf{v}$), so another way of stating the same law is

$$\mathbf{F} = \frac{d\mathbf{p}}{dt} = \frac{d\,(m\mathbf{v})}{dt} = m\frac{d\mathbf{v}}{dt}. \qquad (2.4.3)$$

The SI unit of force is the **newton**, symbol N, which is the force required to give a mass of 1 kg an acceleration of $1\,\mathrm{m\,s^{-2}}$.

Problem 2.1. A 20 kg object has an acceleration of $5\,\mathrm{m\,s^{-2}}$. What is the force acting on the object?

Using (2.4.2) $F = ma$

$$F = 20 \times 5 = 100\,\mathrm{N}.$$

Problem 2.2. A 900 kg car goes from $10\,\mathrm{m\,s^{-1}}$ to $20\,\mathrm{m\,s^{-1}}$ in 5 s. What force is acting on it?

We need one of the equations of uniform motion (1.10.15) $v = u + at$, which we rearrange to give

$$a = \frac{v - u}{t}$$

$$a = \frac{20 - 10}{5} = 2\,\mathrm{m\,s^{-2}}$$

and can then substitute into (2.4.2) $F = ma$

$$F = 900 \times 2 = 1800\,\mathrm{N}.$$

2.4.3 Newton's third law

States that when one object exerts a force on another object, the second object exerts an equal force in the opposite direction on the first object. Another formulation of this law is that for every action there is an equal and opposite reaction, hence the alternative name for this law as the **action-reaction law**. This law applies irrespective of the motion of the two objects, ie it applies if the objects are stationary, moving with uniform velocity or accelerating. Here are some examples of Newton's third law:

- If you press a stone with your finger, the stone also presses back onto your finger with an equal and opposite force.

- If a car is accelerating, its wheels exert a force on the road, which also exerts an equal and opposite force on the wheels. From Newton's second law, the road's force on the car accelerates the car. The car's force on the road also accelerates the Earth, but only by an infinitesimal amount because the mass of the car is a tiny fraction of the mass of the Earth.

- If a small car and a massive truck have a head-on collision, the force exerted by the car on the truck will be the same as the force exerted by the truck on the car. However, as the mass of the car is much less than the mass of the truck, the deceleration of the car will be much greater than the deceleration of the truck.

- Drop a pencil and it falls to the ground because the Earth exerts a gravitational force on it. The pencil will also exert an equal and opposite force on the Earth. Although the forces are equal, the resulting accelerations will be very different, ie the Earth's acceleration toward the pencil will be much less than the pencil's acceleration toward the Earth. Therefore, the Earth's displacement toward the pencil will be much less than the pencil's displacement toward the Earth.

2.5 Newton's law of universal gravitation

2.5.1 Statement of the law

Newton's law of universal gravitation – his famous inverse square law – states that **point masses** m_1 and m_2, distance r apart, attract each other with a force equal to

$$F = G\frac{m_1 m_2}{r^2}, \tag{2.5.1}$$

where the quantities m_1 and m_2 represent the **gravitational mass** of each body, and G is the gravitational constant ($G = 6.673 \times 10^{-11}\,\mathrm{N\,m^2\,kg^{-2}}$). We can also write the law as a vector equation, where the force m_1 exerts on m_2 is

$$\mathbf{F} = -\frac{Gm_1m_2}{\mathbf{r}^2}\hat{\mathbf{r}}, \tag{2.5.2}$$

where $\hat{\mathbf{r}}$ is a unit vector in the radial direction from m_1 to m_2, and the minus sign indicates that gravity is a force of attraction (ie acts in the opposite direction to the unit vector). The law refers to point masses. However, it can also be shown that spherically symmetric objects behave gravitationally as if all their mass is concentrated at a central point. Most sizeable celestial objects are approximately spherically symmetric, and therefore can be treated as central point masses.

In order to appreciate the power of this law, it is useful to introduce a little historical context. We start our whistle-stop tour of the quest to understand celestial motion with Nicolaus Copernicus (1473–1543), the Renaissance astronomer, who proposed the heliocentric (and heretical – the Catholic Church weren't at all happy) theory that the Sun, not the Earth, was the centre of the universe. However, proving that the planets orbited the Sun was no easy task. We next introduce an eccentric Danish nobleman called Tycho Brahe (1546–1601) who built an observatory on an island in the strait between Denmark and Sweden and proceeded to make detailed observations of the movements of the planets. Brahe – who actually didn't believe the Earth orbited the Sun – accumulated an enormous amount of accurate astronomical measurements. After falling out with the King of Denmark, Brahe moved to Prague, was appointed Imperial Mathematician by the Austrian Emperor Rudolph and, with a team of assistants, proceeded to try to make sense of all his data. One of those assistants – Johannes Kepler (1571–1630) – was appointed Imperial Mathematician after Brahe's death. It was Kepler who eventually proposed what are now known as Kepler's laws of planetary motion. These are:

- The planets move in elliptical orbits, with the Sun at one focus. (You can draw an ellipse using two pins, a loop of string and a pen – see Figure 2.3. Each pin is a focus – the plural is foci – of the ellipse.)

- A line that connects a planet to the Sun sweeps out equal areas in equal times.

- The square of the period (the period is the time for one complete orbit) of any planet is proportional to the cube of the semi-major axis (ie the longest radius) of its orbit.

Kepler's laws are observational laws. They accurately correlate to Brahe's detailed measurements but contain no underlying mathematical explanation as to how or why gravity behaves as it does. Fast forward to Newton's law of universal gravitation, which is also an observational law, with no underlying explanation as to *why* things move under the influence of gravity, but does answer the *how* question – namely, objects attract each other according to the inverse square law. Furthermore, it can be shown that all three of Kepler's laws are a consequence of Newton's law of gravitation, ie Kepler's laws can be mathematically *derived* from Newton's. Thanks to Newton, there was at last a rigorous theory of planetary motion. Although an approximation to Einstein's general theory of relativity, this law remains an extremely accurate description of non-relativistic situations, our Solar System for example.

Because general relativity is a theory of gravitation it's worth spending a little time taking a closer look at the towering scientific achievement it replaced – Newton's inverse square law – which has been called 'the greatest generalization achieved by the human mind' (Feynman [7]). First, we'll look at how Newton derived his gravitational law, then we'll explore, through somewhat lengthy calculations, how the law can be used to plot the motion of an orbiting planet.

Figure 2.3: Foci of an ellipse.

2.5.2 How Newton derived his law of universal gravitation

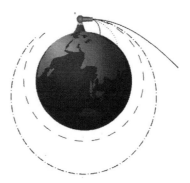

Figure 2.4: Newton's falling cannonball.

We start with Galileo, who discovered that a falling body accelerates uniformly, ie its speed increases at a constant rate. Furthermore, he found that all falling bodies accelerate at the same rate (ignoring air resistance), something Galileo may have discovered by dropping objects off the Leaning Tower of Pisa. He also found that this uniform acceleration is the same in the vertical direction whether you drop the object or throw it horizontally. Drop a stone and it will fall 5 m in the first second. Fire a bullet horizontally from a height of 5 m and it will (again, ignoring the effects of air resistance) takes the same time (1 second) to fall to the ground as a bullet simply dropped from the same height.

Then came Newton, who wondered what would happen if you fired a very fast cannonball from the top of a very tall mountain. Figure 2.4 illustrates Newton's thought experiment, showing the various paths such a cannonball will follow depending on its initial speed.

After 1 second the cannonball would have fallen 5 m But, of course, because of the curvature of the Earth, the ground curves away downwards from a horizontal line. This means that if you fire the cannonball fast enough, after 1 second it will eventually have travelled sufficient distance so that the surface of the Earth has also fallen away 5 m from the horizontal. What does that mean? It means the cannonball though still falling is moving so fast that it cannot catch up with the 'falling' curvature of the ground's surface, ie it is now orbiting the Earth.

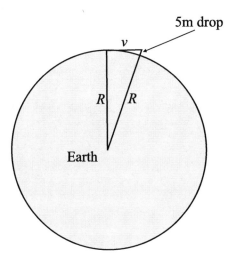

Figure 2.5: Cannonball in orbit.

What is the minimum speed the cannonball needs for it to go into orbit? We can find out using Pythagoras' theorem (1.14.1) as shown in Figure 2.5, where v is the distance the cannonball needs to travel in 1 second.

The radius of the Earth is $6400\,\text{km} = 6,400,000\,\text{m}$. From Pythagoras' theorem (1.14.1) we can therefore say that

$$(R + 5)^2 = v^2 + R^2$$
$$R^2 + 25 + 10R = v^2 + R^2.$$

25 m is a lot smaller than $R = 6,400,000\,\text{m}$, so we'll ignore it, giving

$$v^2 = 10R = 10 \times 6,400,000 = 64,000,000$$

$$v = 8000\,\text{m}.$$

So the cannonball needs to travel 8000 m in 1 second (or about 5 miles per second or about 18,000 mph) in order to enter into a low Earth orbit.

Newton, being a genius, and possibly after watching an apple fall from a tree, then looked at the Moon and wondered if that was also 'falling' around the Earth attracted by the same gravitational force that was acting on his imaginary cannonball. He therefore needed to determine by how much the Moon 'falls' beneath the horizontal in 1 second. If the answer was 5 m then obviously the gravitational force acting on the Moon is the same as that acting on an object falling on Earth. Newton calculated that the Moon actually 'falls' not 5 m, but 1.37 mm (0.00137 m or about one sixteenth of an inch) below a straight line trajectory in 1 second. So the Moon's acceleration towards the Earth is much smaller than the cannonball's. Newton suggested this was because the force acting on the Moon was much less than on the cannonball. But how much less?

Assuming that both the cannonball and the Moon are accelerating towards the centre of the Earth, the ratio of the Moon's acceleration to the cannonball's is 1.37/5000, which is approximately equal to 1/3600.

The radius of the Moon's orbit (384,000 km) is about 60 times the radius of the cannonball's (which is more or less equal to the radius of the Earth). $1/(60)^2 = 1/3600$, which led Newton to assert his famous inverse square law, that the force of gravitational attraction between two bodies is inversely proportional to the square of the distance separating them. Or, as we have seen in (2.5.1)

$$F = G\frac{m_1 m_2}{r^2}.$$

Incidentally, the minimum speed the cannonball needs to be fired at in order to *escape* the Earth's (or any other spherical body's) gravitational field is known as the **escape speed** v_e and is given by

$$v_e = \sqrt{\frac{2GM}{R}}, \tag{2.5.3}$$

where M is the mass and R is the radius of the Earth, planet, star, etc (and air resistance and the planet's rotation is ignored). An object launched from the surface of the Earth with an initial speed v_e will not fall back, or even go into orbit, but will 'slip the surly bonds of Earth', as the poet John Gillespie Magee, Jr. wrote, and leave our planet forever. Escape speed refers to the initial speed of a non-propelled object such as our hypothetical simple cannonball. The concept of escape speed does not apply to powered craft such as a rocket, which could escape the Earth's gravitational pull at a snail's pace provided it had sufficient fuel. If the rocket turns off its engines, it will then need to be travelling at the escape speed in order to leave the Earth's orbit.

Note, the term is escape *speed* not *velocity*, because it refers to an object fired in any direction (as long as it doesn't hit the planet of course). The escape speed of an object on the surface of the Earth is about 11.2 km s^{-1}, much faster than any cannonball, or even rifle bullet.

Also, note that the escape speed does not depend on the mass of the object trying to do the escaping.

2.5.3 Gravitational acceleration

If an object is freely falling in a gravitational field, Newton's second law (2.4.1) tells us that the object must be subject to a force, the magnitude and direction of which is given by

$$\mathbf{F} = M\mathbf{a}.$$

The magnitude and direction of the force on the freely falling object is also given by Newton's law of gravitation (2.5.2)

$$\mathbf{F} = -\frac{Gm_1 m_2}{\mathbf{r}^2}\hat{\mathbf{r}}.$$

Experiment shows that the inertial mass M equals the gravitational mass m to an accuracy of at least one part in 10^{11}. We can therefore say

$$\mathbf{F} = m_1\mathbf{a} = -\frac{Gm_1m_2}{r^2}\hat{\mathbf{r}}$$

$$\mathbf{a} = -\frac{GM}{r^2}\hat{\mathbf{r}}, \tag{2.5.4}$$

where M is the mass of the gravitating body, Earth for example, and the minus sign indicating the acceleration decreases as r increases. We have shown the fundamental result that the acceleration of a body due to gravity does not depend on the mass of that body: feathers fall as fast as bricks (ignoring, of course, the effects of air resistance). Equation (2.5.4) is the vector description of gravitational acceleration. In scalar form we can write (2.5.4) as

$$a = G\frac{M}{r^2}. \tag{2.5.5}$$

Problem 2.3. Calculate the acceleration due to gravity (a) on the surface of the Earth, and (b) 100 km above the surface of the Earth.

(a) Assume that the mass of the Earth is 5.98×10^{24} kg, the radius of the Earth is 6,378,100 m, and the gravitational constant $G = 6.673 \times 10^{-11}\,\mathrm{N\,m^2\,kg^{-2}}$.

Gravitational acceleration is given by (2.5.5)

$$a = \frac{GM}{r^2},$$

where G is the gravitational constant, M is the mass of the gravitating body (Earth in this case), and r is the distance from the centre of M to the accelerating body.

$$a = \frac{6.673 \times 10^{-11} \times 5.98 \times 10^{24}}{(6,378,100)^2} = 9.81\,\mathrm{m\,s^{-2}}.$$

(b) For 100 km above the surface of the Earth we need to increase the radius by 100,000 m, and

$$a = \frac{6.673 \times 10^{-11} \times 5.98 \times 10^{24}}{(6,478,100)^2} = 9.51\,\mathrm{m\,s^{-2}}.$$

2.5.4 How the law predicts elliptical orbits

As already noted, Newton's laws of motion and his law of universal gravitation can be used to derive all of Kepler's laws. The mathematics is hard going, and we won't pursue it here. Nevertheless, it is instructive to see how Newtonian mechanics can be used in a quantitative manner to accurately predict the motions of the planets. The orbital path we end up with will be an ellipse, though we won't formally have proved that. However, even without a rigorous mathematical proof, it's still rather magical to take a couple of simple equations and use them to produce an accurate graphical representation of a planet's journey around the Sun. This example is from Feynman [8]. Our plan is to numerically calculate the planet's changing position, based on its velocity and acceleration,

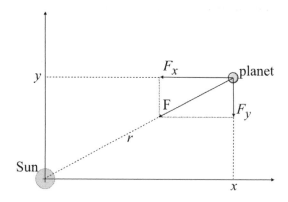

Figure 2.6: Force of gravity on a planet.

as derived from Newton's law of universal gravitation and second law. We assume, for the sake of simplicity, that the Sun is stationary.

First, we consider the gravitational forces on a planet at a distance r from the Sun, as shown in Figure 2.6. Newton's law of universal gravitation tells us there is a force $\mathbf{F}$ directed toward the Sun with a magnitude

$$|\mathbf{F}| = \frac{Gm_sm_p}{r^2}, \tag{2.5.6}$$

where m_s is the mass of the Sun and m_p is the mass of the planet. The force $\mathbf{F}$ can be resolved into horizontal and vertical components F_x and F_y and, because we are dealing with similar triangles, we can say

$$\frac{F_x}{|\mathbf{F}|} = \frac{-x}{r},$$

and therefore, using (2.5.6),

$$F_x = \frac{-|\mathbf{F}|\,x}{r} = \frac{-Gm_sm_px}{r^3},$$

the minus sign showing that negative F_x corresponds to positive x. Similarly, with regard to the F_y component of $\mathbf{F}$, we can say

$$F_y = \frac{-|\mathbf{F}|\,y}{r} = \frac{-Gm_sm_py}{r^3}.$$

Pythagoras also tells us

$$r = \sqrt{x^2 + y^2}. \tag{2.5.7}$$

We also know, from Newton's second law, that when any force acts on an object the component of that force in the x direction will equal the mass of the object multiplied by the change in velocity (ie the acceleration a_x) of the object in the x direction. Similarly, we find acceleration a_y in the y direction. Therefore, we can say

$$m_p\frac{dv_x}{dt} = \frac{-Gm_sm_px}{r^3}$$

$$\frac{dv_x}{dt} = a_x = \frac{-Gm_sx}{r^3},$$

which tells us the acceleration of the planet in the x direction, and

$$\frac{dv_y}{dt} = a_y = \frac{-Gm_s y}{r^3},$$

which tells us the acceleration of the planet in the y direction. We can also assume, for simplicity's sake, that the constant $Gm_s = 1$, so

$$\frac{dv_x}{dt} = a_x = \frac{-x}{r^3} \qquad (2.5.8)$$

and

$$\frac{dv_y}{dt} = a_y = \frac{-y}{r^3}. \qquad (2.5.9)$$

We assume that the initial position of the planet at time $t = 0.00$ is at $x = 0.500$ and $y = 0.000$. We also assume that the initial velocity in the x direction $v_x = 0.000$ and in the y direction $v_y = 1.630$ as shown in Figure 2.7. Different initial assumptions would give us different orbital curves, not all of which would be stable closed orbits.

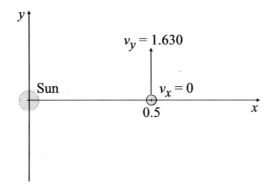

Figure 2.7: Planet's position and velocity components.

Our aim is to calculate the planet's position, velocity and acceleration a at regular time intervals (denoted by the Greek letter epsilon ϵ). Obviously, the smaller we make the time interval ϵ, the more accurate will be our predicted orbital curve. Given the planet's position, we can find its instantaneous acceleration using (2.5.7), (2.5.8) and (2.5.9). We'll do this at time $t = 0.0, 0.1, 0.2, 0.3$, etc, meaning we are using time intervals of $\epsilon = 0.1$. Velocity is a little more tricky as, for the best approximation, we need to find an average velocity at a time midway between known position times. We therefore calculate the velocity at $t = 0.05, 0.15, 0.25, 0.35$, etc (ie at $t + \epsilon/2$) and to do this we use the following equation to find the approximate average velocity for the period $t - \epsilon/2$ to $t + \epsilon/2$.

$$v(t + \epsilon/2) = v(t - \epsilon/2) + \epsilon a(t). \qquad (2.5.10)$$

As you can see, all we are doing is multiplying the acceleration at time t by the time interval ϵ to give us the change in velocity in that time interval. We then add that change in velocity to the velocity at $t - \epsilon/2$ to get a reasonable approximation for the velocity at time $t + \epsilon/2$.

For our initial calculation of velocity at $t = 0.05$ we use a slightly amended version of (2.5.10)

$$v\left(\epsilon/2\right) = v\left(0\right) + \left(\epsilon/2\right) a\left(0t\right). \tag{2.5.11}$$

We calculate the approximate position of the planet at $t + \epsilon$ by adding the previous position at t to the change in position found by multiplying the time interval ϵ by the velocity at $t + \epsilon/2$, using the equation

$$x\left(t + \epsilon\right) = x\left(t\right) + \epsilon v\left(t + \epsilon/2\right). \tag{2.5.12}$$

A spreadsheet is ideal for this type of laborious but repetitive calculation. We start with the initial values at $t = 0.00$

$$x\left(0\right) = 0.500 \quad y\left(0\right) = 0.000$$
$$v_x\left(0\right) = 0.000 \quad v_y\left(0\right) = +1.630$$

Using (2.5.7), (2.5.8) and (2.5.9) we find

$$r\left(0\right) = 0.500 \quad 1/r^3\left(0\right) = 8.000$$
$$a_x\left(0\right) = -4.000 \quad a_y\left(0\right) = 0.000$$

We now calculate the velocities $v_x\left(0.05\right)$ and $v_y\left(0.05\right)$ using (2.5.11)

$$v_x\left(0.05\right) = 0 - 0.05 \times 4.000 = -0.200,$$

$$v_y\left(0.05\right) = 1.630 + 0.05 \times 0 = 1.630.$$

Next we find the planet's position at $t = 0.1$ using (2.5.12)

$$x\left(0.1\right) = 0.500 + 0.1 \times \left(-0.20\right) = 0.480,$$

$$y\left(0.1\right) = 0.0 + 0.1 \times \left(1.630\right) = 0.163.$$

And repeat the above calculations

$$r\left(0.1\right) = \sqrt{0.480^2 + 0.163^2} = 0.507,$$
$$1/r^3 = 7.673,$$
$$a_x\left(0.1\right) = -0.480 \times 7.673 = -3.683,$$
$$a_y\left(0.1\right) = -0.163 \times 7.673 = -1.251,$$
$$v_x\left(1.5\right) = -0.200 + \left(0.1\right) \times \left(-3.683\right) = -0.568,$$
$$v_y\left(1.5\right) = 1.630 + \left(0.1\right) \times \left(-1.251\right) = 1.505.$$

The first few results, from $t = 0.00$ to $t = 0.45$ are shown in Table 2.1.

A plot of the results for a complete orbit, from $t = 0.00$ to $t = 4.10$, is shown in Figure 2.8. The curve shows the movement of the planet around the Sun. Each little square marker shows the planet's position at successive $\epsilon = 0.1$ time intervals.

t	x	v_x	a_x	y	v_y	a_y	r	$1/r^3$
0.000	0.500	0.000	-4.000	0.000	1.630	0.000	0.500	8.000
0.050		-0.200			1.630			
0.100	0.480		-3.685	0.163		-1.251	0.507	7.677
0.150		-0.568			1.505			
0.200	0.423		-2.897	0.313		-2.146	0.527	6.847
0.250		-0.858			1.290			
0.300	0.337		-1.958	0.443		-2.569	0.556	5.805
0.350		-1.054			1.033			
0.400	0.232		-1.112	0.546		-2.617	0.593	4.794
0.450		-1.165			0.772			

Table 2.1: Calculation of planetary motion from $t = 0.00$ to $t = 0.45$.

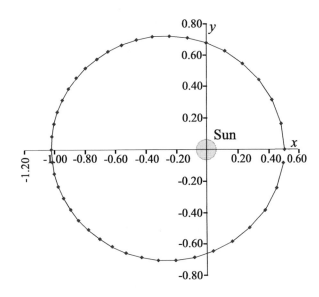

Figure 2.8: Calculated orbit of a planet around the Sun.

Although we haven't rigorously proved it, the orbit does indeed describe an ellipse, with the Sun at one focus. Notice also how the planet moves faster when nearer the Sun, which correlates to Kepler's second law – a line that connects a planet to the Sun sweeps out equal areas in equal times. By plotting orbits with different periods we could also show Kepler's third law – the square of the period of any planet is proportional to the cube of the semi-major axis of its orbit.

And this method doesn't just work for one planet. As long as we know their initial positions, we can calculate the gravitational interactions between the planets (and the Sun) and predict their orbital motion. Feynman [8] finishes his chapter on Newton's laws of dynamics by saying,

> 'We began this chapter not knowing how to calculate even the motion of a mass on a spring. Now, armed with the tremendous power of Newton's laws, we can not only calculate such simple motions, but also, given only a machine to handle the arithmetic, even the tremendously complex motions of the planets, to as high a degree of precision as we wish!'

2.5.5 The gravitational field

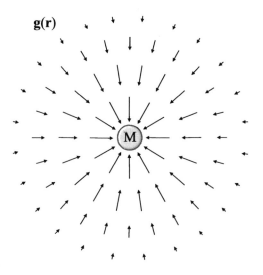

Figure 2.9: Gravitational field around a spherical mass.

The gravitational field $\mathbf{g}\,(\mathbf{r})$ at any point $\mathbf{r}$ in space is defined as the gravitational force felt by a unit mass placed at point $\mathbf{r}$. It is a vector field because the gravitational force is moving the unit mass inwards in a radial direction, and is found by taking the vector equation of Newton's law of universal gravitation (2.5.2)

$$\mathbf{F} = -\frac{Gm_1m_2}{\mathbf{r}^2}\hat{\mathbf{r}} \tag{2.5.13}$$

and substituting 1 (a unit mass) for one of the two masses to give

$$\mathbf{g}\,(\mathbf{r}) = -G\frac{M}{|\mathbf{r}|^2}\hat{\mathbf{r}}, \tag{2.5.14}$$

where M is the mass of the gravitating body, Earth for example.

We can plot (2.5.14) to show the gravitational field around a uniform spherical mass (see Figure 2.9).

We met the idea of the divergence of a vector field (remember sources and sinks) in Section 1.13.3. It can be shown that the divergence of the gravitational field at any point is given by

$$\nabla \cdot \mathbf{g} = -4\pi G \rho, \tag{2.5.15}$$

where ρ (the Greek letter rho) is the mass density at that point. Evidently, if the point is in empty space, $\rho = 0$, and the divergence $\nabla \cdot \mathbf{g} = 0$.

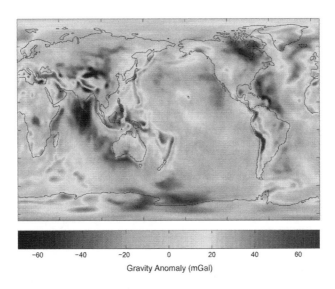

Figure 2.10: Earth's gravity as measured by the GRACE mission.

Because the Earth isn't a uniformly dense, perfect sphere, precisely calculating its gravitational field is not straightforward. Since its launch in 2002, the NASA/German Aerospace Centre GRACE (Gravity Recovery and Climate Experiment) mission has been accurately mapping the Earth's gravitational field (see Figure 2.10). According to the NASA website, 'the gravity variations that GRACE studies include: changes due to surface and deep currents in the ocean; run-off and ground water storage on land masses; exchanges between ice sheets or glaciers and the oceans; and variations of mass within the Earth.'

2.5.6 Tidal forces

Forces associated with the variation of a gravitational field are known as **tidal forces**. We can illustrate the effects of these forces by imagining an observer in a huge room suspended above a planet, as shown in Figure 2.11. Notice how the gravitational field varies both vertically and horizontally throughout the room. Travelling the length of the room would be a decidedly odd experience as the direction of the gravitational field would vary from straight down to almost horizontal.

Another way to visualise tidal forces is to imagine dropping a handful of small pebbles down a very deep mineshaft. We assume that the pebbles are sufficiently small that we can ignore any

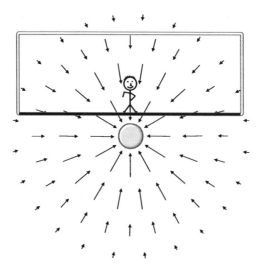

Figure 2.11: Tidal forces.

mutually attractive gravitational forces between them, and that they start their descent neatly grouped together into a round ball. As they fall, tidal forces stretch the ball vertically and squeeze it horizontally as shown in Figure 2.12.

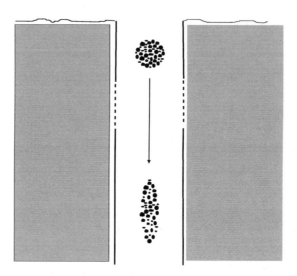

Figure 2.12: Tidal forces on pebbles falling down a very deep mineshaft.

Gravitational field variation in a room-sized volume on Earth is extremely small. Even ascending

100 km above the Earth's surface only results in a decrease in gravitational acceleration of $9.81 - 9.51 = 0.30 \, \mathrm{m \, s^{-2}}$ (Problem 2.3). The most noticeable tidal effects on the Earth are, of course, the ocean's tides, caused primarily by variations in the Moon's gravitational field. However, in the vicinity of a sufficiently dense mass, such as a black hole, tidal forces can be enormous, as we'll see in Chapter 10.

2.5.7 Gravitational potential energy

If a force acting on an object makes the object move in the direction of the force, the force is said to do **work** on the object. Work equals force multiplied by the distance through which the force acts. The SI unit of work (and energy) is the **joule**, symbol J, which is the work done (or energy expended) in applying a force of 1 newton through a distance of 1 metre. We can speak of positive and negative work depending on the direction of the force and object. If I lift a brick off the floor I am doing positive work on the brick (the force I exert is in the same direction as the movement of the brick). Gravity, on the other hand, is doing negative work on the brick (the gravitational force pulling the brick down is in the opposite direction to the upward movement of the brick). However, if I then drop the brick, gravity is doing positive work on the brick (ie the gravitational force is in the same direction as the movement of the falling brick). Gravity is an example of a **conservative force**, where the work done by the force in moving an object from one point to another depends only on the initial and final position of the object, and is independent of the path taken. The negative work done by gravity as I raise the brick off the floor is therefore exactly equal to the positive work done by gravity as the brick falls back down to the floor. Another example of a conservative force is the force exerted by a (perfect) spring. Friction, on the other hand, is an example of a non-conservative force.

There are two types of mechanical energy: **potential energy** and **kinetic energy**. Potential energy is stored energy, the energy an object has because of its position or configuration; a stretched spring, for example, has potential energy. We give an object potential energy by performing work on it against a conservative force. Gravitational potential energy is the energy an object possesses because of its position in a gravitational field. We can give an object gravitational potential energy by lifting it against the force of gravity. If we then release the object it will fall, and the potential energy will be converted to kinetic energy (the energy an object possesses due to its motion). The concept of potential energy is only meaningful in a conservative force field. In such a field, the work done against a conservative force, such as gravity, is stored as potential energy, which is recoverable. If I throw a ball upwards, the work done by gravity is negative, and the gravitational field transforms the kinetic energy of the ball to potential energy. When the ball starts to fall, the work done by gravity is positive and the gravitational field transforms the potential energy of the ball to kinetic energy. Whether the ball is rising or falling, its change in potential energy ΔPE_g is equal to the negative of the work done on it W_g by the force of gravity, which we can write as $\Delta PE_g = -W_g$.

Potential energy is a relative not an absolute quantity. It's the difference in potential energy that matters not the absolute value at a given point. My raised brick has a certain potential energy with reference to the floor, but a greater potential energy with reference to the ground (if I'm on top of a multi-story building). We choose a zero reference point convenient for the particular problem we are trying to solve. This is where things can become confusing, because when looking at potential energy far away from the Earth's surface the convention is that the potential energy of an object in a gravitational field is taken to be zero when r equals infinity. This makes sense because the gravitational force approaches zero as r approaches infinity. However, it does mean, somewhat strangely, that the object's potential energy will then decrease negatively as it approaches the gravitational source (a quantity that decreases from zero must be decreasing negatively). Remember, it's the difference in potential energy that's important. Following this convention, the gravitational potential energy of an object is then defined as the negative of work done in bringing the object to that point from infinity.

We know that gravitational force is given by (2.5.1) $F = G\frac{m_1 m_2}{r^2}$.

Recalling that work equals force multiplied by distance, the work done by the force of gravity in moving an object through a gravitational field from infinity to a separation r is given by

$$W_g = \int_\infty^r -G\frac{m_1 m_2}{r'^2} dr' = G\frac{m_1 m_2}{r}, \tag{2.5.16}$$

where m_1 and m_2 are the masses of the gravitational source (a planet, for example) and the object being moved through the gravitational field. The minus sign is necessary because the gravitational force increases as r decreases.

Because potential energy is the negative of the work done by gravity, the potential energy of an object in a gravitational field is

$$PE_g = -G\frac{m_1 m_2}{r}. \tag{2.5.17}$$

More generally, we can say that if we move an object through the planet's gravitational field, from a distance r_1 outwards to a distance r_2 (both distances measured from the planet's centre) gravity does work on the object equal to

$$W_g = G\frac{m_1 m_2}{r_2} - G\frac{m_1 m_2}{r_1} = Gm_1 m_2 \left(\frac{1}{r_2} - \frac{1}{r_1}\right) \tag{2.5.18}$$

and the change in gravitational potential energy is

$$\Delta PE_g = -W_g = -Gm_1 m_2 \left(\frac{1}{r_2} - \frac{1}{r_1}\right). \tag{2.5.19}$$

This equation is true for any change in distance r_1 to r_2.

If that change in distance h is very small compared to the planet's radius we can use the force (ma) times distance (h) approximation

$$PE_g = mah.$$

Note that we are now assuming that the object's potential energy equals zero on the planet's surface, not at infinity. Near the surface of the Earth, the acceleration due to gravity is $g = 9.81\,\mathrm{m\,s^{-2}}$, and the above equation becomes

$$PE_g = mgh. \tag{2.5.20}$$

Problem 2.4. A 2 kg rock is thrown vertically into the air. What gravitational potential energy does it possess at its highest point 3 m above the ground?

Using (2.5.20), $PE_g = mgh = 2 \times 9.81 \times 3 = 58.86\,\mathrm{J}$.

Although the calculation would be more complicated (involving the gravitational constant and the radius and mass of the earth), we would of course obtain the same result using equation (2.5.19).

2.5.7.1 The gravitational potential field

The gravitational potential (ϕ) at a point in a gravitational field is defined as the work done in bringing a unit mass to that point from infinity. In other words, ϕ is the gravitational potential energy of a unit mass in a gravitational field. Using (2.5.17) $PE_g = -G\frac{m_1 m_2}{r}$ and letting one of the masses equal 1 kg we can therefore define the gravitational potential field, which is a scalar field, as

$$\phi = \frac{-GM}{r}. \tag{2.5.21}$$

To be precise, 2.5.21 is only true for a point at a distance r from a point mass M or at a distance r from the centre of and *outside* a uniform sphere of mass M. So if the sphere has radius R, then 2.5.21 is true if $r \geq R$.

For a point *inside* a uniform spherical mass, ie for $r < R$, things become a bit more complicated, but it can be shown that the gravitational potential is given by

$$\phi = \frac{-GM}{2R^3}\left(3R^2 - r^2\right). \tag{2.5.22}$$

Substitute $r = R$ into (2.5.22) and we get back to (2.5.21).

2.5.7.2 Gravitational field and gravitational potential field

Equation (1.13.3) tells us that the magnitude $|\mathbf{r}|$ of the position vector $\mathbf{r}$ is given by

$$r = |\mathbf{r}| = \sqrt{x^2 + y^2 + z^2}. \tag{2.5.23}$$

In Cartesian coordinates we can therefore rewrite the gravitational potential field (2.5.21) as

$$\phi = \frac{-GM}{r} = \frac{-GM}{\sqrt{x^2 + y^2 + z^2}}.$$

Recall from Section 1.13.2 that the gradient (denoted by ∇) of a scalar field ϕ is given by (1.13.6)

$$\nabla\phi = \left(\frac{\partial\phi}{\partial x}, \frac{\partial\phi}{\partial y}, \frac{\partial\phi}{\partial z}\right).$$

We can apply the differential operator ∇ once to find the gradient of the gravitational potential field:

$$\nabla\phi = -GM\left(\frac{\partial}{\partial x}\frac{1}{\sqrt{x^2 + y^2 + z^2}}, \frac{\partial}{\partial y}\frac{1}{\sqrt{x^2 + y^2 + z^2}}, \frac{\partial}{\partial z}\frac{1}{\sqrt{x^2 + y^2 + z^2}}\right).$$

We can find the partial derivatives by hand or, more easily, using a suitable online calculus calculator. For example, using the WolframAlpha Calculus and Analysis Calculator [33], type 'd/dx (1/(x^2 + y^2 + z^2)^(1/2)), d/dy (1/(x^2 + y^2 + z^2)^(1/2)), d/dz (1/(x^2 + y^2 + z^2)^(1/2))' (omit the quotes) into the input box to obtain

$$\nabla\phi = -GM\left(\frac{-x}{(x^2 + y^2 + z^2)^{3/2}}, \frac{-y}{(x^2 + y^2 + z^2)^{3/2}}, \frac{-z}{(x^2 + y^2 + z^2)^{3/2}}\right)$$

$$= GM \left(\frac{x}{\left(x^2 + y^2 + z^2\right)^{3/2}}, \frac{y}{\left(x^2 + y^2 + z^2\right)^{3/2}}, \frac{z}{\left(x^2 + y^2 + z^2\right)^{3/2}} \right)$$

$$= GM \left(\frac{(x, y, z)}{\left(x^2 + y^2 + z^2\right)^{3/2}} \right). \tag{2.5.24}$$

As $\sqrt{x^2 + y^2 + z^2} = r = |\mathbf{r}|$, then $\left(x^2 + y^2 + z^2\right)^{3/2} = r^3 = |\mathbf{r}^3|$ and we can write (2.5.24) as

$$\nabla \phi = \frac{GM}{|\mathbf{r}^3|} (x, y, z). \tag{2.5.25}$$

By definition, $\boldsymbol{r} = (x, y, z) = |\mathbf{r}|\hat{\mathbf{r}}$. The (x, y, z) term here denotes an ordinary Cartesian vector (Cartesian because we're using derivatives with respect to x, y and z). The basis vectors are implied (ie we're smart enough to know they're really there) so we don't bother to show them. Sometimes angled brackets are used to denote a Cartesian vector, eg $\langle x, y, z \rangle$.

So we can rewrite (2.5.25) as

$$\nabla \phi = \frac{GM|\mathbf{r}|\hat{\mathbf{r}}}{|\mathbf{r}^3|}$$

and cancel to give

$$\nabla \phi = \frac{GM}{|\mathbf{r}^2|} \hat{\mathbf{r}}. \tag{2.5.26}$$

But the right-hand side of (2.5.26) equals the gravitational field (2.5.14)

$$\mathbf{g}\left(\mathbf{r}\right) = -G \frac{M}{|\mathbf{r}|^2} \hat{\mathbf{r}}.$$

So we have found an equation that relates the gravitational potential field (ϕ – potential energy per unit mass, a scalar quantity) to the gravitational field ($\mathbf{g}$ – force per unit mass, a vector quantity):

$$\mathbf{g} = -\nabla \phi = -\left(\frac{\partial \phi}{\partial x}, \frac{\partial \phi}{\partial y}, \frac{\partial \phi}{\partial z} \right). \tag{2.5.27}$$

2.5.7.3 Poisson's equation

We've seen that the divergence of the gravitational field is given by (2.5.15)

$$\nabla \cdot \mathbf{g} = -4\pi G \rho,$$

where ρ is the mass density at that point. If we substitute (2.5.27) into this equation we obtain

$$\nabla \cdot \nabla \phi = \nabla^2 \phi = 4\pi G \rho \tag{2.5.28}$$

or, in terms of Cartesian coordinates,

$$\nabla^2 \phi = \left(\frac{\partial^2 \phi}{\partial x^2} + \frac{\partial^2 \phi}{\partial y^2} + \frac{\partial^2 \phi}{\partial z^2} \right) = 4\pi G \rho,$$

which is known as **Poisson's equation**. If we have a chunk of matter (a planet for example) then Poisson's equation tells us the relationship between the gravitational potential ϕ at a point within that chunk and the mass density ρ at that point. Lambourne ([17]) refers to Poisson's equation as providing, 'The essential summary of Newtonian gravitation in terms of a differential equation ... It is entirely equivalent to Newton's inverse square law, but has the advantage that it is a differential equation for a scalar quantity that may be straightforward to solve.'

Along with (2.5.27) $\mathbf{g} = -\nabla\phi$, Poisson's equation is one of the fundamental field equations of Newtonian gravitation and one we'll be returning to when looking at the Newtonian limit of general relativity.

Note that if the point is outside of the mass, then $\rho = 0$ and Poisson's equation becomes

$$\nabla^2\phi = 0,$$

known as **Laplace's equation**. For example, if we consider the Earth's gravitational field, every point inside the Earth will be described by $\nabla^2\phi = 4\pi G\rho$, and every point outside the Earth will be described by $\nabla^2\phi = 0$.

Problem 2.5. Prove that the divergence of the gradient of the gravitational potential field in empty space equals zero.

In other words, we need to prove that $\nabla^2\phi = 0$, where $\phi = \frac{-GM}{r}$. Recalling the terminology from Section 1.13.3, we say we want to find the Laplacian (1.13.9) of $\phi = \frac{-GM}{r}$, ie we need to apply the differential operator ∇ twice. As we did above, we can rewrite ϕ in Cartesian coordinates as

$$\phi = \frac{-GM}{r} = \frac{-GM}{\sqrt{x^2 + y^2 + z^2}}.$$

Equation (2.5.24) gives us the result of applying ∇ once:

$$\nabla\phi = GM\left(\frac{x, y, z}{(x^2 + y^2 + z^2)^{3/2}}\right).$$

Then, we can apply the differential operator ∇ a second time:

$$\nabla^2\phi = GM\left(\frac{\partial}{\partial x} + \frac{\partial}{\partial y} + \frac{\partial}{\partial z}\right)\left(\frac{x, y, z}{(x^2 + y^2 + z^2)^{3/2}}\right).$$

And then, by hand or using a suitable online calculus calculator ([33], for example), find the partial derivatives, giving

$$\nabla^2\phi = GM\left(\frac{2x^2 - y^2 - z^2}{(x^2 + y^2 + z^2)^{5/2}} + \frac{2y^2 - x^2 - z^2}{(x^2 + y^2 + z^2)^{5/2}} + \frac{2z^2 - x^2 - y^2}{(x^2 + y^2 + z^2)^{5/2}}\right),$$

and then (almost magically, as the x, y, z terms on the top line cancel out) we obtain

$$\nabla^2\phi = 0.$$

3 Special relativity

Common sense is the collection of prejudices acquired by age eighteen.

ALBERT EINSTEIN

3.1 Introduction

First, let's note what we don't see in Newton's equations. The dog that doesn't bark is c, the speed of light. In contrast, as will become clear, c is absolutely central to Einstein's theory of special relativity. Furthermore, Newtonian mechanics assumes time and space to be unrelated absolutes. Special relativity (which applies only in the absence of gravity) unites these two quantities into a single fluid entity called spacetime. Both time and spatial distance in spacetime are no longer absolute, depending instead on an observer's relative velocity. Special relativity insists, however, that the laws of physics are independent of the uniform motion of any observer.

Foster and Nightingale [9] summarise the importance of special relativity, stating the theory:

> 'Gives a satisfactory description of all physical phenomena (when allied with quantum theory), with the exception of gravitation. It is of importance in the realm of high relative velocities, and is checked out by experiments performed every day, particularly in high-energy physics.'

Our key to understanding special relativity is seeing how events in spacetime are measured in different frames of reference. What do we mean by frames of reference?

Say, for example, I am on a train and have wired up two light bulbs – one at one end of the train and one at the other. The bulbs are connected so that when I flick a switch they both come on simultaneously. Using a ruler, I can measure the distance between the light bulbs and I can flick my switch and watch both bulbs light up. My description of what's happening regarding the light bulbs consists of two simple quantities: their distance apart and the time interval (zero, according to my observation) between them coming on. Now imagine a trackside observer with highly sophisticated equipment for measuring these two same quantities. If the train is moving relative to this observer will her measurements be the same as mine, ie will she record the same distance separating the bulbs, and the same zero time interval between them lighting up, as I do? No, says the theory of special relativity, she most definitely won't. That discrepancy between the measurements in two reference frames is what we are attempting to understand.

Spacetime diagrams are a useful, geometrical tool for visualising some of the basic properties of special relativity. We can then build on these insights by moving to a more algebraic approach, introducing the Lorentz transformations and working towards an understanding of how special relativity reformulates the laws of mechanics. But first we need to introduce a few essential concepts. These are:

- time

- spacetime

- events

- frames of reference

- inertial frames

- coordinate transformations

- the Galilean transformations

- the two postulates of special relativity.

3.2 Basic concepts

3.2.1 Time

We've seen that Newtonian mechanics assumes time to be independent of physical phenomena and the same for all observers. As we'll find out, this is not the case for special relativity, which forces us to abandon many of our cherished 'common sense' ideas about the meaning of time. Simultaneity, for example, is no longer always absolute – two events separated in space may occur at the same time for one observer but at different times for another.

When we speak about time in special relativity we aren't of course only talking about the reading of a clock. Instead we are referring to a deeper, more fundamental idea of time in the sense of the intrinsic rate of natural processes. Muons, a type of tiny sub-atomic particle, are produced in great numbers by cosmic ray interaction with the top of the Earth's atmosphere at a height of about 10 km. They have such a small lifetime (about two millionths of a second) that few should survive their journey to the Earth's surface. The reason that most do survive is because, travelling at close to the speed of light, they are affected by relativistic time dilation – the muon's 'internal clock' (whatever that is) runs more slowly in other words. In terms of time, and the merest scrap of existence that is the muon, you really can't get much more intrinsic than that!

3.2.2 Spacetime

In Newtonian mechanics, events are described using three-dimensional Euclidean space plus an independent scale of absolute time. In both special and general relativity space and time are fused together into a single four-dimensional entity (or continuum) known as **spacetime**. Spacetime in special relativity is flat (parallel lines do not meet). It is therefore easiest to describe it using Cartesian (x, y, z) coordinates plus a time (t) coordinate. Although we are using Cartesian coordinates, we have now moved away from the familiar Euclidean space we've been using so far. Yes, spacetime is flat (parallel lines never meet, for example), but it isn't Euclidean for the simple reason that

Figure 3.1: Hermann Minkowski (1864 -1909).

distances between points in spacetime are described using a non-Euclidean metric. In contrast, the spacetime of general relativity is curved, not flat.

The spacetime of special relativity is called **Minkowski space** or **Minkowski spacetime**, after the German mathematician Hermann Minkowski (Figure 3.1), Einstein's mathematics professor at Zurich polytechnic, who in 1908 introduced spacetime to the world in a public lecture with the famous line:

> 'The views of space and time which I wish to lay before you have sprung from the soil of experimental physics, and therein lies their strength. They are radical. Henceforth space by itself, and time by itself, are doomed to fade away into mere shadows, and only a kind of union of the two will preserve an independent reality.'

3.2.3 Events

We are interested in things that happen in spacetime called **events**. An event is something that happens instantaneously at a single point in spacetime, such as a light flashing, or a point on a moving object passing another point. All events in spacetime are defined using the four coordinates t, x, y, z.

Imagine a particle moving through spacetime. We can think of the particle's progress as a succession of events. If we link all those events together we would have a line representing the particle's progress through spacetime. That line is called the particle's **world-line**.

3.2.4 Frames of reference

Special relativity addresses how observers moving relative to each other measure events in spacetime. The coordinate system from which each observer takes measurements is called a **frame of reference** or **frame**. We are using nice, simple Cartesian (x, y, z) coordinates, so we could imagine our frames of reference to be a series of infinitely large Cartesian coordinate systems zipping about through spacetime. Let's call one of those frames S. We can determine the spatial position of any event in S by using the x, y, z coordinates. But we also need describe the time an event happens. We therefore imagine that our frame S is full of an infinite number of regularly spaced clocks, all of which are synchronised and run at the same rate. To find when an event happened we simply consult the clock adjacent to it.

This may seem an odd, laboured sort of way to measure time. Why not just imagine an observer sitting in a comfy chair, seeing a distant event and noting the time by consulting an adjacent super-precise clock or wristwatch. The trouble with that scenario is that it doesn't tell us when the event happened, only when the observer saw the event happen, which isn't necessarily the same thing. Alpha Centauri, the nearest star to Earth, is 4.4 light-years distant. If it blew up today we wouldn't know for 4.4 years. In order to understand spacetime we have to assume that we know exactly when events happen. That's the reason we fill our reference frame with synchronised clocks.

It's often helpful to dispense with the notion of a human observer physically making measurements in a reference frame. Instead we can define a reference frame as simply a coordinate system, where each event is located in spacetime by three spatial x, y, z coordinates and one time t coordinate.

3.2.5 Inertial frames

Special relativity is particularly concerned with uniformly moving frames of reference, known as inertial frames. We've already met inertial frames when looking at Newtonian mechanics (Section 2.3) and saw that objects in them obey Newton's first law, ie the object will remain at rest or in uniform motion in a straight line unless acted upon by an external force. Inertial frames in special relativity are known as **Lorentz frames**.

So are Lorentz frames the same as Newtonian inertial frames? Only in the sense that they are both uniformly moving frames where objects obey Newton's first law. They differ fundamentally in how they deal with gravity – special relativity and Lorentz frames are concerned with the behaviour of objects and light rays in the absence of gravitation. Newtonian inertial frames, on the other hand, can include gravity, treating it as just another force. Lorentz frames can only be precisely constructed in flat spacetime, in other words, spacetime that isn't curved by the proximity of mass-energy. One reason for this is because gravitational time dilation (the phenomenon that clocks run slower in a gravitational field – see Section 9.4.2) makes it impossible to globally synchronise clocks. See the Section 'Non-existence of Lorentz frame at rest on Earth' in Schutz [28] for a more detailed discussion of this topic.

However, although an exact global inertial frame cannot be constructed in a gravitational field, for many purposes a local frame of reference on Earth is a useful approximation to an inertial frame.

Later, when discussing general relativity and the equivalence principle, we'll see that a freely falling frame in a gravitational field is actually a locally inertial frame.

3.2.6 Coordinate transformations

We need to compare the measurements of observers in relative motion. Say we have an observer O measuring the time and space coordinates of an event, such as a light flashing on and off, in frame S.

Let's just be clear what we mean here. The flashing light is just an event happening in spacetime. We could describe that event using an infinite number of reference frames, but we have chosen one that we call S. The observer in frame S uses his Cartesian coordinates and synchronised clocks to measure the event and define its position by assigning it four spacetime coordinates t, x, y, z. Another observer O' (the little mark is called a prime, so we say, 'O prime') in a different frame S' that is moving with a constant velocity relative to S, measures the same event with his Cartesian coordinates and synchronised clocks, and assigns it four coordinates t', x', y', z'. Unless the frames coincide, t, x, y, z will not equal t', x', y', z'. But, because they are both nice, simple Cartesian coordinate systems, we would expect there to be a reasonably straightforward set of equations allowing us to relate the coordinates t, x, y, z to the coordinates t', x', y', z'. This set of equations is known as a **coordinate transformation**.

So, what is the correct coordinate transformation for two observers in uniform relative motion? Before special relativity the answer would have been a wonderfully simple set of equations known as the **Galilean transformations**, which we now look at.

3.2.7 The Galilean transformations

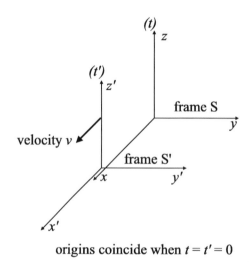

origins coincide when $t = t' = 0$

Figure 3.2: Two frames of reference in standard configuration.

Two frames, S and S', can of course be orientated in an infinite number of ways. To make our lives easier, we arrange our frames in what is known as **standard configuration** (see Figure 3.2), where:

- The x, y, z axes of frame S are parallel to the x', y', z' axes of frame S'.

- The origin of frame S' moves along the x axis of frame S with velocity v as measured in S.

- The origins of frames S and S' coincide when time $t = 0$ in frame S and when time $t' = 0$ in frame S'.

Using the standard configuration is just a convenient way of setting up any two Cartesian coordinate-based moving frames.

As the origins of the two frames coincide when $t = t' = 0$ the distance travelled by the origin of frame S' is equal to vt. Therefore, if the event happens at distance x in the S frame it will happen at distance $x' = x - vt$ in the S' frame.

The Galilean transformations are therefore given by

$$t' = t, \qquad (3.2.1)$$

$$x' = x - vt,$$

$$y' = y,$$

$$z' = z.$$

Problem 3.1. A penny is on a railway track 60 m past the end of a station platform. A train passes through the station towards the penny at a constant velocity of 100 km/hour. A passenger on the train zeros his stopwatch exactly when the rear of the train passes the end of the platform. How far from the rear of the train will the penny be after (a) 1 second, and (b) 10 seconds?

The railway track is the S frame, with its origin at the end of the platform. The penny's position is therefore at $x = 60$. The train is the S' frame with its origin at the rear of the train and travelling at $v = 100$ km/hour. We need to find x' – the distance from the rear of the train to the penny when $t' = t = 1$, and $t' = t = 10$. We can ignore all other coordinates.

(a) First, we convert v to metres per second:

$$v = \frac{100 \times 1000}{3600} = 27.8 \, \mathrm{m\,s^{-1}}.$$

Using (3.2.1):

$$x' = x - vt$$

$$x' = 60 - (27.8 \times 1) = 32.2 \, \mathrm{m}.$$

The penny is 32.2 m *in front of* the rear of the train after 1 second.

(b)

$$x' = 60 - (27.8 \times 10) = -218 \, \mathrm{m}.$$

The penny is 218 m *behind* the rear of the train after ten seconds, ie the rear of the train has passed the penny.

Note that the times will be the same for an observer on the platform because of the Galilean assumption that $t' = t$.

_____

Now let's assume that an object is moving with a velocity u with respect to frame S. What is w, the object's velocity with respect to frame S'?

You may be able to guess the answer. If I'm driving a car at 70 mph and the car chasing me is doing 80 mph, then that car has a relative velocity w to me of $w = 80 - 70 = 10$ mph. If the car behind me is only doing 50 mph, then it has a relative velocity of $w = 50 - 70 = -20$ mph, in effect it is moving backwards away from me, hence the minus sign.

We can formalise these calculations using the Galilean transformation equations (3.2.1)

$$x' = x - vt,$$

$x = ut$, so we can write

$$x' = ut - vt = (u - v)\,t = (u - v)\,t',$$

and the velocity of the object with respect to frame S$'$ is given by

$$w = \frac{x'}{t'} = u - v$$

or

$$u = w + v, \tag{3.2.2}$$

which is the well known **law of Galilean addition of velocities**.

This is the common sense assumption that says, for example, that if I'm on bike pedalling at 15mph relative to an observer standing by the road, and I throw a ball straight ahead of me at a velocity of 50 mph, the velocity of the ball with respect to the stationary observer equals $15 + 50 = 65$ mph.

The Galilean transformations and consequent law of addition of velocities work beautifully in our everyday world of trains, bikes and balls, none of which travel at anything near the speed of light. But what if I'm again cycling at 15 mph and I now shine a torch straight ahead of me. What is the velocity of the light beam with respect to the roadside observer. The Galilean addition of velocities law tells us the answer is the speed of light plus 15 mph.

'NO, IT'S NOT', said Einstein. In fact he said that no matter how fast you are travelling, any inertial observer (ie an observer in an inertial frame) would measure the same value c for the speed of light, where $c = 3 \times 10^8 \, \mathrm{m\,s^{-1}}$ in a vacuum. In short, Einstein suggested that the speed of light with respect to any inertial frame is independent of the motion of the light source.

Abandoning long established Galilean assumptions about coordinate transformations and the addition of velocities means abandoning many of our most cherished notions concerning the nature of time and space. Einstein was prepared to take such a drastic step because the alternative hypothesis, that the speed of light was not constant, was less tenable. In particular, he was relying on the work of Scottish physicist James Clerk Maxwell, who in 1865 published *A Dynamical Theory of the Electromagnetic Field*. This paper set out the four fundamental and eminently successful equations governing the behaviour of electric and magnetic fields. These are known as Maxwell's equations and are seen as one of the crowning triumphs of nineteenth century physics.

Maxwell proposed that light is an electromagnetic phenomenon with a speed that is both constant in all directions and independent of the motion of the light source. But this result is inconsistent with the Galilean transformation equations, which insist that velocity is not an invariant quantity. In other words, Maxwell's equations are not invariant under the Galilean transformations. Maxwell, like many physicists of his day, thought the solution might lie with an invisible, unknown stuff called the **ether**, which fills all space and is the medium through which light travels, much as sound waves travel through air or ripples through water.

According to this theory, light would travel with its correct, Maxwell equation-derived speed only with respect to the stationary ether, known as the ether frame. Observers moving with respect to the ether could then use the Galilean transformations to find the speed of light relative to them. It was a clever solution except (a) no one could find any trace of the ether (and never have done), and

(b) sophisticated attempts (the most famous being the Michelson-Morley experiment in 1887) to confirm that light travels with different speeds depending on the relative velocity of the observer all returned a constant result for the speed of light.

Plus there was another problem, which seems to have troubled Einstein the most. This was to do with the idea of Galilean relativity (see Section 2.3), which states there is no preferred frame of reference for describing the laws of mechanics. Einstein trusted this principle to such an extent that he eventually extended it to cover all the laws of physics and used it as one of the foundations of special relativity. Unfortunately, the notion of a stationary ether was incompatible with Galilean relativity. This is because if light only travelled with its 'correct' speed c with respect to the ether frame, then this frame would constitute an absolute frame of reference. An absolute frame is a preferred frame. An inertial observer would then, by using the Galilean transformations, be able to calculate (assuming that they could ever get the experiments to work!) the relative velocity of their own frame of reference. Inertial frames could then be distinguished from each other, and bye-bye Galilean relativity.

Either Maxwell's theory or the Galilean transformations had to be wrong. Although various physicists tried to 'fix' Maxwell's equations, Einstein eventually came to the conclusion that the problem lay with the Galilean transformations. These, he showed, were actually special 'low velocity solutions' for a more general set of coordinate transformation equations called the Lorentz transformations.

Einstein's solution – the theory of special relativity – was to reject the need for an ether and accept Maxwell's constant speed of light at face value. The latter requirement meant reformulating the 'common sense' assumptions about time and space that subtly underpin the Galilean transformations.

3.2.8 The two postulates of special relativity

Einstein based his theory of special relativity on two postulates, or fundamental assumptions, about the way the universe works.

1. The principle of relativity – the laws of physics are the same in any inertial frame of reference, regardless of position or velocity.

2. The constancy of the speed of light in a vacuum – the speed of light in vacuum has the same value $c = 3 \times 10^8 \, \mathrm{m\,s^{-1}}$ in all inertial frames of reference.

The first postulate, the extension of Galilean relativity to all the laws of physics, isn't too hard to accept and understand. The second postulate is the bombshell. This is the one whose implications fly against our everyday assumptions about time and space.

We begin our exploration of special relativity by looking at spacetime diagrams.

3.3 Spacetime diagrams

3.3.1 Units and notation

How do we visualise four-dimensional spacetime? We don't, but we do take heart from the words [14] of British theoretical physicist and cosmologist Stephen Hawking, who wrote:

'It is impossible to imagine a four-dimensional space. I personally find it hard enough to visualise three-dimensional space! However, it is easy to draw diagrams of two-dimensional spaces, such as the surface of the Earth.'

And so we make much use of spacetime diagrams (also known as Minkowski diagrams), where instead of an x and y axis there is (usually) a vertical time axis and a horizontal spatial x axis.

Special relativity is predicated on the assumption that light travels at a constant speed for any inertial observer. We therefore need to be able to draw the path of a light ray on our spacetime diagram. Such a ray could start at any point on the x axis and travel either towards increasing or decreasing values of x (we can imagine physically standing on a particular point of the x axis and shining a beam of light either to our right or to our left along the axis).

The snag with trying to draw the path of a light ray on a spacetime diagram is, because light travels very, very fast, if we used SI units of seconds and metres for our two axes, a line representing that path would be as near to horizontal as makes no difference.

To get around this problem, we multiply the time in seconds by the speed of light in metres per second and use this quantity ct as the units for the vertical time axis. One unit of ct is the speed of light c $\left(3 \times 10^8 \,\mathrm{m\,s^{-1}}\right)$ multiplied by the time t $\left((1/3) \times 10^{-8}\,\mathrm{s}\right)$ that it takes light to travel 1 metre, which equals 1.

By using ct units of time we have defined the speed of light c as being equal to 1 and can now draw the path of a light ray as a line with a slope of $45°$, pointing either to the left or to the right depending on its direction of travel.

Remember that the t in ct still refers to the SI time unit of seconds. But one unit of ct doesn't refer to 1 second but to $(1/3) \times 10^{-8}$ seconds, a much smaller quantity. As distance equals speed multiplied by time, using ct units of time effectively means we are measuring time in metres.

To summarise these new 'natural' units and how we convert them to standard SI units:

$$3 \times 10^8 \,\mathrm{m\,s^{-1}} = 1,$$

meaning we have defined the speed of light as being equal to 1.

$$1\,\mathrm{s} = 3 \times 10^8 \,\mathrm{m},$$

$$1\,\mathrm{m} = \frac{1}{3 \times 10^8}\,\mathrm{s}.$$

Using the metre as a unit of both time and distance is a convenient and common practice in relativity.

A single point in Minkowski space is called an event and has four components (ct, x, y, z). In index notation (see Section 1.15) we can describe the components of an event as $x^\mu = (ct, x, y, z) = \left(x^0, x^1, x^2, x^3\right)$ where $x^0 = ct, x^1 = x, x^2 = y, x^3 = z$. x^μ is also sometimes called the particle's four-position. A line in Minkowski space represents a particle's movement through spacetime – a bit like how a vapour trail shows the path of a jet aeroplane – and is called the particle's world-line.

3.3.2 Spacetime diagrams – basic concepts

Figure 3.3 shows a two-dimensional slice of spacetime for an observer O in an inertial frame of reference S. A single point in this space is an event, something that happens instantaneously at a

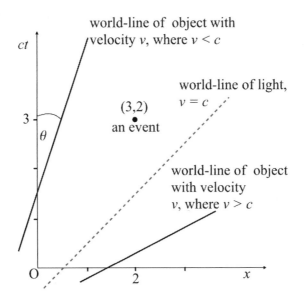

Figure 3.3: A simple spacetime diagram for an observer in frame S.

certain value of x. For example the point $(3, 2)$ describes a point in spacetime with time coordinate $ct = 3$ and spatial coordinate $x = 2$.

(Remember, we are using metre units of time, so if $ct = 3$, then $t = 3 \div (3 \times 10^8) = 10^{-8}$ seconds.)

Because we are using ct time units, a $45°$ straight line shows the path of a light ray. We can draw an infinite number of such lines, each representing a light ray starting at a different value of x when $ct = 0$. An object travelling at a constant velocity less than c will have a straight world-line at an angle greater than $45°$ to the x axis.

If an object is travelling with a constant velocity v then that velocity will equal distance travelled divided by time taken and is given by

$$v = \frac{x}{t}. \tag{3.3.1}$$

The object will have a straight world-line with an angle θ to the ct axis, where

$$\tan \theta = \frac{x}{ct}. \tag{3.3.2}$$

Substituting (3.3.1) into (3.3.2) gives

$$\tan \theta = \frac{v}{c}. \tag{3.3.3}$$

Something travelling with a constant velocity significantly slower than light will be shown by a straight line very close to the vertical ct axis.

Problem 3.2. The world-line of a particle moving with constant velocity makes an angle of $10°$ with the ct axis. How fast is the particle travelling?

Equation (3.3.3) tells us that

$$\tan\theta = \frac{v}{c} = \tan(10°) = 0.18,$$

therefore

$$v = c\tan\theta = c\tan(10°) = 0.18c.$$

The particle is travelling at $0.18c = 0.18 \times 3 \times 10^8 = 5.4 \times 10^7 \, \mathrm{m\,s^{-1}}$.

Problem 3.3. What angle would the world-line of a rifle bullet travelling with a velocity of $1000 \, \mathrm{m\,s^{-1}}$ make with the ct axis?

Let the rifle bullet's world-line make an angle θ with the ct axis. The time t taken for the bullet to travel 1 metre equals $1/1000 = 10^{-3}\,\mathrm{s}$. Equation (3.3.3) tells us

$$\tan\theta = \frac{v}{c} = \frac{10^3}{(3\times10^8)} = \frac{1}{3\times10^5}.$$

If we know the tangent we can find the angle by using an inverse tan calculator:

$$\theta = 0.00019°.$$

Which is a very small angle. To visualise how small, think of a right angled triangle with an adjacent side of $300{,}000\,\mathrm{m} = 300\,\mathrm{km} \approx 186$ miles and an opposite side of 1 metre! And that's for an object – a rifle bullet – that in our everyday world is moving very fast indeed.

3.3.3 Adding a second observer

How would we show another inertial frame S′ belonging to a second observer O′? First, we make life simpler for ourselves and assume that both frames are in standard configuration (see Section 3.2.7) and that frame S′ is moving with a constant velocity v relative to frame S.

We can start by drawing the ct' time axis for frame S′. But how do we draw this axis in our diagram?

First, because we are using frames in standard configuration, we know that the origins of frames S and S′ coincide when time $ct = 0$ in frame S and when time $ct' = 0$ in frame S′. Therefore, the ct' axis must pass through the origin of frame S.

Second, consider every possible event that can happen in frame S′ when the spatial coordinate x' equals zero. All these points, joined together, will form the ct' axis. But the point $x' = 0$ is moving along the x axis with a velocity v (it must be as the frame S′ is moving with that velocity). We can therefore draw the ct' axis as the world-line of the moving point $x' = 0$. Equation (3.3.1) tells us that the ct' axis will make an angle θ with the vertical ct axis of frame S, as shown in Figure 3.4.

We now need to draw the x' axis of frame S′. This axis can be drawn by connecting all the events with the time coordinate $ct' = 0$. But how to do this?

As with the ct' axis, we know the x' axis must pass through the origin of frame S because we are in standard configuration.

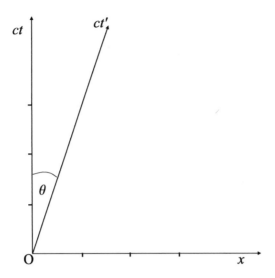

Figure 3.4: The ct' time axis of a second frame S$'$.

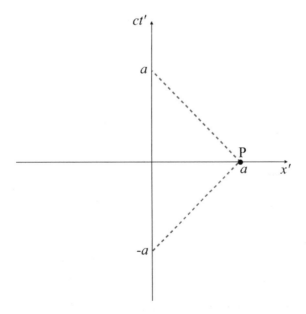

Figure 3.5: The path of a reflected light ray.

Recall that one of the postulates of special relativity is that the speed of light c is the same for all inertial observers. This means that any light ray will always be drawn with a 45° slope for any inertial frame. Figure 3.5 shows the spacetime diagram of observer O' (with axes ct' and x'). The broken line represents a light ray emitted at $x' = 0$, $ct' = -a$, striking a mirror at $x' = a$, $ct' = 0$ and reflected back to $x' = 0$, $ct' = a$. We call the point (or event) P where $x' = a$, $ct' = 0$. As a varies, so will the position of point P. In fact, because of the constancy of the speed of light, point P (where ct' always equals zero) traces out the x' axis for different values of a.

We use this property of point P to define the position of the x'axis in Figure 3.6. This time we are back to showing the reference frame S of observer O, who records a light ray (the broken line) emitted at point Q ($x' = 0$, $ct' = -a$), striking a mirror at point P ($x' = a$, $ct' = 0$), and reflected back to point R ($x' = 0$, $ct' = a$). The broken lines from points Q and R must be at 45° to the x axis. Where these lines intersect must be at point P, the position of the mirror at $x' = a$, $ct' = 0$. Therefore, a straight line drawn through P and the origin must define the x' axis.

The triangle ORP is an isosceles triangle (two sides of equal length, and therefore two equal angles) bisected by a line at 45° to the x axis and passing through O. Therefore, the x' axis must make the same angle with the x axis as the ct' axis makes with the ct axis, ie $\theta = \alpha$.

We can therefore say that the equation of the x'axis is

$$ct = (\tan \alpha)\, x = \left(\frac{v}{c}\right) x \qquad (3.3.4)$$

and the equation of the ct'axis is

$$ct = \frac{x}{(\tan \alpha)} = \left(\frac{c}{v}\right) x. \qquad (3.3.5)$$

Figure 3.7 shows the frames S and S' from the point of view of O (who is observing from frame S) where frame S' (on which O' is sitting) is moving to the right.

The same physical situation is shown in Figure 3.8, but this time from the point of view of O', who sees O moving to the left.

Now, this might seem a little confusing. We have an inertial frame S with coordinate axes x and ct, and we have another inertial frame S' (moving with uniform velocity v relative to S) with its own coordinate axes x' and ct', all drawn on the same diagram. How do we now describe the coordinates of a particular event?

It's straightforward. All we need do is construct coordinate lines parallel to the respective axes. Take, for example, event A shown in Figure 3.9. We read off A's coordinates with respect to the x and ct axes normally, as we would on any Cartesian grid. These are x_1 and ct_1 respectively. We then construct a line passing through A and parallel to the x' axis. Where this line crosses the ct' axis gives us the ct' coordinate – ct'_1. Similarly, we then construct a line passing through A and parallel to the ct' axis. Where this line crosses the x' axis give us the x' coordinate – x'_1.

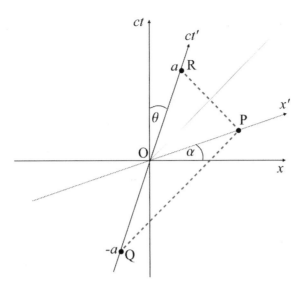

Figure 3.6: The x' axis of a second frame S'.

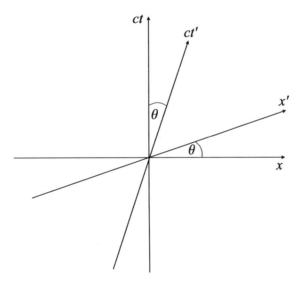

Figure 3.7: Spacetime diagram for observer O.

It's here worth noting the obvious point that the x' and x axes do not coincide. They would if we were using the Galilean transformations (3.2.1), which are of course based on Galilean/Newtonian assumptions of absolute time and space. In that case we would still draw a sloping t' axis (where $t' = t$), but the x' and x axes would be on the same line (where $x' = x - vt$). The second postulate (the constancy of the speed of light) necessitates a sloping x' axis and destroys all assumptions of absolute time and space.

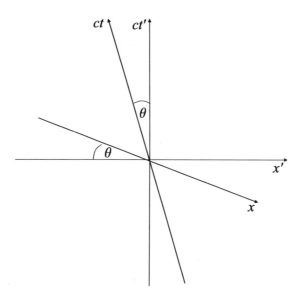

Figure 3.8: Spacetime diagram for observer O'.

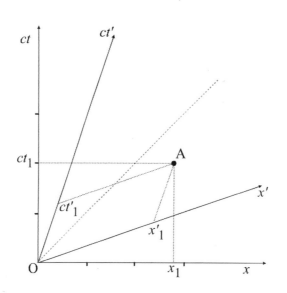

Figure 3.9: Reading the coordinates of an event.

Depending on the accuracy of our drawing and assuming that we knew how to calibrate the ct' and x' axes, we could estimate the coordinates of event A as measured by the observers on frames S and S'. Taking measurements from graphs isn't of course very accurate. Later, we discuss the Lorentz transformations, which allow us to algebraically calculate the measurements of observers in different inertial frames, including calibrating the ct' and x' axes.

3.3.4 Simultaneity and causality

We can now demonstrate the surprising result that the simultaneity of events can depend on the observer's reference frame. Figure 3.10 shows four events – A, B, C, D. In what order do they happen?

Observer O in inertial frame S will see the events happening (as time ct increases) separately in the order A, C, D, B. However, observer O′ in inertial frame S′ will use the ct' axis to record his time and will first see A and D occurring simultaneously and then see C and B occurring simultaneously. Just by the fact that the two observers don't share a common time axis, two events that are simultaneous for one observer cannot be simultaneous for the other.

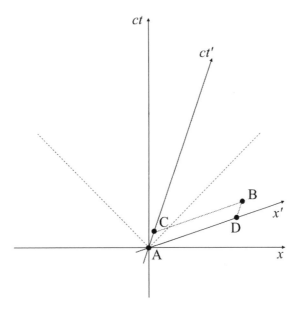

Figure 3.10: Relativity of simultaneity.

This phenomenon is known as the **relativity of simultaneity**. If two events occur at the same time at the same point in space, all observers agree they occurred simultaneously. If the events are separated in space then whether they are simultaneous or not depends on the reference frame of the observer.

We can also see from Figure 3.10 that not only will the two observers disagree about which events are simultaneous, but in the case of C and D they will also not even see the events occurring in the same order. Now this result is even stranger than the relativity of simultaneity because it appears to overthrow the fundamental notion of **causality**. Event X can only cause event Y if X occurs before Y. Say that X is the event of me dropping a book and Y is the event of the book hitting the floor. If an observer cannot agree with me which event happened first we could end up with the bizarre situation of them seeing the book hitting the floor before I've dropped it!

Fortunately, one of the consequences of special relativity is that, in an observer's inertial frame, no information signal or material object can travel faster than light. What that means is that although observers may disagree about the order of two events, they will not disagree about the order of two events that can be linked by a light signal.

(The 'in an observer's inertial frame' condition is crucial. In an expanding universe, as we'll see when we look at cosmology, galaxies can move away from us at the speed greater than that of light. However, this motion is not in any observer's inertial frame as it is space itself that is expanding.)

In Figure 3.11 we have introduced the **lightcone**, formed by rotating the light rays (the two dotted lines) passing through A around the ct axis (the y axis shows the lightcone occupies two spatial dimensions: x and y). Assuming that all four events still lie directly above the x axis, we now see that the only event (out of B, C and D) that could be linked to A by a signal not moving faster than the speed of light is C. This is because only C is 'inside' the lightcone. In other words, because no signal can travel faster than light, the only event that could be caused by A is C. Events, such as A and C, that lie within each other's lightcones are said to be **causally related**. If two events are causally related in one inertial frame they are causally related in all inertial frames.

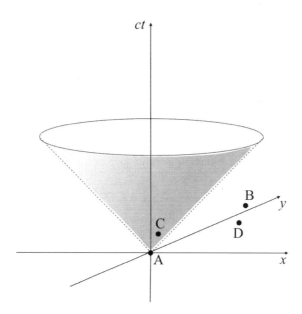

Figure 3.11: The lightcone.

If any event, such as C, is inside A's lightcone, it is possible to draw a ct' axis through A (an event at the origin) and C. This means there is some inertial frame where A and C occur at the same place but at different times.

Conversely, only a signal travelling faster that light (a line at less than 45°to the x axis) could link events A to B, A to D and C to D. So these events are not causally related. As there is no possible causal link between C and D, for any frame of reference, my book will never hit the floor before it is dropped.

If any event, such as B, is outside A's lightcone, it is possible to draw an x' axis through A (an event at the origin) and B. This means there is some inertial frame where A and B occur at the same time but at different places.

The lightcone (see Figure 3.12) is a very useful concept in relativity, showing in three dimensions (one of time, two of space) the location of those events that may have a causal relationship with an event occurring now at the origin. The sides of the lightcone are formed from light rays passing through the origin. The cone below the origin, represents past events that may have *caused* the event at the origin. The cone above the origin contains events that may have *been caused* by the event at the origin. A plane passing through the origin and the x and y axes represents the present.

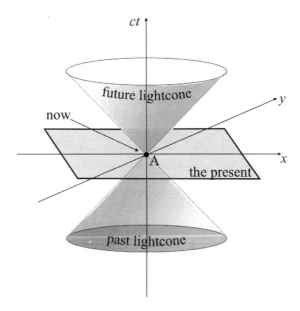

Figure 3.12: Past, present and future events.

3.3.5 Invariance of the interval

Observers O and O$'$ will describe the position of the four events A, B, C, D shown in Figure 3.10 using different coordinate values. However, they will agree on the spacetime separation between any of the two events, where the separation or interval Δs^2 is given by

$$\Delta s^2 = c^2 \Delta t^2 - \Delta x^2 - \Delta y^2 - \Delta z^2. \tag{3.3.6}$$

This theorem is known as the **invariance of the interval** and we'll prove it, using the Lorentz transformations, in Section 3.4.9. We say the interval is invariant because any two different inertial observers will calculate Δs^2 and obtain the same answer.

Because the interval Δs^2 may be positive, negative or zero in value, it's best to think of it as a single symbol rather than the square of something. We don't really want to go down the road of taking square roots of negative numbers.

Problem 3.4. Two events occur at $(ct_1, x_1, y_1, z_1) = (3, 7, 0, 0)$ and $(ct_2, x_2, y_2 z_2)$ $= (5, 5, 0, 0)$. What is their spacetime interval?

$\Delta ct = 5 - 3 = 2 \, \text{m}$, $\Delta x = 5 - 7 = -2 \, \text{m}$.

The spacetime interval (3.3.6) is given by

$$\Delta s^2 = c^2 \Delta t^2 - \Delta x^2 - \Delta y^2 - \Delta z^2$$

$$\Delta s^2 = (2 \, \text{m})^2 - (-2 \, \text{m})^2 = 0.$$

Depending on the values of ct, x, y, z the spacetime separation Δs^2 may be positive, negative or zero.

Figure 3.13 illustrates the three ways that an event A at the origin may be related to other events in spacetime:

- Time-like interval – where $\Delta s^2 > 0$, describes events within A's lightcone. These events are causally related to A, and there will be some inertial frame where A and C occur at the same place but at different times.

- Space-like interval – where $\Delta s^2 < 0$, describes events outside A's lightcone. These events are not causally related to A, and there will be some inertial frame where A and C occur at the same time but at different places.

- Light-like interval – where $\Delta s^2 = 0$, describes events on A's lightcone. These events are causally related to A, but they can only be linked to A by a light signal.

Schutz [28] refers to the invariance of the interval as 'probably the most important theorem of special relativity.'

On our spacetime diagram, where we ignore the y and z axes, (3.3.6) reduces to

$$\Delta s^2 = c^2 \Delta t^2 - \Delta x^2, \tag{3.3.7}$$

this being the equation observers O and O' would use to describe the separation of any of the four events A, B, C, D in Figure 3.10.

3.3.6 Invariant hyperbolae

Now we've learned that the interval Δs^2 is invariant for any inertial observer, we can see how to calibrate the x' and ct' axes of frame S' on a spacetime diagram. Consider the following equations:

$$c^2 t^2 - x^2 = -a^2, \tag{3.3.8}$$

$$c^2 t^2 - x^2 = b^2. \tag{3.3.9}$$

Both these define curves known as **hyperbolae** (singular – **hyperbola**) on the spacetime diagram of observer O. These are shown in Figure 3.14.

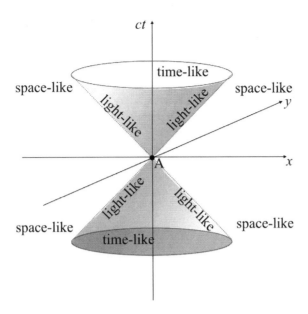

Figure 3.13: Spacetime separations.

(A hyperbola is a type of conic section: one of the family of curves you get if you slice through a cone. The other conic sections are the circle, ellipse and parabola, which we met when drawing simple functions in Section 1.5.1.)

There is an obvious similarity between the two equations (3.3.8) and (3.3.9), and the spacetime interval equation (3.3.7) $\Delta s^2 = c^2 \Delta t^2 - \Delta x^2$. Recalling that space-like intervals are where $\Delta s^2 < 0$ and time-like intervals are where $\Delta s^2 > 0$, we can see that if we let $\Delta s^2 = -a^2$ we have defined a space-like interval, and if we let $\Delta s^2 = b^2$ we have defined a time-like interval. What this means is that each hyperbola shown in Figure 3.14 joins up all the events that have an equal spacetime separation from an event at the origin. The two hyperbolae passing through $x = -a$ and $x = a$ join all the events with a space-like interval $-a^2$; the hyperbolae passing through $ct = -b$ and $ct = b$ join up all the events with a time-like interval b^2.

Figure 3.15 shows a frame S′ moving with uniform relative motion with respect to a frame S. We are trying to calibrate the ct' and x' axes of frame S′. Consider the invariant hyperbola $c^2 t^2 - x^2 = b^2$ passing through events A and B. Event A is on the ct axis, and therefore $x = 0$. If we let event A occur at $ct = 1$, then b^2, the time-like separation of A from the origin, must also equal 1. Because the hyperbola joins up every event with the same spacetime interval, event B on the ct' axis (where $x' = 0$) must also occur when $ct' = 1$.

Similarly, the invariant hyperbola $c^2 t^2 - x^2 = -a^2$, passing through events C and D, connects events with the same space-like separation of $-a^2$. If we let C occur at $x = 1$, then $ct = 0$ and $-a^2$ must equal -1. Event D, on the same invariant hyperbola, must also occur at $x' = 1$ (where $ct' = 0$).

Notice that D appears to be further from the origin than C, even though they have the same spacetime interval. Similarly, B and A have the same spacetime separation from the origin even though B looks to be further away. These confusions arise from the fact that we are trying to

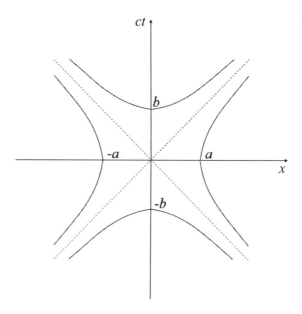

Figure 3.14: Invariant hyperbolae.

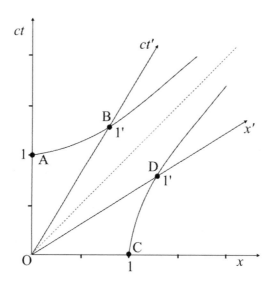

Figure 3.15: Calibrating the ct' and x' axes.

represent non-Euclidean spacetime on a flat Euclidean surface. What do we mean by non-Euclidean? We mean the distance between any two points is not given by the Euclidean line element $ds^2 =$

$dt^2 + dx^2$, but by the non-Euclidean line element $ds^2 = dt^2 - dx^2$. The minus sign makes all the difference.

We can use invariant hyperbolae to illustrate two important physical implications of the invariance of the interval: time dilation and length contraction.

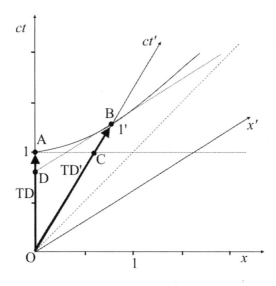

Figure 3.16: Time dilation.

Time dilation   Figure 3.16 shows an invariant hyperbola passing through the ct axis at $ct = 1$, and the ct' axis at $ct' = 1$. The horizontal dotted line passing through events A and C is a line of simultaneity for observer O, meaning all events on that line have the same time value of $ct = 1$. The sloping dotted line passing through B and D (actually a tangent to the hyperbola at B) is a line of simultaneity for observer O', meaning all events on that line have the same time value of $ct' = 1$. What do the two observers measure?

- Observer O' measures event C occurring at a time $ct' < 1$ on his ct' axis. However, observer O measures the same event occurring at time $ct = 1$ on his ct axis. From the point of view of O, the clocks on frame S' belonging to O' are running slow. The black arrow **TD** is the time dilation observed by O.

- Observer O measures event D occurring at a time $ct < 1$ on his ct axis. However, observer O' measures the same event occurring at time $ct' = 1$ on his ct' axis. From the point of view of O', the clocks on frame S belonging to O are running slow. The black arrow **TD'** is the time dilation observed by O'.

The remarkable result that both observers measure each other's clocks to be running slow! This effect is known as **time dilation**.

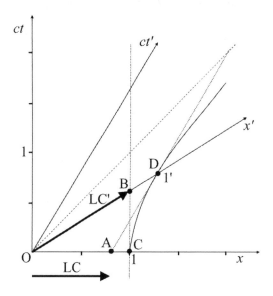

Figure 3.17: Length contraction.

Length contraction   Figure 3.17 shows an invariant hyperbola passing through the x axis at $x = 1$, and the x' axis at $x' = 1$. The vertical dotted line passing through events B and C is a line that has a constant value of $x = 1$ for observer O. The sloping dotted line (actually a tangent to the hyperbola at D) passing through A and D is parallel to the ct' axis, and therefore has a constant value of $x' = 1$ for observer O'. What do the two observers measure?

- Observer O' measures distance OD as $x' = 1$ on his x' axis. Point A will also have the same value $x' = 1$ for O' because it is on the line AD. However, observer O measures the same distance as OA < 1 on his x axis. From the point of view of O, the distance OD $= 1$ has contracted to OD < 1. The black arrow **LC** is the length contraction observed by O.

- Observer O measures distance OC as $x = 1$ on his x axis. Point B will also have the same value $x = 1$ for O because it is on the line BC. However, observer O' measures the same distance as OB < 1 on his x' axis. From the point of view of O', the distance OC $= 1$ has contracted to OC < 1. The black arrow **LC'** is the length contraction observed by O'.

Both observers measure moving objects to shrink in the direction of motion, a phenomenon known as **length contraction** or **Lorentz contraction**.

It's essential to realise that both time dilation and length contraction are genuine, observable effects, not an optical illusion caused by faulty clocks, dodgy rulers, or the time it takes light to travel from a moving event to an observer. An astronaut in a rocket flying past the Earth near the speed of light would measure the planet as being squashed along her direction of travel (see Figure 3.18). However, *measurement* is not the same as *visual appearance*. What an observer of a rapidly moving object actually sees or photographs depends not only on the mathematics of length contraction, but also on the time lag of photons from different parts of the object arriving at her eyes or camera, a phenomenon known as the **Terrell** or **Penrose–Terrell effect**.

Figure 3.18: A *measured* squashed Earth means you are moving close to the speed of light.

Of course, we don't notice these phenomena in our everyday world because we don't travel at relative speeds anything near the speed of light. As Michio Kaku [16] whimsically expresses it:

'For everyday velocities, Newton's laws are perfectly fine. This is the fundamental reason why it took over two hundred years to discover the first correction to Newton's laws. But now imagine the speed of light is only 20 miles per hour. If a car were to go down the street, it might look compressed in the direction of motion, being squeezed like an accordion down to perhaps 1 inch in length, for example, although its height would remain the same. Because the passengers in the car are compressed down to 1 inch, we might expect them to yell and scream as their bones are crushed. In fact, the passengers see nothing wrong, since everything inside the car, including the atoms in their bodies, is squeezed as well.

As the car slows down to a stop, it would slowly expand from about 1 inch to about 10 feet, and the passengers would walk out as if nothing happened. Who is really compressed? You or the car? According to relativity, you cannot tell, since the concept of length has no absolute meaning.'

3.4 The Lorentz transformations

Having gained some insight into the strange nature of spacetime from looking at spacetime diagrams, we need to develop a precise algebraic formulation of how coordinates change for different inertial observers. We do this using a set of equations called the **Lorentz transformations**, first derived by the Dutch physicist Hendrik Lorentz (1853–1929).

Let's reiterate the point of what we are now doing. An event t, x, y, z in spacetime is described

by an observer O sitting on an inertial frame S using his coordinate system. Another observer O'
in an inertial frame S' describes the same event as t', x', y', z' using his coordinate system. The
fundamental question is: how are the coordinates t, x, y, z and t', x', y', z' related? As usual, we are
assuming that both frames are in standard configuration.

We've seen that if we assume the validity of the second postulate (the speed of light c is constant for
all inertial observers) we cannot use the Galilean transformations to compare one set of coordinates
to another. Instead we use the Lorentz transformations. We'll derive the Lorentz transformations
later (in Section 3.4.1). For now we'll simply give them.

We start with the **Lorentz factor**, denoted by the Greek letter gamma γ, where

$$\gamma = \frac{1}{\sqrt{1 - (v/c)^2}}, \tag{3.4.1}$$

v is the relative velocity of the two frames and c is, as ever, the speed of light.

We can see by the structure of this equation that if v is equal to or greater than c, we end up with
the square root of zero or of a negative number. The expression for γ then becomes meaningless,
suggesting that c is the maximum possible speed in nature. We are therefore only considering
situations where v is less than c.

Now we see that, providing $v \neq 0$, the Lorentz factor will always be greater than 1 (because $v < c$,
and therefore $\sqrt{1 - (v/c)^2} < 1$). For normal everyday speeds it will be very, very close to 1 because
$(v/c)^2 \approx 0$. Figure 3.19 shows how the Lorentz factor changes with velocity v.

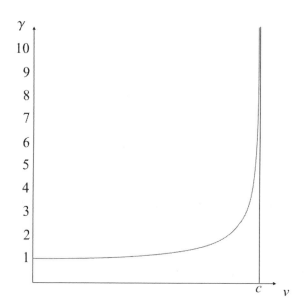

Figure 3.19: Lorentz factor as a function of velocity.

Some values of the Lorentz factor for various values of v/c are:

v/c	γ
0.000	1.000
0.100	1.005
0.300	1.048
0.500	1.155
0.700	1.400
0.900	2.294
0.990	7.089
0.999	22.366

Problem 3.5. What is the value of γ for two inertial frames moving with a relative velocity of $1000\,\mathrm{m\,s^{-1}}$?

Using (3.4.1)

$$\gamma = \frac{1}{\sqrt{1 - (v/c)^2}} = \frac{1}{\sqrt{1 - (1000/(3 \times 10^8))^2}}$$

$$\gamma = 1.0000000000056.$$

This velocity, incidentally, is the assumed velocity of a rifle bullet we used in Problem (3.3). As you can see, the Lorentz factor for two frames moving with a relative velocity equal to the speed of a rifle bullet is still exceedingly close to 1.

The Lorentz factor plugs into the Lorentz transformations to give the values of the coordinates t', x', y', z'.

The Lorentz transformations are:

$$t' = \gamma\left(t - \frac{vx}{c^2}\right), \tag{3.4.2}$$

$$x' = \gamma\left(x - vt\right),$$

$$y' = y,$$

$$z' = z.$$

The inverse Lorentz transformations (to give the values of the coordinates t, x, y, z) are:

$$t = \gamma\left(t' + \frac{vx'}{c^2}\right), \tag{3.4.3}$$

$$x = \gamma\left(x' + vt'\right),$$

$$y = y',$$

$$z = z'.$$

Note that when $v \ll c$ the Lorentz transformations approximate to the Galilean transformations (3.2.1):

$$t' = t,$$

$$x' = x - vt,$$

$$y' = y,$$
$$z' = z.$$

In Section 3.3.3 we drew the coordinates ct' and x' of a second inertial frame S' on the spacetime diagram of frame S. A useful way of looking at the Lorentz transformations is that they allow us to calibrate the axes of a second frame on a spacetime diagram.

Problem 3.6. An observer in frame S records an event X occurring at coordinates ($ct = 3\,\text{m}$, $x = 4\,\text{m}$, $y = 0\,\text{m}$, $z = 0\,\text{m}$). Another observer O' in frame S' is moving with velocity $v = 3c/4$ in the positive x direction. What are the coordinates of event X as described by O'?

We first need to calculate the Lorentz factor (3.4.1), which equals

$$\gamma = \frac{1}{\sqrt{1 - (v/c)^2}} = \frac{1}{\sqrt{1 - (3/4)^2}} = \frac{4}{\sqrt{7}}.$$

Using the Lorentz transformations (3.4.2) we can find the t', x', y', z' coordinates. We are using ct units of time, so we need to multiply $t' = \gamma\left(t - \frac{vx}{c^2}\right)$ by c to give

$$ct' = c\gamma\left(t - \frac{vx}{c^2}\right) = \frac{4}{\sqrt{7}}\left(ct - \frac{cvx}{c^2}\right) = \frac{4}{\sqrt{7}}\left(3 - \frac{3c^2 \times 4}{4c^2}\right) = \frac{4}{\sqrt{7}}(3 - 3) = 0\,\text{m}$$

$$x' = \gamma(x - vt) = \frac{4}{\sqrt{7}}\left(4 - \frac{3}{4}ct\right) = \frac{4}{\sqrt{7}}\left(4 - \frac{3 \times 3}{4}\right) = \frac{4}{\sqrt{7}}\left(\frac{7}{4}\right) = \sqrt{7}\,\text{m}$$

$$y' = y = 0\,\text{m}$$

$$z' = z = 0\,\text{m}.$$

Recall that because we are using ct units of time, all coordinates are expressed in units of distance, ie metres (m).

3.4.1 Deriving the Lorentz transformations

We can now derive the Lorentz transformations. Remember, we are trying to find the relationship between the coordinates of an event in two inertial frames S and S', with S' moving with a velocity v with respect to S. As usual, we are assuming that both frames are in standard configuration (so we can ignore the y and z coordinates and concentrate on the t and x coordinates.

We first assume that general equations relating the t and x coordinates in the S frame to the t' and x' coordinates in the S' frame will be of the form

$$t' = a_0 + a_1 t + a_2 x + a_3 t^2 + a_4 x^2 + \cdots$$

$$x' = b_0 + b_1 x + b_2 t + b_3 x^2 + b_4 t^2 \cdots$$

the dots represent higher powers of x and t. The a_i and b_i terms are unknown constants.

Newton's first law holds in all inertial frames, meaning the frames cannot be accelerating, and therefore we assume a linear relationship and omit the squared and higher powered terms to give

$$t' = a_0 + a_1 t + a_2 x,$$

$$x' = b_0 + b_1 x + b_2 t.$$

We now need to find the unknown constants $a_0, a_1, a_2, b_0, b_1, b_2$.

Because the frames are in standard configuration we know that when the origins coincide $t' = t = 0$ and $x' = x = 0$. Therefore, it must be that $a_0 = b_0 = 0$ and we can now say

$$t' = a_1 t + a_2 x, \tag{3.4.4}$$

$$x' = b_1 x + b_2 t. \tag{3.4.5}$$

After time t, the origin of frame S' (given by $x' = 0$) will have moved a distance $x = vt$ in frame S. We can therefore rewrite (3.4.5) as

$$0 = b_1 vt + b_2 t,$$

which can be rewritten as

$$b_2 = -b_1 v.$$

Next we divide (3.4.5) by (3.4.4) and substitute the above value of b_2 to give

$$\frac{x'}{t'} = \frac{b_1 x - b_1 vt}{a_1 t + a_2 x}. \tag{3.4.6}$$

After time t', the origin of frame S (given by $x = 0$) will be at $x' = -vt'$ in frame S'. We can therefore rewrite (3.4.6) as

$$\frac{-vt'}{t'} = \frac{-b_1 vt}{a_1 t},$$

where we can cancel t', t and v to find $a_1 = b_1$, which we can substitute into (3.4.6) and then divide top and bottom of the right-hand side of (3.4.6) by t to get

$$\frac{x'}{t'} = \frac{b_1 (x/t) - vb_1}{b_1 + a_2 (x/t)}. \tag{3.4.7}$$

Now recall that one of the postulates of special relativity is that the speed of light in a vacuum has the same value c in all inertial frames of reference. This means that if a beam of light is emitted in the positive direction along the x axis from the origin $(ct = 0, x = 0)$ it will have a speed of $c = x'/t'$ as well as $c = x/t$, which we can substitute into (3.4.7) to get

$$c = \frac{b_1 c - vb_1}{b_1 + a_2 c},$$

which rearranges to

$$cb_1 + c^2 a_2 = b_1 c - vb_1$$

giving (because we know $a_1 = b_1$)

$$a_2 = \frac{-vb_1}{c^2} = \frac{-va_1}{c^2}.$$

We can now rewrite (3.4.4) $t' = a_1 t + a_2 x$ to give

$$t' = a_1 \left(t - vx/c^2 \right). \tag{3.4.8}$$

Looking at (3.4.5) $x' = b_1 x + b_2 t$, we recall that $b_1 = a_1$ and $b_2 = -b_1 v = -a_1 v$ and we can say

$$x' = a_1 \left(x - vt \right). \tag{3.4.9}$$

We now need to write (3.4.8) and (3.4.9) in terms of t and x. In effect this means we have changed from being an observer on frame S looking at frame S', to being an observer on frame S' looking at frame S. Frame S is therefore now moving away from us along the negative x' axis so we need to replace v by $-v$ to get

$$t = a_1 \left(t' + vx'/c^2 \right), \tag{3.4.10}$$

$$x = a_1 \left(x' + vt' \right). \tag{3.4.11}$$

We are almost there. We now substitute (3.4.10) and (3.4.11) into (3.4.9) to get

$$x' = a_1 \left(a_1 \left(x' + vt' \right) - va_1 \left(t' + vx'/c^2 \right) \right) = a_1 \left(a_1 x' + a_1 vt' - va_1 t' - v^2 a_1 x'/c^2 \right)$$

$$x' = \left(a_1 \right)^2 x' - v^2 \left(a_1 \right)^2 x'/c^2$$

$$1 = \left(a_1 \right)^2 \left(1 - v^2/c^2 \right)$$

$$a_1 = \frac{1}{\sqrt{1 - \left(v/c \right)^2}}.$$

Bingo! We have now shown that the constant a_1 is the Lorentz factor (3.4.1) γ and can substitute this into (3.4.8), (3.4.9), (3.4.10) and (3.4.11) to obtain the Lorentz transformations (3.4.2) and inverse Lorentz transformations (3.4.3).

3.4.2 Lorentz transformation matrix

Another way of expressing the Lorentz transformations is in matrix form:

$$\begin{pmatrix} ct' \\ x' \\ y' \\ z' \end{pmatrix} = \begin{pmatrix} \gamma & -\gamma v/c & 0 & 0 \\ -\gamma v/c & \gamma & 0 & 0 \\ 0 & 0 & 1 & 0 \\ 0 & 0 & 0 & 1 \end{pmatrix} \begin{pmatrix} ct \\ x \\ y \\ z \end{pmatrix}, \tag{3.4.12}$$

where the 4×4 matrix on the right-hand side is known as the **Lorentz transformation matrix**. Recalling the rule for matrix multiplication (from Section 1.12.1) we see that

$$ct' = (\gamma \times ct) + \left(\frac{-\gamma v}{c} \times x \right) + (0 \times y) + (0 \times z)$$

$$ct' = \gamma \left(ct - \frac{vx}{c} \right)$$

and dividing by c gives

$$t' = \gamma \left(t - \frac{vx}{c^2} \right),$$

which is the Lorentz transformation for t'. Similarly, we find x', y' and z' by multiplying the two right-hand side matrices. We'll be looking at transformation matrices in greater detail when we discuss vectors and tensors in general relativity.

We can write (3.4.12) in a more compact form, using index notation (see Section 1.15) as

$$[x'^{\mu}] = [\Lambda^{\mu}_{\nu}][x^{\nu}], \tag{3.4.13}$$

where $[x'^{\mu}]$ represents the column vector on the left-hand side of (3.4.12), $[\Lambda^{\mu}_{\nu}]$ is the Lorentz transformation matrix, and $[x^{\nu}]$ is the column vector to the right of that matrix. Λ is the Greek letter Lambda, by the way. The indices μ and ν (the Greek letters mu and nu) take the values 0 to 3, so the components of $[x'^{\mu}]$ are

$$\left(x'^{0}, x'^{1}, x'^{2}, x'^{3}\right) = (ct', x', y', z')$$

and the components of $[x^{\nu}]$ are

$$\left(x^{0}, x^{1}, x^{2}, x^{3}\right) = (ct, x, y, z).$$

The components of the transformation matrix are

$$[\Lambda^{\mu}_{\nu}] = \begin{pmatrix} \Lambda^0_0 & \Lambda^0_1 & \Lambda^0_2 & \Lambda^0_3 \\ \Lambda^1_0 & \Lambda^1_1 & \Lambda^1_2 & \Lambda^1_3 \\ \Lambda^2_0 & \Lambda^2_1 & \Lambda^2_2 & \Lambda^2_3 \\ \Lambda^3_0 & \Lambda^3_1 & \Lambda^3_2 & \Lambda^3_3 \end{pmatrix},$$

where the μ index refers to the μth row and the ν index refers to νth column. So, for example, $\Lambda^0_0 = \gamma$, $\Lambda^3_3 = 1$, $\Lambda^0_1 = -\gamma v/c$, etc.

The quantity $[x^{\mu}]$ is known as the **four-position** (our first example of a **four-vector**) as it describes an event or position in spacetime using the four components (ct, x, y, z).

Strictly speaking, the brackets in (3.4.13) mean we are referring to matrices. (From now on we won't bother to use this clunky bracket notation for single index vectors.) We can rewrite this equation in terms of the components of these matrices as

$$x'^{\mu} = \Lambda^{\mu}_{\nu} x^{\nu} \tag{3.4.14}$$

using the Einstein summation convention (see Section 1.15). For example, if $\mu = 3$, then (3.4.14) becomes

$$x'^{3} = \Lambda^3_0 x^0 + \Lambda^3_1 x^1 + \Lambda^3_2 x^2 + \Lambda^3_3 x^3$$

$$= (0 \times ct) + (0 \times x) + (0 \times y) + (1 \times z)$$

$$x'^{3} = z.$$

We'll be making much use of this shorthand index notation from now on.

3.4.3 A second observer revisited

In Section 3.3.3 we used the constancy of the speed of light c to construct the coordinate lines ct' and x' of a second inertial frame S' on the spacetime diagram of frame S. We can now describe these axes algebraically using the Lorentz transformations.

131

The Lorentz transformations for ct' is ((3.4.2) multiplied by c)

$$ct' = c\gamma \left(t - \frac{vx}{c^2} \right).$$ (3.4.15)

We want to find the equation of the x' axis, which is the line where $ct' = 0$. Equation (3.4.15) then becomes

$$0 = c\gamma \left(t - \frac{vx}{c^2} \right)$$

giving

$$ct = \left(\frac{v}{c} \right) x,$$ (3.4.16)

which is the equation of the x' axis.

Similarly, the Lorentz transformation for x' is

$$x' = \gamma \left(x - vt \right).$$

We want to find the equation of the ct' axis, which is the line where $x' = 0$. We can therefore write

$$0 = \gamma \left(x - vt \right)$$

giving

$$t = \frac{x}{v}.$$

Multiplying by c gives

$$ct = \left(\frac{c}{v} \right) x,$$ (3.4.17)

which is the equation of the ct' axis.

Thankfully, (3.4.16) and (3.4.17) are the same line equations as (3.3.4) and (3.3.5) that we earlier showed geometrically.

Figure 3.20 shows the lines $ct = \left(\frac{v}{c} \right) x$ and $ct = \left(\frac{c}{v} \right) x$, which are the equations of the x' and ct' axes of frame S$'$.

3.4.4 Interval transformation rules

The Lorentz transformations (3.4.2) and (3.4.3) tell us how events (ie single points in spacetime) transform from one frame to another. If we have two events in spacetime there will be a difference between the corresponding time and spatial coordinates. As we've already noted, these differences $\Delta t, \Delta x, \Delta y, \Delta z$ are called intervals.

For example, if we had two events $(t_1, x_1) = (1, 3)$ and $(t_2, x_2) = (5, 4)$ then the time interval $\Delta t = 5 - 1 = 4$, and the spatial interval $\Delta x = 4 - 3 = 1$.

It is often useful to see how spacetime intervals transform. We derive the interval transformation rules as follows.

For two events labelled 1 and 2, the Lorentz transformations (3.4.2) are

$$t_1{}' = \gamma \left(t_1 - \frac{vx_1}{c^2} \right),$$ (3.4.18)

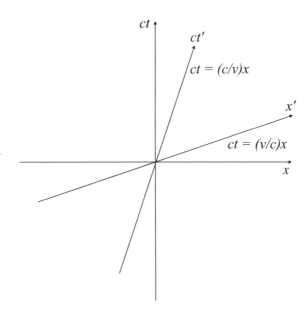

Figure 3.20: Finding the coordinate lines for a second observer using the Lorentz transformations.

$$x'_1 = \gamma\left(x_1 - vt_1\right),$$
$$y'_1 = y_1,$$
$$z'_1 = z_1$$

and

$$t'_2 = \gamma\left(t_2 - \frac{vx_2}{c^2}\right), \tag{3.4.19}$$
$$x'_2 = \gamma\left(x_2 - vt_2\right),$$
$$y'_2 = y_2,$$
$$z'_2 = z_2.$$

If we find $t_2' - t_1'$ and $x_2' - x_1'$ etc we get these transformation rules for intervals

$$\Delta t' = \gamma\left(\Delta t - v\Delta x/c^2\right), \tag{3.4.20}$$
$$\Delta x' = \gamma\left(\Delta x - v\Delta t\right), \tag{3.4.21}$$
$$\Delta y' = \Delta y,$$
$$\Delta z' = \Delta z,$$

where $\Delta t = t_2 - t_1$, $\Delta x = x_2 - x_1$, $\Delta y = y_2 - y_1$ and $\Delta z = z_2 - z_1$.

Similarly, using the inverse Lorentz transformations (3.4.3), we find

$$\Delta t = \gamma\left(\Delta t' + v\Delta x'/c^2\right), \tag{3.4.22}$$
$$\Delta x = \gamma\left(\Delta x' + v\Delta t'\right), \tag{3.4.23}$$
$$\Delta y = \Delta y',$$
$$\Delta z = \Delta z'.$$

3.4.5 Proper time and time dilation

In Section 3.3.6, when examining the properties of invariant hyperbolae, we mentioned the counter-intuitive phenomenon of time dilation. We now take a closer look at time dilation using the Lorentz transformations.

Up to now we've been using coordinate time, which refers to the time measured by a distant observer using that observer's own clock. As we've seen, each inertial frame has its own coordinate time (the different time axes on a spacetime diagram). Coordinate time varies from observer to observer. But what about the time measured by an observer using their own clock? That is an invariant measure of time known as **proper time**, which we can use to calculate time dilation.

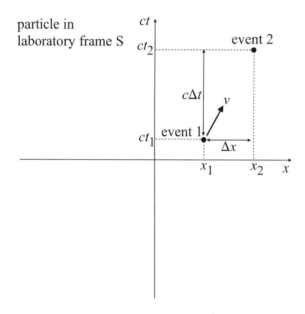

Figure 3.21: Particle moving in the laboratory frame S.

Say we are in a laboratory investigating sub-atomic particles that are moving with a constant velocity v in the positive x direction – see Figure 3.21. We assume that the particle we are studying is short lived and is created at a point in spacetime we can call event 1 before decaying at another point labelled event 2.

We are the observers O in what we can refer to as the laboratory frame S. If, in the laboratory frame, event 1 occurs at (t_1, x_1) and event 2 at (t_2, x_2) then we measure the lifetime of the particle to be $\Delta t = t_2 - t_1$ (which we'll call ΔT) and the distance the particle travels to be $\Delta x = x_2 - x_1$.

But there is another obvious frame of reference to consider and that is for an observer O′ in the frame S′ moving with the particle. This is called the particle's **rest frame**.

You can visualise this scenario by imagining you are driving a car, and there's a bag of shopping next to you on the passenger seat. The bag of shopping is equivalent to the particle, you are the observer O′, and the car is the bag's rest frame S′. People standing by the road will see the bag of shopping zooming past them at 70 mph or whatever. But for you, the bag is stationary, quietly keeping you company on your way back home from the supermarket.

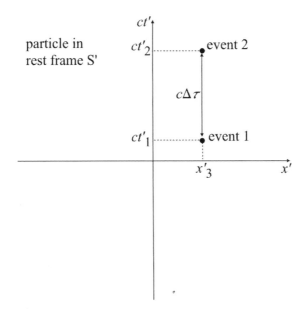

Figure 3.22: Particle in its rest frame S'.

Figure 3.22 shows the particle in its rest frame S'. Just as the bag of shopping is stationary relative to you, the driver, so according to observer O' the particle doesn't move spatially between being created and decaying. If, in the rest frame, event 1 occurs at (t'_1, x'_3) and event 2 at (t'_2, x'_3) then observer O' will measure the lifetime of the particle to be $\Delta t' = t'_2 - t'_1$. As the particle doesn't move in its rest frame $\Delta x' = 0$.

In special relativity, proper time is the time measured by an observer in their own rest frame. Think of 'proper' in terms of a time that is the 'property' of the observer, not as a synonym for 'correct'. Proper time is denoted by the Greek letter tau τ. In our example, we could imagine an accurate clock strapped to the particle. That's hard to visualise, so it's probably more helpful to think about the particle's natural 'internal clock', the unknown mechanism that eventually says to the particle (just as it says to all decrepit living things), 'Time's up I'm afraid.' The time recorded by that internal clock is the particle's proper time.

For our particle, the proper time interval between event 1 and event 2 is therefore given by

$$\Delta \tau = \Delta t' = t'_2 - t'_1.$$

So now we have two measures of the particle's lifetime: the coordinate time $\Delta T = \Delta t = t_2 - t_1$ we measured using our instruments in the laboratory, and the proper time $\Delta \tau = \Delta t' = t'_2 - t'_1$ as measured by the particle's own internal clock. We can relate the two using the interval transformation rules (3.4.22)

$$\Delta t = \gamma \left(\Delta t' + v \Delta x'/c^2 \right).$$

Knowing $\Delta x' = 0$, we obtain

$$\Delta T = \gamma \Delta \tau \tag{3.4.24}$$

or, in terms of proper time,

$$\Delta \tau = \frac{\Delta T}{\gamma}. \tag{3.4.25}$$

Another way of seeing what is going on here is that a process that takes a certain proper time ($\Delta \tau$ – measured by definition in its own rest frame) has a longer duration (ΔT) measured by another observer moving relative to the rest frame, ie *moving clocks run slow.*

This is the phenomenon of time dilation that we met earlier in Section 3.3.6.

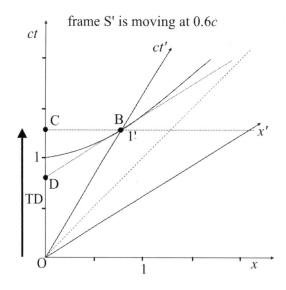

Figure 3.23: Time dilation with frame S' moving at $0.6c$.

Figure 3.23 is a simplified version of the spacetime diagram we used in Section 3.3.6, but where we now stipulate that frame S' is moving at $0.6c$.

The black arrow **TD** is the time dilation ΔT observed by O of event $B = 1$ as measured by O' along his ct'axis. How do we calculate the length of the black arrow **TD**?

Using (3.4.24), where the proper time of event B measured by O' is $\Delta \tau = 1$ we find

$$\Delta T = \gamma \Delta \tau$$

$$= \frac{\Delta \tau}{\sqrt{1 - (v/c)^2}} = \frac{1}{\sqrt{1 - (0.6/1)^2}} = \frac{1}{\sqrt{0.64}}$$

$$\Delta T = 1.25.$$

Problem 3.7. A muon lives, on average, for $\Delta\tau = 2.2\,\mu$s in its own rest frame. If a muon is travelling with speed $v = 0.995c$ relative to an observer on Earth, what is its lifetime as measured by that observer?

First, we calculate the Lorentz factor (3.4.1), which equals

$$\gamma = \frac{1}{\sqrt{1 - (v/c)^2}} = \frac{1}{\sqrt{1 - (0.995)^2}} = 10.013.$$

Time dilation, given by (3.4.24) is

$$\Delta T = \gamma\Delta\tau$$

$$= 10.013 \times 2.2$$

$$= 22.03\,\mu\text{s}.$$

Problem 3.8. An astronaut travelling in a rocket at $0.9c$ measures her heart rate at 70 beats per minute. What will her heart rate be as measured by an observer back on Earth?

The proper time interval $\Delta\tau$ between heartbeats equals

$$\frac{60}{70} = 0.86\,\text{seconds/beat}.$$

Time dilation, given by (3.4.24), is

$$\Delta T = \gamma\Delta\tau$$

$$= \left(\frac{1}{\sqrt{1 - (v/c)^2}}\right)\Delta\tau$$

$$= \left(\frac{1}{\sqrt{1 - (0.9/1)^2}}\right)0.86 = \left(\frac{1}{0.44}\right)0.86 = 1.95\,\text{seconds/beat}$$

$$= \frac{1}{1.95} = 0.51\,\text{beats/second} = 0.51 \times 60\,\text{beats/minute}$$

$$= 30.6\,\text{beats/minute}.$$

3.4.6 Length contraction

We've already used spacetime diagrams to explore the phenomenon of length contraction. We now take a closer look at length contraction using the Lorentz transformations.

Figure 3.24 shows a rod moving lengthways along the x axis with velocity v in an inertial laboratory frame S. The length L of the rod can be measured by assuming that one end of the rod corresponds to an event 1 (t, x_1), and the other end to an event 2 (t, x_2). In other words, we are measuring the distances x at the same time ct, meaning $\Delta t = 0$. The length of the rod will then be given by

$$L = \Delta x = x_2 - x_1.$$

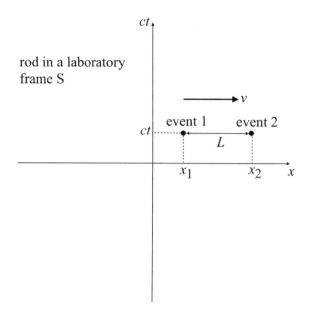

Figure 3.24: Rod in a laboratory frame.

Now, just as we did for time dilation, consider the rod in a rest frame S′, where the rod is lying stationary along the x' axis, as shown in Figure 3.25. We know that events 1 and 2 will still occur at the ends of the rod, but we don't know the time these events occur. If, in the rest frame, event 1 occurs at (t'_1, x'_1) and event 2 at (t'_2, x'_2) then observer O′ will measure the length of the rod L_p (known as the rod's **proper length**) to be

$$L_p = \Delta x' = x'_2 - x'_1.$$

We don't know t'_1 and t'_2, therefore we don't know $\Delta t' = t'_2 - t'_1$. But that isn't a problem as we can use the interval transformation equation (3.4.21), which doesn't include $\Delta t'$

$$\Delta x' = \gamma \left(\Delta x - v \Delta t \right).$$

Substituting our expressions for L, L_p and $\Delta t = 0$ we obtain

$$L_p = \gamma L \tag{3.4.26}$$

or

$$L = \frac{L_p}{\gamma}. \tag{3.4.27}$$

Figure 3.26 shows the same spacetime diagram we used in Section 3.3.6, but where we now stipulate that frame S′ is moving at $0.6c$.

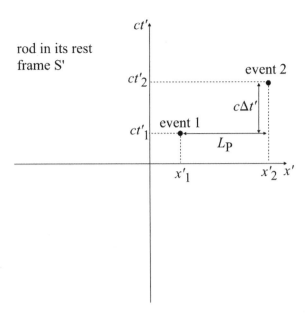

Figure 3.25: Rod in its rest frame.

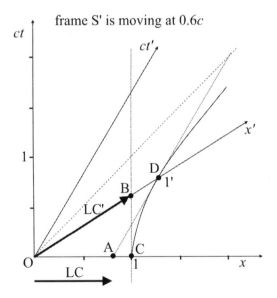

Figure 3.26: Length contraction with frame S' moving at 0.6c.

Recall that the black arrow **LC** is the length contraction L observed by O of a length OD $= 1$ as measured by O' along his x' axis. How do we calculate the length of the black arrow **LC**?

Using (3.4.27), where $L_p = 1$ we find

$$L = \frac{L_p}{\gamma}$$

$$= \frac{1}{\left(\frac{1}{\sqrt{1-(v/c)^2}} \right)} = \sqrt{1 - (0.6/1)^2} = \sqrt{0.64}$$

$$L = 0.8.$$

The same calculation would give us the same result for the length of the black arrow **LC'** representing the length contraction L observed by O' of a length OC $= 1$ measured by O along his x axis.

3.4.7 The end of simultaneity

We saw when looking at spacetime diagrams that the simultaneity of events can depend on the observer's reference frame. We can show this algebraically using the interval transformation rule (3.4.20)

$$\Delta t' = \gamma \left(\Delta t - v \Delta x/c^2 \right).$$

If two events occur simultaneously in frame S, then $\Delta t = 0$ and the above equation becomes

$$\Delta t' = -\gamma \left(v \Delta x/c^2 \right).$$

As long as they don't occur at the same point (ie $\Delta x \neq 0$) then we can say

$$\Delta t' = -\gamma \frac{vL}{c^2},$$

where L is the distance Δx. This equation is an algebraic formulation of the relativity of simultaneity. We can see that for very low speeds, where $v \ll c$, then $\Delta t' \approx 0$.

3.4.8 Velocity transformations

Figure 3.27 shows two inertial frames in standard configuration with an object moving with velocity v_x along the x axis of frame S. What velocity v' will the object be moving according to an observer in frame S'?

We've met the Galilean answer to this earlier (3.2.2) in the context of me driving a car at 70 mph and being followed by another vehicle doing 80 mph. According to the Galilean transformations, the car behind me has a relative velocity to my vehicle of $80 - 70 = 10$ mph. For the example shown in 3.27 the Galilean answer would be

$$v' = v_x - v. \tag{3.4.28}$$

In the above car example, $v_x = 80$ mph, $v = 70$ mph, and $v' = 10$ mph.

Now let's tackle the problem using the Lorentz transformations.

We can use the two interval transformation rules 3.4.20

$$\Delta t' = \gamma \left(\Delta t - v \Delta x/c^2 \right)$$

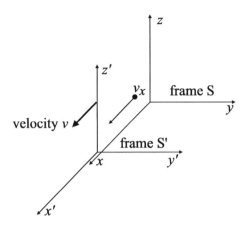

Figure 3.27: Velocity transformation.

and 3.4.21

$$\Delta x' = \gamma \left(\Delta x - v \Delta t \right)$$

to describe two events that occur on the x axis of frame S. If we divide the second equation by the first we obtain

$$\frac{\Delta x'}{\Delta t'} = \frac{\gamma \left(\Delta x - v \Delta t \right)}{\gamma \left(\Delta t - v \Delta x / c^2 \right)} = \frac{\left(\Delta x - v \Delta t \right)}{\left(\Delta t - v \Delta x / c^2 \right)}.$$

Next we divide the top and bottom expressions on the right-hand side by Δt to give

$$\frac{\left(\Delta x / \Delta t - v \right)}{\left(1 - \left(\Delta x / \Delta t \right) v / c^2 \right)}. \tag{3.4.29}$$

If we bring the two events on the x axis closer and closer together, eventually – as Δx and Δt approach 0 – the quantities $\Delta x' / \Delta t'$ and $\Delta x / \Delta t$ become the instantaneous velocities v'_x and v_x of an object moving through the two events. Equation (3.4.29) then becomes

$$v'_x = \frac{v_x - v}{1 - v_x v / c^2}. \tag{3.4.30}$$

What does this equation tell us?

First, if v_x and v are very small compared to the speed of light, then $v_x v / c^2 \approx 0$ and (3.4.30) reduces to the Galilean velocity equation (3.4.28) $v' = v_x - v$ that we met at the start of this section.

Second, let's consider what happens if the object is now a light ray moving in the opposite direction to frame S' (ie $v_x = -c$) as shown in Figure 3.28. How fast does an observer on frame S' measure the light ray?

Now, this is the sixty four thousand dollar question, because the second postulate of special relativity states that the speed of light in vacuum has the same value $c = 3 \times 10^8 \, \mathrm{m\,s^{-1}}$ in all inertial frames of reference. So we know that according to the second postulate the answer should be $-c$. But what does the velocity transformation equation (3.4.30) give us?

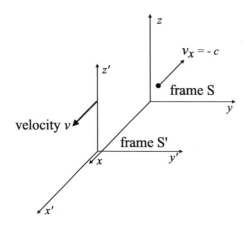

Figure 3.28: Light ray moving in opposite direction to frame S′.

We start by plugging in $v_x = -c$ to get

$$v'_x = \frac{-c - v}{1 - (-c)\, v/c^2}$$

and juggle this around a bit to get

$$v'_x \left(1 + v/c\right) = -\left(c + v\right)$$

and then

$$\frac{v'_x}{c} \left(c + v\right) = -\left(c + v\right)$$

and thus

$$v'_x = -c,$$

which is exactly as required by the second postulate. And as expected, we can also see that $v'_x = -c$ is independent of the relative motion v of frames S and S′.

Problem 3.9. Two rockets A and B are moving in opposite directions, A at $0.75c$, B at $0.85c$ with respect to an observer on Earth (see Figure 3.29). How fast does an observer on rocket A measure B to be travelling?

We let frame S be Earth, frame S′ be rocket A travelling at $v = 0.75c$, and $v_x = -0.85c$ be the velocity of rocket B. We plug these values into (3.4.30) to solve for v'_x, the velocity of B relative to A.

$$v'_x = \frac{v_x - v}{1 - v_x v/c^2}$$

$$= \frac{-0.85c - 0.75c}{1 - (-0.85c \times 0.75c/c^2)} = \frac{-1.6c}{1.638}$$

$$v'_x = -0.977c,$$

the minus sign indicates that rocket B is moving in the opposite direction to rocket A.

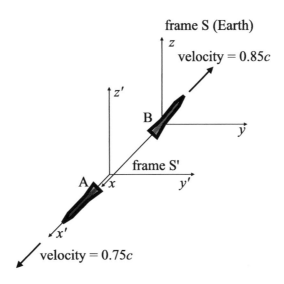

Figure 3.29: Two rockets.

3.4.9 Invariance of the interval – again

We mentioned when looking at spacetime diagrams in Section 3.3.5 that the separation of two events in spacetime is the same for all inertial observers. We'll now prove that theorem.

In Section 1.14 we saw that the separation between two points x_1, y_1, z_1 and x_2, y_2, z_2 in three-dimensional Euclidean space is given by 1.14.4

$$\Delta l^2 = \Delta x^2 + \Delta y^2 + \Delta z^2,$$

where $\Delta x = x_2 - x_1, \Delta y = y_2 - y_1$ and $\Delta z = z_2 - z_1$.

Recall from Section 1.14.1 that 1.14.4 actually defines the geometry of the space, in the sense that if 1.14.4 is true for every interval, then the space must be Euclidean. This means that we can describe the two points x_1, y_1, z_1 and x_2, y_2, z_2 using any other Cartesian coordinate system, for example

$$\Delta l'^2 = \Delta x'^2 + \Delta y'^2 + \Delta z'^2,$$

and we would find that

$$\Delta l^2 = \Delta l'^2.$$

The distance between any two points in Euclidean space is the same irrespective of the coordinate system used to describe the points.

Similarly, as we have seen, there is an invariant interval in four-dimensional spacetime. It is similar to 1.14.4 except it includes both positive and negative signs. The spacetime interval Δs^2 is given by (3.3.6)

$$\Delta s^2 = c^2 \Delta t^2 - \Delta x^2 - \Delta y^2 - \Delta z^2.$$

Crucially, this spacetime separation is invariant under Lorentz transformations, which means it is true for all inertial reference frames.

We can prove 3.3.6 using the interval transformation rules from Section 3.4.4

$$\Delta t' = \gamma \left(\Delta t - v\Delta x/c^2\right), \; \Delta x' = \gamma \left(\Delta x - v\Delta t\right), \; \Delta y' = \Delta y, \; \Delta z_1' = \Delta z_1,$$

$$\Delta t = \gamma \left(\Delta t' + v\Delta x'/c^2\right), \; \Delta x = \gamma \left(\Delta x' + v\Delta t'\right), \; \Delta y = \Delta y', \; \Delta z = \Delta z'.$$

Meaning we only need to show that

$$c^2\Delta t^2 - \Delta x^2 = c^2\Delta t'^2 - \Delta x'^2,$$

where $\Delta t \neq \Delta t'$ and $\Delta x \neq \Delta x'$.

The algebra is more straightforward if we write the interval transformation rules in terms of $\beta = v/c$ and $\gamma = 1/\sqrt{1 - \beta^2}$ to give

$$c\Delta t' = \gamma \left(c\Delta t - \beta\Delta x\right), \; \Delta x' = \gamma \left(\Delta x - \beta c\Delta t\right), \; \Delta y' = \Delta y, \; \Delta z_1' = \Delta z_1,$$

$$c\Delta t = \gamma \left(c\Delta t' + \beta\Delta x'\right), \; \Delta x = \gamma \left(\Delta x' + \beta c\Delta t'\right), \; \Delta y = \Delta y', \; \Delta z = \Delta z'$$

and we see that

$$
\begin{aligned}
c^2\Delta t'^2 - \Delta x'^2 &= \gamma^2 \left(c\Delta t - \beta\Delta x\right)^2 - \gamma^2 \left(\Delta x - \beta c\Delta t\right)^2 \\
&= \gamma^2 \left(c^2\Delta t^2 - 2\beta c\Delta t\Delta x + \beta^2\Delta x^2\right) - \gamma^2 \left(\Delta x^2 - 2\beta c\Delta t\Delta x + \beta^2 c^2\Delta t^2\right) \\
&= \gamma^2 \left(c^2\Delta t^2 - 2\beta c\Delta t\Delta x + \beta^2\Delta x^2 - \Delta x^2 + 2\beta c\Delta t\Delta x - \beta^2 c^2\Delta t^2\right) \\
&= \gamma^2 \left(c^2\Delta t^2 + \beta^2\Delta x^2 - \Delta x^2 - \beta^2 c^2\Delta t^2\right) \\
&= \gamma^2 \left(c^2\Delta t^2 \left(1 - \beta^2\right) + \Delta x^2 \left(\beta^2 - 1\right)\right) \\
&= \gamma^2 \left(c^2\Delta t^2 \left(1 - \beta^2\right) - \Delta x^2 \left(1 - \beta^2\right)\right) \\
&= c^2\Delta t^2 - \Delta x^2,
\end{aligned}
$$

because

$$\gamma^2 \left(1 - \beta^2\right) = \left(\frac{1}{1 - \beta^2}\right) \left(1 - \beta^2\right) = 1.$$

We have therefore shown (eventually!) that

$$c^2\Delta t^2 - \Delta x^2 = c^2\Delta t'^2 - \Delta x'^2.$$

And, as noted, because in the standard configuration we are using, $y = y'$ and $z = z'$ we have also shown that

$$c^2\Delta t^2 - \Delta x^2 - \Delta y^2 - \Delta z^2 = c^2\Delta t'^2 - \Delta x'^2 - \Delta y'^2 - \Delta z'^2,$$

meaning we have proven the invariance of the spacetime interval under Lorentz transformations.

The importance of this proof is as follows. Say we have observers O and O′ travelling in inertial frames S and S′ and measuring two events in spacetime. The observers will not agree on the time and distance separating the events, but they will agree on the spacetime separation of the events, ie they will agree that

$$\Delta s^2 = c^2\Delta t^2 - \Delta x^2 - \Delta y^2 - \Delta z^2 = c^2\Delta t'^2 - \Delta x'^2 - \Delta y'^2 - \Delta z'^2 = \Delta s'^2$$

ie

$$\Delta s^2 = \Delta s'^2.$$

3.4.10 The interval and proper time

Recall from Section 3.4.5 that proper time is the time measured by an observer in their own rest frame, ie the time between two events as measured in a frame where the events are in the same position. Equation (3.4.25)

$$\Delta\tau = \frac{\Delta T}{\gamma}$$

relates the proper time $\Delta\tau$ between two such events and the coordinate time ΔT of another observer who is in uniform motion relative to the events. We used the example of a short-lived particle, where proper time $\Delta\tau$ is the lifetime of the particle measured in its own rest frame, and the coordinate time ΔT is the lifetime of the particle measured by an observer in a laboratory frame.

The invariance of the interval gives us a means of expressing proper time in terms of the spacetime interval Δs^2.

First, consider the particle's rest frame. The spacetime interval between any two positions of the particle in such a frame is given by

$$\Delta s^2 = c^2\Delta t^2 - \Delta x^2 - \Delta y^2 - \Delta z^2$$

$$\Delta s^2 = c^2\Delta t^2 - \Delta 0^2 - \Delta 0^2 - \Delta 0^2 = c^2\Delta t^2.$$

But as proper time $\Delta\tau$, by definition, is the time measured by an observer in their own rest frame, we can say $\Delta\tau = \Delta t$, and therefore

$$\Delta s^2 = c^2\Delta t^2 = c^2\Delta\tau^2. \tag{3.4.31}$$

However, the spacetime separation of events is an invariant quantity, ie is measured the same for all inertial observers. We can therefore say that (3.4.31) tells us not only the proper time between events occurring at the same position, but also applies to time separated events measured from any frame, ie

$$c^2\Delta\tau^2 = \Delta s^2 = c^2\Delta t^2 - \Delta x^2 - \Delta y^2 - \Delta z^2. \tag{3.4.32}$$

Problem 3.10. You fancy a holiday on your favourite planet in the Andromeda galaxy, but you don't want to age more than 1 year on the journey there. If the planet is 2 million light-years away, approximately how fast do you need to travel for you to age just 1 year on your trip?

We assume that there is an inertial frame all the way from the Earth to your destination, ie special relativity applies. We'll call this frame the Earth frame. Your spaceship is moving with constant velocity through the Earth frame so (3.4.32) tells us your proper time, which we want to equal 1 year:

$$c^2\Delta\tau^2 = \Delta s^2 = c^2\Delta t^2 - \Delta x^2 - \Delta y^2 - \Delta z^2.$$

Now assume that $c = 1$ (so our final velocity will be expressed as a fraction of c) to give

$$\Delta\tau^2 = \Delta t^2 - \Delta x^2 - \Delta y^2 - \Delta z^2,$$

which we can rearrange to give

$$\Delta\tau^2 = \Delta t^2\left(1 - \frac{\left(\Delta x^2 + \Delta y^2 + \Delta z^2\right)}{\Delta t^2}\right).$$

The $\frac{(\Delta x^2 + \Delta y^2 + \Delta z^2)}{\Delta t^2}$ term is actually the square of your velocity as measured by an observer in the Earth frame, so we can say

$$\Delta \tau^2 = \Delta t^2 \left(1 - v^2\right)$$
$$= \Delta t^2 \left(1 - v\right)\left(1 + v\right). \tag{3.4.33}$$

We can now assume that your velocity is going to have to be very close to the speed of light, ie $v \approx c = 1$, so we say

$$(1 + v) \approx (1 + 1) = 2.$$

Therefore, (3.4.33) becomes

$$\Delta \tau^2 = 2\Delta t^2 \left(1 - v\right)$$

or

$$v = 1 - \frac{\Delta \tau^2}{2\Delta t^2}. \tag{3.4.34}$$

As you are travelling very close to the speed of light, we assume that Δt equals the time it takes a ray of light to make the trip as measured in the Earth frame, ie we assume that $\Delta t = 2$ million years. We want $\Delta \tau$ to equal 1, so (3.4.34) becomes

$$v = 1 - \frac{1^2}{2 \times \left(2 \times 10^6\right)^2} = 1 - \frac{1}{8 \times 10^{12}}$$

$$v = 1 - \left(1.25 \times 10^{-13}\right)$$

$$v = 0.999999999999875c.$$

Meaning, you need a very fast spaceship to only age 1 year on the journey.

3.5 The Minkowski metric

As we saw when looking at Euclidean geometry in Section 1.14.1, the metric is a function that defines the distance between two points in a particular space. Once we know the metric of a space, we know (theoretically, at least) everything about the geometry of the space, which is why the metric is of fundamental importance.

To refresh our memory, recall that the Euclidean metric (1.14.6) is

$$[g_{ij}] = \begin{pmatrix} 1 & 0 & 0 \\ 0 & 1 & 0 \\ 0 & 0 & 1 \end{pmatrix}.$$

We've just met one form of the function that defines the distance between two points in spacetime – it's the spacetime interval Δs^2 given by (3.3.6)

$$\Delta s^2 = c^2 \Delta t^2 - \Delta x^2 - \Delta y^2 - \Delta z^2.$$

If we make the intervals Δs, Δt, Δx, Δy and Δz infinitesimally small, we end up with coordinate differentials ds, cdt, dx, dy, dz (we met these in Section 1.14.1), which we can use to define the Minkowski line element

$$ds^2 = c^2 dt^2 - dx^2 - dy^2 - dz^2. \tag{3.5.1}$$

Just as we did when defining the Euclidean metric, we use the metric coefficients of the Minkowski line element to define the Minkowski metric, which is denoted by $\eta_{\mu\nu}$ (η is the Greek letter eta).

Looking at 3.5.1, we can see there are $1 \times c^2 dt^2$, $-1 \times dx^2$, $-1 \times dy^2$ and $-1 \times dz^2$ terms, so the metric coefficients are $+1, -1, -1, -1$, which we can arrange into a 4×4 matrix

$$[\eta_{\mu\nu}] = \begin{pmatrix} \eta_{00} & \eta_{01} & \eta_{02} & \eta_{03} \\ \eta_{10} & \eta_{11} & \eta_{12} & \eta_{13} \\ \eta_{20} & \eta_{21} & \eta_{22} & \eta_{23} \\ \eta_{30} & \eta_{31} & \eta_{32} & \eta_{33} \end{pmatrix} = \begin{pmatrix} 1 & 0 & 0 & 0 \\ 0 & -1 & 0 & 0 \\ 0 & 0 & -1 & 0 \\ 0 & 0 & 0 & -1 \end{pmatrix}. \tag{3.5.2}$$

This matrix simply tells us how to multiply the differentials cdt, dx, dy, dz to obtain the line element (3.5.1). We can see this by writing the matrix as a table

$$[\eta_{\mu\nu}] = \begin{array}{c|cccc} - & cdt & dx & dy & dz \\ cdt & 1 & 0 & 0 & 0 \\ dx & 0 & -1 & 0 & 0 \\ dy & 0 & 0 & -1 & 0 \\ dz & 0 & 0 & 0 & -1 \end{array}$$

and seeing that $cdt \times cdt \times 1 = c^2 dt^2$, $dx \times dx \times -1 = -dx^2$, $dy \times dy \times -1 = -dy^2$, $dz \times dz \times -1 = -dz^2$ with all the other products of cdt, dx, dy, dz equalling zero.

The indices μ, ν after the η symbol identify the elements of the matrix by reference to its rows and columns. The convention is that the metric coefficients run from 0 to 3. So $\eta_{00} = 1$, $\eta_{11} = -1$, $\eta_{22} = -1$ and $\eta_{13} = 0$, etc.

We can rewrite the Minkowski line element (3.5.1) more concisely using index notation and the Einstein summation convention as

$$ds^2 = \eta_{\mu\nu} dx^\mu dx^\nu, \tag{3.5.3}$$

where dx^μ and dx^ν are the coordinate differentials cdt, dx, dy, dz. Using similar notation we can rewrite the spacetime interval Δs^2 (3.3.6) as

$$\Delta s^2 = \eta_{\mu\nu} \Delta x^\mu \Delta x^\nu, \tag{3.5.4}$$

where Δx^μ and Δx^ν represent the four spacetime components (ct, x, y, z).

In Section 3.4.10 we saw that proper time could be expressed in terms of the spacetime separation

$$c^2 \Delta\tau^2 = \Delta s^2 = c^2 \Delta t^2 - \Delta x^2 - \Delta y^2 - \Delta z^2.$$

Using index notation we can therefore express proper time in terms of the metric tensor as

$$c^2 (\Delta\tau)^2 = \eta_{\mu\nu} \Delta x^\mu \Delta x^\nu,$$

and using the line element (3.5.3)

$$c^2 d\tau^2 = \eta_{\mu\nu} dx^\mu dx^\nu. \tag{3.5.5}$$

The configuration of the plus and minus signs in the metric tensor is called the **metric signature**. In this book we use the metric signature $+ - - -$. Some textbooks use the opposite metric signature $- + ++$, meaning they would write the Minkowski metric as

$$[\eta_{\mu\nu}] = \begin{pmatrix} -1 & 0 & 0 & 0 \\ 0 & 1 & 0 & 0 \\ 0 & 0 & 1 & 0 \\ 0 & 0 & 0 & 1 \end{pmatrix}.$$

It doesn't matter which convention is used as long as you are consistent with the signs when doing calculations.

All the above has assumed we are using Cartesian coordinates. Of course, we can describe events in spacetime using any coordinate system, it's just that Cartesian coordinates are often the most straightforward. In spherical coordinates (see Section 1.9.3) the Minkowski line element is given by

$$ds^2 = c^2dt^2 - dr^2 - r^2d\theta^2 - r^2\sin^2\theta d\phi^2. \tag{3.5.6}$$

3.6 Mechanics in special relativity

3.6.1 Conservation laws

We first need to mention a fundamental concept in physics known as the **conservation laws**. Briefly, these state that certain properties of an isolated physical system will be conserved no matter how that system changes. The conservation of energy and the conservation of momentum are the two conservation laws that we'll encounter.

A pendulum is a good example of the conservation of a form of energy known as mechanical energy. When the ball of the pendulum is at its highest point it is momentarily stationary and has maximum potential energy and zero kinetic energy. When the ball is swinging through its lowest point, it has zero potential energy and maximum kinetic energy. But at all points through the swing of the pendulum the sum of the potential and kinetic energies is constant, ie the mechanical energy is conserved.

3.6.2 Invariance

A quantity is an invariant in special relativity if it has either the same value or the same form in all inertial frames. Invariants with the 'same value' refer to invariant physical quantities (we've seen several already) including:

- The speed of light in a vacuum c.

- The spacetime separation (3.3.6) $\Delta s^2 = c^2\Delta t^2 - \Delta x^2 - \Delta y^2 - \Delta z^2$.

- The proper time (3.4.32) $c^2\Delta\tau^2 = c^2\Delta t^2 - \Delta x^2 - \Delta y^2 - \Delta z^2$.

Another very useful invariant physical quantity is **rest mass** m, which is the mass of an object or particle in its rest frame. Whenever we refer to mass from now on we'll be referring to rest mass, also known as invariant or proper mass.

The relationships between various invariant physical quantities can be described using equations. If these equations are invariant (ie have the 'same form' in all inertial frames) they are called **form-invariant** or **covariant** (confusingly, this is not the same usage of covariant as in the 'covariant vectors' we meet later). We know that the assumption that the laws of physics must take the same form in all inertial frames (the principle of relativity) is one of the postulates of special relativity. We'll now look at some form-invariant laws of mechanics.

3.6.3 Four-velocity

We saw in Section 1.11 how parametric equations using a single parameter can be used to define a curve through space such as the path of a ball (we used the example of a ball thrown over a cliff in Cartesian (x, y) coordinates using the variable $(t - \text{time})$ as a parameter).

The path of a particle moving in ordinary three-dimensional Euclidean space can be described using three functions of t (time), one for x, one for y and one for z. The three functions $x = f(t)$, $y = f(t)$, $z = f(t)$ are called parametric equations and give a vector whose components represent the object's spatial velocity (or three-velocity) in the x, y, z directions. The spatial velocity of the particle is a tangent vector to the path (ie the vector points along the path) and has components $\left(\frac{dx}{dt}, \frac{dy}{dt}, \frac{dz}{dt} \right)$. Spatial velocities are frame-dependent and therefore not invariant under Lorentz transformations.

However, there is a type of velocity vector in special relativity that is form-invariant, and this is called the **four-velocity**.

Consider the velocity of a particle moving along a world-line in four-dimensional spacetime. As we have seen, a clock fastened to the particle will measure the particle's proper time (τ, which we know is invariant), and therefore it makes sense to use τ as the parameter along the path. The four-velocity of a particle is the rate of change of its four-position with respect to proper time. As with three-velocity, four-velocity $\vec{U}$ is a tangent vector to the particle's world-line and is defined as having components

$$U^\mu = \frac{dx^\mu}{d\tau} = \left(c\frac{dt}{d\tau}, \frac{dx}{d\tau}, \frac{dy}{d\tau}, \frac{dz}{d\tau} \right). \tag{3.6.1}$$

Four-velocity, having a time component as well as three spatial components, is a type of four-vector, a crucial form-invariant quantity in special relativity. We'll be looking at four-vectors in much greater detail later.

To determine the components of the four-velocity recall that (3.4.24) gives us coordinate time ΔT in relation to proper time $\Delta \tau$

$$\Delta T = \gamma \Delta \tau.$$

As we are using ct units of time we can rewrite this as

$$x^0 = ct = c\gamma\tau.$$

Taking the derivative with respect to proper time gives

$$U^0 = \frac{dx^0}{d\tau} = c\gamma.$$

We can use the chain rule (1.10.4) to find the spatial components of U^μ ($\mu = i = 1, 2, 3$)

$$U^i = \frac{dx^i}{d\tau} = \frac{dx^i}{dx^0}\frac{dx^0}{d\tau} = \frac{dx^i}{dx^0}c\gamma = \frac{dx^i}{d(ct)}c\gamma = \frac{dx^i}{cdt}c\gamma = \frac{dx^i}{dt}\gamma.$$

$\frac{dx^i}{dt}$ is the particle's ordinary spatial velocity, which is a vector $\mathbf{v}$ with components $v = \frac{dx^1}{dt}, \frac{dx^2}{dt}, \frac{dx^3}{dt} = v_x, v_y, v_z$. The particle's four-velocity is therefore given by

$$U^\mu = \left(U^0, U^1, U^2, U^3 \right) = \frac{dx^\mu}{d\tau} = (c\gamma, \gamma\mathbf{v}) = \gamma(c, \mathbf{v}). \tag{3.6.2}$$

In special relativity the scalar product of two four-vectors $\vec{A}$ and $\vec{B}$ is defined using the Minkowski metric as

$$\vec{A} \cdot \vec{B} = \eta_{\mu\nu} A^\mu B^\nu = A^0 B^0 - A^1 B^1 - A^2 B^2 - A^3 B^3, \tag{3.6.3}$$

and which is invariant under Lorentz transformations.

So the scalar product of the four-velocity is given by

$$\vec{U} \cdot \vec{U} = \eta_{\mu\nu} U^\mu U^\nu = \gamma^2 c^2 - \gamma^2 \left((v_x)^2 + (v_y)^2 + (v_z)^2 \right)$$

$$= \gamma^2 \left(c^2 - v^2 \right).$$

But as

$$\gamma^2 = \left(\frac{1}{\sqrt{1 - (v/c)^2}} \right)^2 = \frac{1}{1 - \frac{v^2}{c^2}} = \frac{c^2}{c^2 - v^2},$$

we find that

$$\eta_{\mu\nu} U^\mu U^\nu = c^2, \tag{3.6.4}$$

which is obviously an invariant.

3.6.4 Relativistic momentum

In Newtonian mechanics, the momentum (which we'll call $\mathbf{p}_{Newtonian}$) of a particle equals the particle's mass m multiplied by its ordinary spatial velocity $\mathbf{v}$:

$$\mathbf{p}_{Newtonian} = m\mathbf{v},$$

with spatial velocity having the components $\left(\frac{dx}{dt}, \frac{dy}{dt}, \frac{dz}{dt} \right)$. Providing no external forces act on a Newtonian system (where speeds are much less than the speed of light) momentum is conserved. For example, if a particle of mass m_A moving with velocity v_A collides with a particle of mass m_B moving with velocity v_B, and after the collision the particle of mass m_A moves with velocity u_A and the particle of mass m_B moves with velocity u_B then

$$m_A v_A + m_B v_B = m_A u_A + m_B u_B. \tag{3.6.5}$$

In special relativity velocities transform in complicated ways between different inertial frames, and therefore we can't use the Newtonian conservation of momentum law (3.6.5). Instead we need to introduce the notion of **relativistic momentum**. To do this we use proper time τ instead of coordinate time t and define relativistic momentum $\mathbf{p}$ as

$$\mathbf{p} = (p_x, p_y, p_z) = m \left(\frac{dx}{d\tau}, \frac{dy}{d\tau}, \frac{dz}{d\tau} \right). \tag{3.6.6}$$

Because (3.4.25) gives us coordinate time ΔT in relation to proper time $\Delta \tau$

$$\Delta \tau = \frac{\Delta T}{\gamma},$$

we can express $\mathbf{p}$ in terms of coordinate time as

$$\mathbf{p} = m\gamma \left(\frac{dx}{dt}, \frac{dy}{dt}, \frac{dz}{dt} \right) = m\gamma\mathbf{v}, \tag{3.6.7}$$

which will happily transform between different inertial frames using the Lorentz transformations. Importantly, this means that unlike Newtonian momentum relativistic momentum is conserved in all inertial frames. Notice that at slow speeds ($\mathbf{v} \ll c$) $\gamma \approx 1$ and the relativistic momentum $\mathbf{p}$ approximates to the Newtonian momentum $\mathbf{p}_{Newtonian}$.

3.6.5 Relativistic kinetic energy

The kinetic energy of a particle is the energy it possesses due to its motion. In Newtonian mechanics, the kinetic energy ($KE_{Newtonian}$) of a particle of mass m moving with speed v is defined as the work done to accelerate the particle from rest to that speed v. Work done W equals force F multiplied by the distance through which the force acts ($\Delta s = s_1 - s_0$). Therefore, the work done

$$W = \int_{s_0}^{s_1} F dx.$$

Newton's second law relates force to mass and acceleration

$$F = ma$$

so we can say that work done is given by

$$W = \int_{s_0}^{s_1} ma dx.$$

Acceleration is the rate of change of velocity with respect to time $\frac{dv}{dt}$ and we can substitute this into the above equation to give

$$W = \int_{s_0}^{s_1} m\frac{dv}{dt} dx$$

and using the chain rule $\frac{dv}{dx}\frac{dx}{dt}dx = \frac{dv}{dt}dx$ we can write

$$W = \int_{s_0}^{s_1} m\frac{dv}{dx}\frac{dx}{dt} dx = \int_{v_0}^{v_1} mv dv, \tag{3.6.8}$$

where v_1 is the velocity of the particle at distance s_1, and v_0 is the velocity at distance s_0. We then integrate to find

$$W = \frac{1}{2}m\left(v_1\right)^2 - \frac{1}{2}m\left(v_0\right)^2$$

and because we define kinetic energy as the work done to accelerate the particle from rest to its final velocity v, we know that $v_0 = 0$ and that the kinetic energy is equal to

$$KE_{Newtonian} = \frac{1}{2}mv^2. \tag{3.6.9}$$

Because Newtonian momentum $p = mv$, we can rewrite (3.6.8) as

$$W = \int_{v_0}^{v_1} v\,dp, \tag{3.6.10}$$

where (implied but not stated) p_0 and p_1 are the particle's initial momentum ($= 0$) and final momentum ($= mv_1$). But this now gives us a means to find the **relativistic kinetic energy** (KE_{rel}) by substituting relativistic momentum (3.6.7) instead of Newtonian momentum in (3.6.10)

$$KE_{rel} = \int_{v_0}^{v_1} v\,d\left(\frac{mv}{\sqrt{1 - (v/c)^2}}\right).$$

We can evaluate this integral by first using integration by parts (1.10.17)

$$KE_{rel} = \left[\frac{mv^2}{\sqrt{1 - (v/c)^2}}\right]_{v_0}^{v_1} - \int_{v_0}^{v_1} \frac{mv}{\sqrt{1 - (v/c)^2}}\,dv.$$

We then need to use integration by substitution (1.10.16) on the second term. First, factor out the constant m to give

$$m \int \frac{v}{\sqrt{1 - (v/c)^2}}\,dv.$$

Let $s = 1 - \frac{v^2}{c^2} \Rightarrow ds = -\frac{2v}{c^2}\,dv$

$$= -\frac{c^2 m}{2} \int \frac{1}{\sqrt{s}}\,ds.$$

The integral of $\frac{1}{\sqrt{s}}$ is $2\sqrt{s}$, giving

$$= -c^2 m \sqrt{s} + C.$$

Substitute back for $s = 1 - \frac{v^2}{c^2}$

$$= -c^2 m \sqrt{1 - (v/c)^2} + C,$$

which means we can now write

$$KE_{rel} = \left[\frac{mv^2}{\sqrt{1 - (v/c)^2}} + mc^2\sqrt{1 - (v/c)^2}\right]_{v_0}^{v_1}.$$

Next, multiply top and bottom of the right-hand term by $\sqrt{1 - (v/c)^2}$ to get everything over a common denominator

$$KE_{rel} = \left[\frac{mv^2 + mc^2\left(1 - (v/c)^2\right)}{\sqrt{1 - (v/c)^2}}\right]_{v_0}^{v_1} = \left[\frac{mv^2 + mc^2 - mc^2v^2/c^2}{\sqrt{1 - (v/c)^2}}\right]_{v_0}^{v_1}$$

$$= \left[\frac{mc^2}{\sqrt{1 - (v/c)^2}} \right]^{v_1}_{v_0}$$

$$= \left(\frac{mc^2}{\sqrt{1 - (v_1/c)^2}} - \frac{mc^2}{\sqrt{1 - (v_0/c)^2}} \right).$$

Let $v_1 = v$, and we know that the particle is accelerating from rest, therefore $v_0 = 0$, so we finally arrive at the equation for relativistic kinetic energy of a particle of mass m moving with speed v :

$$KE_{rel} = mc^2 \left(\frac{1}{\sqrt{1 - (v/c)^2}} - 1 \right) = (\gamma - 1)\, mc^2 = \frac{mc^2}{\sqrt{1 - (v/c)^2}} - mc^2, \qquad (3.6.11)$$

where γ is the Lorentz factor (3.4.1).

This looks very different to (3.6.9), the equation for Newtonian kinetic energy. However, using Taylor's theorem (Section 1.10.3) it is possible to expand the Lorentz factor

$$\gamma = \frac{1}{\sqrt{1 - (v/c)^2}} = 1 + \frac{1}{2}\frac{v^2}{c^2} + \frac{3}{8} \left(\frac{v^2}{c^2} \right)^2 + \dots$$

on to infinity. Therefore

$$\cdot\ KE_{rel} = \left[\left(1 + \frac{1}{2}\frac{v^2}{c^2} + \frac{3}{8} \left(\frac{v^2}{c^2} \right)^2 + \dots \right) - 1 \right] mc^2.$$

In Newtonian systems we can assume that $v \ll c$, and therefore ignore the squares and higher terms of $\left(\frac{v^2}{c^2} \right)$ giving

$$KE_{rel} \approx \left[\left(1 + \frac{1}{2}\frac{v^2}{c^2} \right) - 1 \right] mc^2$$

$$KE_{rel} \approx mc^2 + \frac{mc^2 v^2}{2c^2} - mc^2$$

$$KE_{rel} \approx \frac{1}{2}mv^2.$$

So at slow speeds the relativistic kinetic energy approximates to the Newtonian kinetic energy.

3.6.6 Total relativistic energy

If we rearrange the equation for relativistic kinetic energy (3.6.11) we can write

$$E = \gamma mc^2 = \frac{mc^2}{\sqrt{1 - (v/c)^2}} = KE_{rel} + mc^2 \qquad (3.6.12)$$

and now we have an equation that gives the **total relativistic energy** E of a particle in an inertial frame. Total relativistic energy consists of the particle's relativistic kinetic energy plus the second term mc^2, which is the particle's mass energy E_o. It can be shown theoretically and has been verified experimentally that, providing no external forces act, total relativistic energy is conserved in all inertial frames, irrespective of whether mass or kinetic energy are conserved. In high-speed particle collisions, for example, mass, kinetic energy, even the total number of particles may not be conserved, but the total relativistic energy of the system will be.

If the particle is at rest (ie $v = 0$) the Lorentz factor $\gamma = \frac{1}{\sqrt{1-(v/c)^2}}$ reduces to 1 and

$$E = E_0 = mc^2. \tag{3.6.13}$$

This is Einstein's famous **mass-energy equation**, which states that mass and energy are in some sense equivalent, and that even when at rest a particle will still have energy due to its mass. Obviously, c^2 is a big number, so a small amount of mass yields a large amount of energy.

Problem 3.11. Calculate the increase in mass if two 7 kg lumps of clay, each travelling at 1000 mph, collide. Assume an inelastic head-on collision.

Total relativistic energy is given by (3.6.12)

$$E = \frac{mc^2}{\sqrt{1 - (v/c)^2}}.$$

From the conservation of total relativistic energy $E_{before} = E_{after}$.

$v = 1000$ miles per hour $= 447 \, \mathrm{m \, s^{-1}}$

$(v/c)^2 = \left(447/\left(3 \times 10^8\right)\right)^2 = \left(1.49 \times 10^{-6}\right)^2 = 2.22 \times 10^{-12}$

$\sqrt{1 - (v/c)^2} = 0.99999999999889.$

An inelastic head-on collision means the velocity after impact is zero.

Let the mass of each total lump of clay before collision $= m$, and mass of the combined lumps after collision $= M$, then

$$\frac{2mc^2}{\sqrt{1 - (v/c)^2}} = \frac{Mc^2}{1} = Mc^2$$

$$\frac{2m}{\sqrt{1 - (v/c)^2}} = M$$

$$M = \frac{14}{0.99999999999889} = 14.00000000001554.$$

Therefore, the increase in mass

$$= 14.00000000001554 - 14 = 1.55 \times 10^{-11} \, \mathrm{kg}.$$

An exceedingly small amount!

3.6.7 Four-momentum

If we multiply the four-velocity (3.6.1) by the rest mass m of a particle we get another four-vector that goes by the name of **four-momentum**.

$$P^\mu = mU^\mu. \tag{3.6.14}$$

If we recall the definition of the four-velocity (3.6.2)

$$U^\mu = (c\gamma, \gamma\mathbf{v})$$

and then multiply by m we get

$$P^\mu = (mc\gamma, m\gamma\mathbf{v}). \tag{3.6.15}$$

But recall (3.6.12) $E = \gamma mc^2$ is the equation for total relativistic energy, and from (3.6.7) $\mathbf{p} = m\gamma\mathbf{v}$ is the equation for relativistic momentum. We can then see that $mc\gamma$ is the total relativistic energy divided by the speed of light, and can rewrite (3.6.15) as

$$P^\mu = (E/c, \mathbf{p}) = (E/c, p_x, p_y, p_z). \tag{3.6.16}$$

Four-momentum provides a complete description of the total relativistic energy (its time component) and relativistic momentum (its spatial components) of a particle. Schutz [28] summarises this by saying the four-momentum of a particle 'is a vector where the components in some frame give the particle energy and momentum relative to that frame.' As we'll see later, the all important energy-momentum tensor, the right-hand side of the Einstein field equations and the source of spacetime curvature, is actually a measure of the rate of flow per unit area of four-momentum.

3.6.8 Four-force

Newton's second law of motion (2.4.3) says that the force acting on a body equals the body's rate of change of momentum

$$\mathbf{F} = \frac{d\mathbf{p}}{dt}.$$

We can extend this to special relativity and define the four-force as the rate of change of four-momentum

$$F^\mu = \frac{dP^\mu}{d\tau}, \tag{3.6.17}$$

which we use in Section 7.4 to show how free particles move in curved spacetime.

3.6.9 Energy-momentum relation

The scalar product of the four-velocity (3.6.4) is given by

$$\eta_{\mu\nu} U^\mu U^\nu = c^2.$$

So, as four-momentum (3.6.14) is given by $P^\mu = mU^\mu$, we can write

$$\eta_{\mu\nu} P^\mu P^\nu = m^2 \eta_{\mu\nu} U^\mu U^\nu = m^2 c^2.$$

But the scalar product of four-momentum $P^\mu = (E/c, \mathbf{p})$ can also be found directly using (3.6.3)

$$\eta_{\mu\nu} P^\mu P^\nu = E^2/c^2 - (p_x)^2 - (p_y)^2 - (p_z)^2 = E^2/c^2 - p^2.$$

Combining these two expressions for the scalar product of four-momentum we get

$$\frac{E^2}{c^2} - p^2 = m^2 c^2,$$

which rearranges to give

$$E^2 = p^2 c^2 + m^2 c^4. \tag{3.6.18}$$

For a particle in its rest frame (ie its momentum is zero), this reduces to

$$E = mc^2,$$

which as we saw earlier (3.6.13) when looking at total relativistic energy is the famous mass-energy equation. Light (and other electromagnetic radiation) can be thought of as a stream of photons, a type of elementary particle. A photon, which has zero 'rest mass', does have energy and momentum. So if we let $m = 0$ in (3.6.18) we get

$$E = pc,$$

which describes the energy-momentum relation for a photon.

Problem 3.12. The energy of a photon is given by $E = hc/\lambda$, where $h = 6.63 \times 10^{-34}$ J s is Planck's constant, and λ (the Greek letter lambda) is the wavelength. A photon of blue light has a wavelength of $400 \times 10^{-9} m$. What is its momentum?

$E = pc$ and $E = hc/\lambda$ therefore

$$pc = hc/\lambda$$

$$p = h/\lambda$$

$$p = \frac{6.63 \times 10^{-34}}{400 \times 10^{-9}}$$

$$p = 1.66 \times 10^{-27} \, \text{kg m s}^{-1}.$$

4 Introducing the manifold

If only I had the theorems! Then I should find the proofs easily enough.

BERNHARD RIEMANN

4.1 Introduction

Newtonian gravity is incompatible with special relativity because, in Newton's theory, the gravitational force acts instantaneously across any distance. Special relativity does not allow this, instead imposing a natural speed limit – the speed of light. In order to reconcile relativity and gravity, we now need to move from the (flat) Minkowski space of special relativity to the curved spacetime of general relativity. Einstein's great insights in formulating his theory of general relativity were that:

- matter and energy curve spacetime, and

- spacetime can be modelled using a mathematical structure known as a **pseudo-Riemannian manifold**.

Gravity in general relativity therefore ceases to be regarded as a force, but rather as a property of the geometry of spacetime.

To get across this idea of the geometry of a curved space, Misner et al [23] use the example of an ant crawling over the skin of an apple. Determined to walk in as straight a line as possible, the ant carefully measures all its paces to be of equal length. The ant, if it ventures too close to the top of the apple, even though it is trying to walk in a straight line, will find itself being drawn into the dimple where the stalk is. An imaginative ant might say there's a force attracting it towards the base of the stalk. There isn't a force of course. The ant's path is determined not by a force, but by the curvature – the geometry – of the surface of the apple. Misner et al imagine a physics student watching the ant's progress, then taking a knife, carefully removing the track of apple peel along which the ant has walked and laying it on the face of his book. After observing that the track 'ran as straight as a laser beam ... No more economical path could the ant have taken to cover the 10 cm from start to end of that strip of skin', the student reflects that the ant's path forms 'a beautiful geodesic'.

The (flat) Minkowski space of special relativity is to be found in deep space, far from the influence of any gravitational fields. In flat spacetime free particles move in straight lines. In the vicinity of massive (in the general sense of something with mass) objects – planets or stars for example – spacetime is curved. Free particles, including light, will not then move in straight lines, but will instead move along the 'straightest possible' paths dictated by the geometry of curved spacetime, just as the ant follows the straightest possible path dictated by the surface of the apple. These paths are known as **geodesics**, which we can loosely define as the straightest or shortest distance between two points in a curved space. Geodesics on the surface of a sphere, for example, are parts of a great circle, ie a circle with the same diameter as the sphere and whose centre is the centre point of the sphere, the equator for example, or any line of longitude. If the mathematical structure of spacetime is known – ie how it curves – then the geodesics of moving particles (and planets and stars) can be calculated and tested against observation.

In order to describe curved spaces mathematically we use the concept of a **manifold**, loosely defined as a smoothly curved space that is locally flat. A circle, for example, is a one-dimensional manifold: make it big enough and a small segment looks like a straight line. The surface of a sphere is a two-dimensional manifold: small pieces of it look flat. Spacetime, a four-dimensional manifold, is also locally flat, and in small enough regions reduces to the spacetime of special relativity.

We say a manifold is n-dimensional, because each point on the manifold can be specified by n dimensions or coordinates. The line and circle are one-dimensional manifolds: any point on them can be described using just one coordinate (distance along the line for the line, polar angle for the circle, for example). The plane is a two-dimensional manifold, as is the surface of a sphere: any point on the surface can be described using the spherical polar coordinates θ and ϕ. Spacetime, as we have seen, is four-dimensional: one time coordinate plus three spatial coordinates are needed to specify a point in spacetime.

Not all spatial structures are manifolds. A one-dimensional line emerging from a plane isn't a manifold, nor are two cones apex to apex on top of each other. Because bits of these objects are not locally Euclidean they are not classified as manifolds.

4.2 Riemannian manifolds and the metric

In order to be able to model the spacetime of general relativity, a manifold must also have these two additional properties:

- It must be differentiable. This will become abundantly clear when we look at the transformation properties of tensors, which involve hordes of partial derivatives.

- It must be endowed with a **symmetric metric tensor** (denoted by $g_{\mu\nu} = g_{\nu\mu}$) which, as in special relativity, defines the separation of nearby points on the manifold. The form of the metric will change through the manifold according to how spacetime in that region curves. In deep space, for example, far from the influence of matter and energy, spacetime will approximate the Minkowski space of special relativity and $g_{\mu\nu} \approx \eta_{\mu\nu}$. Near a non-rotating black hole, $g_{\mu\nu}$ will approximate the Schwarzschild metric (see Chapter 9 for details). The metric completely defines the curvature of a manifold and is therefore of fundamental importance in relativity.

A manifold that is differentiable and possesses a symmetric metric tensor is known as a **Riemannian manifold**, after the German mathematician Bernhard Riemann (Figure 4.1).

Riemann discovered that the metric contains all the information we need to describe the curvature of the manifold. If we know the metric, we know the space. That is why the metric is crucial to general relativity.

Figure 4.1: Bernhard Riemann (1826–1866).

An n-dimensional Riemannian manifold has a line element given by:

$$dl^2 = g_{ij}dx^i dx^j, \qquad (4.2.1)$$

where g_{ij} is the metric and dx^i and dx^j are the coordinate differentials we met in Section 1.14.1. Equation 4.2.1 may look complicated, but all it means is that the metric g_{ij} determines the coefficients of the line element. We saw a simple example of this in Section 1.14.1 when we met the three-dimensional Euclidean line element (1.14.5) in Cartesian coordinates

$$dl^2 = dx^2 + dy^2 + dz^2,$$

where the Euclidean metric (1.14.6) is

$$[g_{ij}] = \begin{pmatrix} 1 & 0 & 0 \\ 0 & 1 & 0 \\ 0 & 0 & 1 \end{pmatrix}$$

and tells us there is $1 \times dx^2$, $1 \times dy^2$ and $1 \times dz^2$ in the above Euclidean line element.

And we've also seen the Minkowski line element (3.5.1) in Cartesian coordinates given by

$$ds^2 = c^2 dt^2 - dx^2 - dy^2 - dz^2,$$

where the Minkowski metric (3.5.2) is

$$[\eta_{\mu\nu}] = \begin{pmatrix} 1 & 0 & 0 & 0 \\ 0 & -1 & 0 & 0 \\ 0 & 0 & -1 & 0 \\ 0 & 0 & 0 & -1 \end{pmatrix},$$

which tells us there is $1 \times c^2 dt^2$, $-1 \times dx^2$, $-1 \times dy^2$ and $-1 \times dz^2$ in the above Minkowski line element. Recall that the symbol $\eta_{\mu\nu}$ refers specifically to the Minkowski metric.

Incidentally, we know these metrics describe flat space because their metric coefficients are ± 1. However, if we were to choose a weird coordinate system we might well end up with a complicated metric whose coefficients do not equal ± 1. How do we then tell whether that metric defines a flat space? The answer (to get ahead of ourselves a little, we'll be discussing this in the next section) is the Riemann curvature tensor (6.6.1), the absolute acid test for determining whether a manifold is flat or curved. If the Riemann curvature tensor is zero for all points in a particular space then that space is flat. If the tensor does not equal zero at a point, then the space is curved at that point.

When we think of a circle or the surface of a sphere we consider them as existing in our everyday Euclidean space. We can draw a circle on a piece of graph paper, or place a sphere (a ball for example) on a table. We might then compare the curvature of the circle to the straight lines drawn on the graph paper, or compare the roundness of the sphere to the flatness of the table. In mathematical terms we say the circle and sphere are exhibiting **extrinsic curvature** because their curvature is seen in relation to an external space of higher dimension – two dimensions for the paper, three for the table. However, not all manifolds can be thought of as being embedded in an external space in this manner. Often we need to examine a manifold on its own terms, so to speak, without reference to a space of higher dimension. This type of curvature is known as **intrinsic curvature**. We understand intrinsically curved manifolds by examining them directly, using the metric and mathematical tools derived from the metric to analyse their internal structure. In the context of general relativity, we are only interested in the intrinsic curvature of spacetime.

Place a cylinder of rolled up paper on a table and we can see it has extrinsic curvature compared to the flat surface of the table. However, intrinsically the cylinder is flat – the sum of the internal angles of a triangle drawn on its surface equal $180°$. This isn't surprising, as we could unroll our cylinder and end up with a flat sheet of paper.

Like the cylinder, our ball on the table has an extrinsically curved surface. However, the ball's surface is also intrinsically curved. On a sphere lines of longitude start off parallel on the equator but meet at the poles. Also, the internal angles of a triangle on a sphere can all equal $90°$. Unlike a cylinder, the surface of a sphere cannot be flattened without distortion, as mapmakers find when they try to project the surface of the Earth onto a flat map.

At the start of this section we said that spacetime could be modelled using a pseudo-Riemannian manifold. These structures differ from the Riemannian manifolds we have discussed so far in that they allow dl^2 (in (4.2.1)) to be positive, zero or negative, as is the case with spacetime. The distinction, though factually correct, is not relevant to our needs, and we'll carry on loosely using the terms 'space', 'curved space', 'Riemannian manifold' or 'manifold' from now on.

4.3 Surface of a sphere

One of the simplest two-dimensional manifolds is the surface of a sphere.

In Section 1.14.1 we met the line element (1.14.10) and metric (1.14.11) that describes Euclidean three-dimensional space using spherical coordinates. The line element is

$$dl^2 = dr^2 + r^2 d\theta^2 + r^2 \sin^2\theta d\phi^2.$$

If we set the polar coordinate r to be some constant R we lose the dr term (because r is now a constant) and have defined the line element for the surface of a sphere

$$dl^2 = R^2 d\theta^2 + R^2 \sin^2\theta d\phi^2, \tag{4.3.1}$$

which describes a two-dimensional surface using the two polar coordinates (θ, ϕ). By looking at (4.3.1), we can see that the metric for this surface, using coordinates (θ, ϕ), is

$$[g_{ij}] = \begin{pmatrix} g_{\theta\theta} & g_{\theta\phi} \\ g_{\phi\theta} & g_{\phi\phi} \end{pmatrix} = \begin{pmatrix} R^2 & 0 \\ 0 & R^2 \sin^2 \theta \end{pmatrix}, \tag{4.3.2}$$

where $g_{11} = g_{\theta\theta} = R^2$, $g_{22} = g_{\phi\phi} = R^2 \sin^2 \theta$ and $g_{ij} = 0$ for $i \neq j$.

The inverse metric is

$$[g^{ij}] = \begin{pmatrix} \frac{1}{R^2} & 0 \\ 0 & \frac{1}{R^2 \sin^2 \theta} \end{pmatrix}. \tag{4.3.3}$$

For the unit radius sphere $(R = 1)$ these metrics become

$$[g_{ij}] = \begin{pmatrix} 1 & 0 \\ 0 & \sin^2 \theta \end{pmatrix} \tag{4.3.4}$$

and

$$[g^{ij}] = \begin{pmatrix} 1 & 0 \\ 0 & \frac{1}{\sin^2 \theta} \end{pmatrix}. \tag{4.3.5}$$

Problem 4.1. Find the circumference C of a circle generated by sweeping a point at constant angle θ around the 'north pole' on the surface of a sphere of radius R (see Figure 4.2).

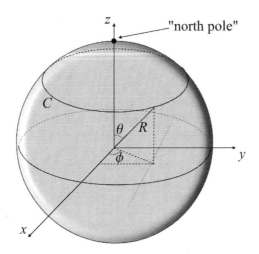

Figure 4.2: Circle on a sphere.

Because θ is constant, $d\theta = 0$ and 1.14.10 becomes

$$dl^2 = R^2 \sin^2 \theta d\phi^2,$$

which is the line element that gives the distance between infinitesimally close points on the circumference C. Taking the square root of both sides gives

$$dl = R \sin \theta d\phi. \tag{4.3.6}$$

As we sweep out our circle around the 'north pole' ϕ changes from 0 to 2π. We integrate 4.3.6 using these limits:

$$C = \int_0^{2\pi} R\sin\theta d\phi = R\sin\theta\,[\phi]_0^{2\pi} = 2\pi R\sin\theta,$$

which is different to the usual equation ($C = 2\pi R$) for the circumference of a circle in two-dimensional Euclidean space. The difference, of course, is because we are working on the curved surface of a sphere, not on a plane.

In Section 6.5 we'll be looking at geodesics ('straightest' line) on the surface of the sphere.

The spacetime manifold is the stage on which the theatre of relativity is played out. We'll now look at the main actors who perform on that stage. These are the mathematical objects known as scalars, contravariant vectors, one-forms and tensors.

5 Scalars, vectors, one-forms and tensors

Someone told me that each equation I included in the book would halve the sales.

STEPHEN HAWKING

5.1 Introduction

The branch of mathematics that is concerned with manifolds and the objects that live on manifolds is called **differential geometry**. In this chapter we look at four of those objects – scalars, contravariant vectors, one-forms (also known as covariant vectors) and tensors. (Actually, scalars, contravariant vectors and one-forms are all different types of tensor, but we'll start off by treating them as separate entities.)

Most authors seem to prefer the term 'one-form' to 'covariant vector', so that's what we'll be using from now on. Also, many authors refer to contravariant vectors simply as vectors, as I've done in the title of this chapter. In fact, we're going to be even sloppier than that and sometimes use 'vector' as a generic term for both contravariant vectors and one-forms. Hopefully, the context should make it clear when we're referring specifically to contravariant vectors and when we're referring to both contravariant vectors and one-forms. Recall that contravariant vector components have an upper index (eg V^a), and one-form components have a lower index (eg V_a). Tensors can have none, one or more indices.

Later on, we'll meet the rules of tensor algebra, including operations such as scaling, where a tensor $T^\lambda{}_\beta$ is multiplied by a scalar S to give a new tensor $X^\lambda{}_\beta$, and contraction, where a tensor $T^\lambda{}_{\beta\lambda}$ is summed over an upper and a lower index to give another tensor T_β. Differential geometry is the theoretical foundation to these rules. However, just as you don't need to be an automotive engineer in order to drive a car, you don't have to know all the underlying mathematics if you want to manipulate tensors – just a working knowledge of the rules of tensor manipulation, which you can more or less learn by rote. So, much of this section is 'under the bonnet' detail – useful but not essential. However, it should give you a deeper understanding of what is going on when we actually start to use tensors in general relativity.

In Section 1.13 we looked at simple vectors representing quantities with magnitude and direction, such as velocity. We can physically draw these vectors as directed line segments – a line with an arrow at one end pointing along the vector's direction. Using Cartesian coordinates, we saw that a vector $\mathbf{V}$ consists of the product of its components (V_x, V_y, V_z), and a set of basis vectors $\hat{e}_x, \hat{e}_y, \hat{e}_z$ (recall the little hat means each basis vector is one unit long) pointing along the x, y, z axes respectively (another way of saying this is that the basis vectors are tangent to the coordinate axes) given by (1.13.1)

$$\mathbf{V} = V_x \hat{e}_x + V_y \hat{e}_y + V_z \hat{e}_z.$$

Because these axes are nice simple straight lines, there's no need for the basis vectors to change direction. This is why Cartesian vectors are so easy to use, because the basis vectors don't change. Bases (plural of basis) of this sort, which are constant, are known as **non-coordinate bases**, ie they don't change with the coordinates.

Spacetime in special relativity, being flat, can also be described using a Cartesian coordinate system. We've seen several examples of four-vectors in special relativity including:

- The four-position $\left(x^\mu = ct, x^i\right)$.

- The four-velocity $\left(U^\mu = \gamma\left(c, \mathbf{v}\right)\right)$.

- The four-momentum $\left(P^\mu = E/c, \mathbf{p}\right)$.

- The four-force $\left(F^\mu = \frac{dP^\mu}{d\tau}\right)$.

Unfortunately, the vectors used in general relativity are not directed line segments stretching from one point to another in space. Instead, each vector is located at a single point in spacetime. In fact, each point in spacetime is itself a vector space and home to an infinite number of vectors. The vector space will be both a **tangent space** (home to those objects known as contravariant vectors), and a **cotangent space** (home to those objects known as one-forms).

Contravariant vectors and one-forms should be thought of as different representations of the same geometrical object. More precisely, they are alternative ways of describing an infinitesimal displacement of a point on a manifold, namely as either:

- a tangent vector to a parameterised curve (aka a contravariant vector), or

- as corresponding to an infinitesimal change in a scalar function (aka a gradient one-form).

Contravariant vectors and one-forms also answer the question: how does a scalar function change along a parameterised curve? The one-form supplies the function, the vector the direction tangent to the curve. Together, at a point, they give the rate of change of the function along the curve.

We'll see later how the metric tensor is used to convert a vector to its corresponding one-form and vice versa.

The reason that the simple vectors in Section 1.13 and the more abstract vectors we are now talking about are both loosely referred to as 'vectors' is that they both obey the rules that define a vector space. In brief, a vector space consists of a group of objects (call the group X, for example) that can be added together and multiplied by a scalar, and the result will be another member of the group X.

Up until now, our indices have referred to particular coordinate systems: x, y, z for Cartesian, r, θ, ϕ for spherical, etc. Differential geometry demands a more abstract use of indices, where they can refer to any permissible coordinate system. Similarly, in general relativity, because we are dealing with curved spacetime, there are no preferred coordinate systems and we need to be able to transform from any one coordinate system to any other (the terminology is that we are using **general coordinates**). So if x^μ are the old coordinates and y^μ are the new coordinates ($\mu = 0, 1, 2, 3$), then any functions linking x^μ and y^μ are permissible as long as (a) the functions are differentiable, and (b) each point in spacetime is uniquely labelled by a set of four numbers. We move freely between different coordinate systems by using the transformation properties (involving partial derivatives of the coordinate functions) of contravariant vectors and one-forms.

Basis contravariant vectors and basis one-forms are also defined in terms of derivatives of the coordinate functions. We don't need to go into much detail, but can just note that basis vectors are tangent to the coordinate curves (along which only one of the coordinates changes), and basis one-forms are gradients of the coordinate surfaces (on which only one of the coordinates remains constant). Bases of this sort (unlike the constant non-coordinate bases of Cartesian coordinates) that change with the coordinates, are known as **coordinate bases**. These are the bases we'll be implicitly using from now on.

However (thankfully!), the transformation properties of the components of vectors and one-forms (and tensors in general) are basis-independent, meaning we usually don't need to worry too much about basis vectors and basis one-forms. Crucially, basis-independence means that if a tensor equation is true in one coordinate system it will be true in all coordinate systems. Because we tend to refer only to the components of vectors, one-forms, etc, we'll continue our habit of loosely referring to a 'vector V^α' or a 'one-form V_α' when, strictly speaking, this notation refers to the components of those objects. Though not totally accurate, this terminology is good enough for our purposes.

Although, in the context of general relativity, we can't meaningfully talk about directed line segments stretching from one point to another in space, we can define an infinitesimal displacement vector in spacetime:

$$d\vec{x} = dx^\mu e_\mu.$$

This is a deceptively simple equation. It is only valid because it includes coordinate basis vectors e_μ which, by definition, give the coordinate differentials dx^μ as the components of $d\vec{x}$. Other, non-coordinate, basis vectors would not generally give such a straightforward (and useful) result. The power of the mathematics that follows is that it implicitly makes use of this definition of $d\vec{x}$ to end up with physical measurable quantities (time, distance, velocity, momentum, etc).

Any contravariant vector or one-form is the product of its components and a basis of some kind. Contravariant four-vectors are often represented by an arrow over the letter (eg $\vec{V}$) so we can say, using the Einstein summation convention, that

$$\vec{V} = V^\alpha e_\alpha, \tag{5.1.1}$$

where V^α and e_α are respectively the components and basis vectors of $\vec{V}$.

One-forms are often represented by a tilde over the letter (eg $\tilde{V}$) so we can say, again using the Einstein summation convention, that

$$\tilde{V} = V_\alpha e^\alpha, \tag{5.1.2}$$

where V_α and e^α are respectively the components and basis one-forms of $\tilde{V}$.

Note that basis vectors (e_α, for example) are themselves vectors, even though they have a lower index. Similarly, basis one-forms (e^α, for example) are one-forms, even though they have an upper index. This convention (a) allows us to write nice balanced tensor equations (such as $\vec{V} = V^\alpha e_\alpha$) with one upper and one lower dummy index, and (b) reflects the fact that basis vectors transform like one-form components and basis one-forms transform like vector components.

At a point on the manifold a contravariant vector acts *linearly* on a one-form (and vice versa) to give a scalar (a real number). This works because the relationship between the basis vectors and basis one-forms is defined (using the Kronecker delta – here used as a tensor, with an upper and lower index) by the equations

$$e^\alpha (e_\beta) = \delta^\alpha_\beta. \tag{5.1.3}$$

So when they act on each other the basis vectors and basis one-forms disappear in a puff of zeros and ones. Because of this reciprocal relationship the basis one-forms e^α are known as the **dual basis** to the basis vectors e_β. Therefore, for any one-form and vector at a point on the manifold

$$\tilde{P}\left(\vec{V}\right) = P_\alpha V^\beta e^\alpha e_\beta = P_\alpha V^\beta \delta^\alpha_\beta = P_\alpha V^\alpha$$

$$= P_0 V^0 + P_1 V^1 + P_2 V^2 + P_3 V^3,$$

which is a scalar. A popular metaphor in the literature is to regard tensors (of which vectors and one-forms are examples) as linear machines with input slots, one for each index. Each lower index slot can be filled with a vector. Each upper index slot can be filled with a one-form. When each slot is filled, out pops a number. A vector and a one-form therefore act together (the process is called 'contraction') to fill each other's slots to produce a number. As we'll see, the scalar product of two vectors is more complicated and requires the intervention of a third tensor – the metric – in order to output a number. (The metric has two lower slots/indices, which are filled by the two vectors.)

Before looking at the transformation properties of contravariant vectors and one-forms we'll see what happens to scalar fields when we change from one coordinate system to another.

5.2 Scalars

In Section 1.13.2 we looked at a scalar field T, where the function $T(x, y, z)$ is the air temperature in a room described by Cartesian coordinates x, y, z. Say we now want to express temperature T in terms of a different coordinate system, spherical polar coordinates for example, with a function $T(r, \theta, \phi)$. How do we do this, and how does T change in this new coordinate system?

We've seen how to express Cartesian coordinates in terms of spherical coordinates using (1.9.2)

$x = r \sin\theta \cos\phi$, $y = r \sin\theta \sin\phi$, $z = r \cos\theta$.

And it wouldn't be too difficult to rearrange these equations to express spherical coordinates in terms of Cartesian coordinates, so that $r = f(x, y, z)$, $\theta = f(x, y, z)$ and $\phi = f(x, y, z)$. However, we don't need to do this as we know that temperature, being a scalar quantity, is an invariant, it doesn't change with a change in coordinate system: if the temperature of the door knob in my room is 15°C when I'm describing the room using Cartesian coordinates, it will be the same temperature if I use spherical or any other coordinate system. If only everything in life was so simple. We can therefore say

$$T(r, \theta, \phi) = T(x, y, z).$$

Note, although the temperature T of the door knob is the same in any coordinate system, the actual values of the coordinates r, θ, ϕ or x, y, z describing the door knob's location will of course be different. Our conclusion is simple:

- Scalars do not change during a coordinate transformation.

Next we look at the transformation properties of contravariant vectors and one-forms, then try to explain, term by term, what these equations actually mean.

5.3 Contravariant vectors and one-forms

We'll start with the transformation equations – stated first, with details to follow.

5.3.1 Contravariant vectors

A contravariant vector is an object with one upper index (also called a superscript or upstairs index) and with components that transform under coordinate transformations as follows:

$$V'^{\alpha} = \frac{\partial x'^{\alpha}}{\partial x^{\beta}} V^{\beta}. \tag{5.3.1}$$

We have here re-introduced the convention first met in Section 3.2.6 that when discussing coordinate transformations the new coordinates are denoted by a prime (eg x'^{α}) and the old coordinates are unprimed (eg x^{β}).

5.3.2 One-forms

A one-form is an object with one lower (or subscript or downstairs) index and with components that transform under coordinate transformations as follows:

$$V'_{\alpha} = \frac{\partial x^{\beta}}{\partial x'^{\alpha}} V_{\beta}. \tag{5.3.2}$$

5.3.3 A little more detail

Both equations are describing coordinate transformations. (5.3.1) tells us how to carry out a coordinate transformation on a contravariant vector, (5.3.2) tells us how to carry out a coordinate transformation on a one-form.

- V^{β} are the components of the contravariant vector $\vec{V}$ expressed in the original x^{β} coordinate system.

- V'^{α} are the components of the same contravariant vector $\vec{V}$ expressed in the new x'^{α} coordinate system.

- V_{β} are the components of the one-form $\tilde{V}$ expressed in the original x^{β} coordinate system.

- V'_{α} are the components of the same one-form $\tilde{V}$ expressed in the new x'^{α} coordinate system.

The terms $\frac{\partial x'^{\alpha}}{\partial x^{\beta}}$ and $\frac{\partial x^{\beta}}{\partial x'^{\alpha}}$ describe how the old and new coordinate systems relate to each other. Specifically, we are taking partial derivatives of the functions linking the coordinates in one system to the coordinates in another system. As we'll see shortly, these terms represent a **transformation matrix**. We multiply the vector in the old x^{β} coordinate system by the correct transformation matrix to get the vector in the new x'^{α} coordinate system.

5.3.4 Examples of one-forms and contravariant vectors

We'll now look at specific examples of one-forms and contravariant vectors.

5.3.4.1 One-forms

Staying with our earlier 'room' example, we'll now consider the components of the gradient ∇T of the scalar field T (1.13.7)

$$\left(\frac{\partial T}{\partial x}, \frac{\partial T}{\partial y}, \frac{\partial T}{\partial z}\right)$$

at a particular point in the room.

- We can now state that the components of the gradient of a scalar field are the components of a one-form.

Let's ask ourselves what would happen if we wanted to convert this gradient one-form to spherical polar coordinates? In that case we would need to change $T(x, y, z)$ into a different function $T(r, \theta, \phi)$, which gives the temperature in terms of spherical polar coordinates r, θ, ϕ. These two functions would not be the same, and therefore we can say

$$\left(\frac{\partial T}{\partial x}, \frac{\partial T}{\partial y}, \frac{\partial T}{\partial z}\right) \neq \left(\frac{\partial T}{\partial r}, \frac{\partial T}{\partial \theta}, \frac{\partial T}{\partial \phi}\right),$$

where $\left(\frac{\partial T}{\partial r}, \frac{\partial T}{\partial \theta}, \frac{\partial T}{\partial \phi}\right)$ are the components of the same one-form, but in spherical coordinates.

Using the primed/unprimed convention for new/original coordinate systems, we'll call the original one-form components $\left(\frac{\partial T}{\partial x}, \frac{\partial T}{\partial y}, \frac{\partial T}{\partial z}\right) = X_\beta$, and the new one-form components $\left(\frac{\partial T}{\partial r}, \frac{\partial T}{\partial \theta}, \frac{\partial T}{\partial \phi}\right) = X'_\alpha$.

(5.3.2) tells us that the transformation equations for a one-form are

$$V'_\alpha = \frac{\partial x^\beta}{\partial x'^\alpha} V_\beta,$$

which we can rewrite for our particular one-form as

$$X'_\alpha = \frac{\partial x^\beta}{\partial x'^\alpha} X_\beta.$$

As mentioned above, the term $\frac{\partial x^\beta}{\partial x'^\alpha}$ represents a transformation matrix, in this case a 3×3 transformation matrix because there are three β coordinates x, y, z and three α coordinates r, θ, ϕ.

So, we can represent our mysterious transformation matrix as

$$\left[\frac{\partial x^\beta}{\partial x'^\alpha}\right] = \begin{bmatrix} ? & ? & ? \\ ? & ? & ? \\ ? & ? & ? \end{bmatrix}.$$

If we knew the elements of this matrix we could multiply it by the one-form X_β to find the new one-form X'_α.

As we'll shortly see, we find the elements of the transformation matrix by taking partial derivatives of the functions relating the coordinates in one system to the coordinates in another system. So the elements of $\frac{\partial x^\beta}{\partial x'^\alpha}$ are the partial derivatives of x^β with respect to x'^α.

For example, the element $\frac{\partial T}{\partial r}$ would be calculated as follows:

$$\frac{\partial T}{\partial r} = \frac{\partial T}{\partial x}\frac{\partial x}{\partial r} + \frac{\partial T}{\partial y}\frac{\partial y}{\partial r} + \frac{\partial T}{\partial z}\frac{\partial z}{\partial r},$$

where x, y, z are all functions of r, θ, ϕ. In a similar fashion we would calculate the other elements $\frac{\partial T}{\partial \theta}$ and $\frac{\partial T}{\partial \phi}$. Don't worry about the details of how we do this for the moment, just try to understand that in order to transform the components of a one-form from one coordinate system to another we need a transformation matrix.

5.3.4.2 Contravariant vectors

We now assume that the air at each point in the room is moving with some velocity which is a function of the three Cartesian coordinates x, y, z.

As we saw in Section 3.6.3 we can use time t to parameterise the path of a moving particle (or air molecule in this case) and thus obtain a vector whose components represent the object's spatial velocity (or three-velocity) in the x, y, z directions. The particle's spatial velocity is then a tangent vector to the path and has components $V^\beta = \left(\frac{dx}{dt}, \frac{dy}{dt}, \frac{dz}{dt}\right)$.

- We can now state that the tangent vector partial derivatives V^β are the components of a contravariant vector.

Again we'll assume that we want to transform the components V^β from Cartesian to spherical polar coordinates, where we'll call it V'^α. (5.3.1) tells us that the transformation equations for a contravariant vector are

$$V'^\alpha = \frac{\partial x'^\alpha}{\partial x^\beta}V^\beta.$$

Again the term $\frac{\partial x'^\alpha}{\partial x^\beta}$ represents a 3×3 transformation matrix, which we need to multiply by V^β in order to calculate V'^α. Notice, however, that $\frac{\partial x'^\alpha}{\partial x^\beta}$ is the inverse of $\frac{\partial x^\beta}{\partial x'^\alpha}$, and therefore the transformation matrix $\frac{\partial x'^\alpha}{\partial x^\beta}$ will be the inverse of the one-form transformation matrix $\frac{\partial x^\beta}{\partial x'^\alpha}$.

As with one-forms, we find the elements of the transformation matrix by taking partial derivatives of the functions relating the coordinates in one system to the coordinates in another system.

As an example of how the components $V^\beta = \left(\frac{dx}{dt}, \frac{dy}{dt}, \frac{dz}{dt}\right)$ transform (details to follow), we note that the first component of V'^α would be calculated as follows:

$$\frac{dr}{dt} = \frac{\partial r}{\partial x}\frac{dx}{dt} + \frac{\partial r}{\partial y}\frac{dy}{dt} + \frac{\partial r}{\partial z}\frac{dz}{dt},$$

where r, θ, ϕ are all functions of x, y, z. In a similar fashion we would calculate the other components $\frac{d\theta}{dt}$ and $\frac{d\phi}{dt}$.

Drawing on the above examples, and in the context of four-dimensional spacetime, we can now give definitions of contravariant vectors and one-forms.

5.3.5 Definition of a contravariant vector

A contravariant vector is a tangent vector to a parameterised curve in spacetime.

If the parameter of the curve is λ, and using a coordinate system x^β the components of the tangent vector $\vec{V}$ are given by

$$V^\beta = \frac{dx^\beta}{d\lambda} = \left(\frac{dx^0}{d\lambda}, \frac{dx^1}{d\lambda}, \frac{dx^2}{d\lambda}, \frac{dx^3}{d\lambda} \right). \tag{5.3.3}$$

A relativistic example of such a curve might be the world-line of a particle moving through spacetime, and an example of a contravariant vector on that curve could then be the particle's four-velocity $\vec{U}$, with components (3.6.1) $U^\mu = \frac{dx^\mu}{d\tau} = \left(c\frac{dt}{d\tau}, \frac{dx}{d\tau}, \frac{dy}{d\tau}, \frac{dz}{d\tau} \right)$. If, for convenience, we use $c = 1$ units, the four-velocity vector components become

$$U^\mu = \left(\frac{dt}{d\tau}, \frac{dx}{d\tau}, \frac{dy}{d\tau}, \frac{dz}{d\tau} \right),$$

where the four-velocity is defined as the rate of change of the particle's four-position (t, x, y, z) with respect to proper time τ, which is the parameter along the curve.

In more advanced texts you may see the vector $\vec{V}$ written as

$$\vec{V} = V^\beta \frac{\partial}{\partial x^\beta},$$

where V^β are the vector's components and the partial derivative operators $\frac{\partial}{\partial x^\beta}$ are the coordinate basis vectors (those things we don't need to worry about). In order to make sense of this formulation, consider an infinitesimal displacement $d\mathrm{P}$ of a point P on the manifold, where P is a function of some coordinate system x^β. If the displacement is along a curve parameterised by λ we can then drop P into $\vec{V} = V^\beta \frac{\partial}{\partial x^\beta}$ to get

$$\vec{V} = V^\beta \frac{\partial \mathrm{P}}{\partial x^\beta} = \frac{dx^\beta}{d\lambda} \frac{\partial \mathrm{P}}{\partial x^\beta} = \frac{d\mathrm{P}}{d\lambda},$$

where $\frac{\partial \mathrm{P}}{\partial x^\beta}$ are the coordinate basis vectors at P. If we define the same point P using another coordinate system, x'^α for example, the components and basis vectors will change, but the vector $\vec{V}$ remains constant.

In this context, $\vec{V}$ is referred to as a **directional derivative operator** – a thing that when supplied with an arbitrary function gives the rate of change of that function, at a point, in the direction of a vector. Drop a function, f for example, into the empty slot of $\vec{V} = V^\beta \frac{\partial}{\partial x^\beta}$ and we get a number – the **directional derivative** of f in the direction of $\vec{V}$.

However, we digress. Vectors as directional derivatives are an abstraction too far for our requirements. We can get along quite happily with the notion that a contravariant vector is simply a tangent vector to a parameterised curve.

5.3.6 Definition of a one-form

Consider a scalar field, where ϕ is a scalar function of x^β ($\beta = 0, 1, 2, 3 = t, x, y, z$, for example) and is invariant under Lorentz transformations (meaning that for each point in spacetime it has the same value in all inertial frames). The gradient components of the scalar field in the coordinate system x^β are given by

$$\frac{\partial \phi}{\partial x^\beta} = \left(\frac{\partial \phi}{\partial x^0}, \frac{\partial \phi}{\partial x^1}, \frac{\partial \phi}{\partial x^2}, \frac{\partial \phi}{\partial x^3} \right) = \left(\frac{\partial \phi}{\partial t}, \frac{\partial \phi}{\partial x}, \frac{\partial \phi}{\partial y}, \frac{\partial \phi}{\partial z} \right). \tag{5.3.4}$$

The partial derivatives $\frac{\partial \phi}{\partial x^\beta}$ are the components of a one-form $\tilde{d}\phi$, known as the **gradient one-form** or **differential one-form** of ϕ.

Formally, the differential one-form $\tilde{d}\phi$ is written as

$$\tilde{d}\phi = \frac{\partial \phi}{\partial x^\beta} \tilde{d}x^\beta,$$

where $\tilde{d}x^\beta$ are the coordinate basis one-forms (more of those things we don't need to worry about). Notice the similarity between this equation and the total differential (equation (1.10.10) – the thing that tells us what happens to a function when some or all of the variables change)

$$df = \frac{\partial f}{\partial w} dw + \frac{\partial f}{\partial x} dx + \frac{\partial f}{\partial y} dy + \frac{\partial f}{\partial z} dz.$$

Hence the name 'differential one-form' for $\tilde{d}\phi$. (Although not all one-forms are differentials of functions, differential one-forms are the only type we are concerned with in this book.)

Recall that on a point on the manifold, because of (5.1.3)

$$e^\alpha (e_\beta) = \delta^\alpha_\beta,$$

a vector acts linearly on *any* one-form to give a scalar (a real number). The differential one-form $\tilde{d}\phi$ is actually *defined* by this relationship, namely it is the thing that acts on a vector $\vec{V}$ to give a number $\vec{V}(\phi)$, ie

$$\tilde{d}\phi \left(\vec{V} \right) = \vec{V}(\phi).$$

For a vector tangent to a curve parameterised by λ, then $\vec{V}(\phi) = \frac{d\phi}{d\lambda}$. Think of $\tilde{d}\phi$ as the infinitesimal change in the function ϕ corresponding to an infinitesimal displacement on the manifold. This change is in an unspecified direction until we ask $\tilde{d}\phi$ to act on a vector $\vec{V}$. We then get a number $\vec{V}(\phi) = \frac{d\phi}{d\lambda}$, the directional derivative we met above, which tells us the rate of change of ϕ in the direction of the vector $\vec{V}$ or, in other words, how ϕ changes along a λ parameterised curve. The equation $\tilde{d}\phi \left(\vec{V} \right) = \vec{V}(\phi)$ is coordinate-independent. However, if we plug in a coordinate function (the function linking one coordinate system to another) $\phi = x^\alpha$ and the basis vectors $\frac{\partial}{\partial x^\beta}$ we obtain

$$\tilde{d}x^\alpha \left(\frac{\partial}{\partial x^\beta} \right) = \frac{\partial x^\alpha}{\partial x^\beta} = \delta^\alpha_\beta,$$

the coordinate basis version of equation (5.1.3)

$$e^\alpha (e_\beta) = \delta^\alpha_\beta.$$

That's why, when using coordinate bases $\tilde{d}x^\alpha$ and $\frac{\partial}{\partial x^\beta}$, a differential one-form $\tilde{d}\phi$ acts on a vector $\vec{V}$ to give a real number.

We are now going to look more closely at how transformation matrices are constructed. Don't worry, YOU DON'T NEED TO DO ANY OF THESE CALCULATIONS WHEN MANIPULATING TENSORS – the maths is just the reason why the rules of tensor algebra actually work.

5.3.7 More on the coordinate transformation of contravariant vectors

Say we want to use a different coordinate system to describe the four-velocity vector $\vec{U}$. The vector itself won't change, but its components ($U^\mu = \frac{dt}{d\tau}, \frac{dx}{d\tau}, \frac{dy}{d\tau}, \frac{dz}{d\tau}$, for example) will. We therefore need to determine how these components change under a coordinate transformation.

Let's call the original coordinate system x^β (where $\beta = 0, 1, 2, 3 = t, x, y, z$) and the new coordinate system y'^α (where $\alpha = 0, 1, 2, 3 = t, r, \theta, \phi$, for example). Using index notation, we therefore rename the original components of $\vec{U}$ as $V^\beta = \left(\frac{dx^0}{d\tau}, \frac{dx^1}{d\tau}, \frac{dx^2}{d\tau}, \frac{dx^3}{d\tau} \right)$, and the new components of $\vec{U}$ as $V'^\alpha = \left(\frac{dy'^0}{d\tau}, \frac{dy'^1}{d\tau}, \frac{dy'^2}{d\tau}, \frac{dy'^3}{d\tau} \right)$. Don't forget that V^β and V'^α represent different components of the same vector $\vec{U}$. The parameter proper time τ, won't change under a coordinate transformation as it doesn't depend on the coordinates, ie it's an invariant. We assume (otherwise none of this works!) that the four y'^α coordinates are nice well behaved functions of the four x^β coordinates, and therefore we can take partial derivatives and use the total derivative (1.10.11) to say

$$\frac{dy'^0}{d\tau} = \frac{\partial y'^0}{\partial x^0} \frac{dx^0}{d\tau} + \frac{\partial y'^0}{\partial x^1} \frac{dx^1}{d\tau} + \frac{\partial y'^0}{\partial x^2} \frac{dx^2}{d\tau} + \frac{\partial y'^0}{\partial x^3} \frac{dx^3}{d\tau}.$$

But the $\frac{dx^\beta}{d\tau}$ terms ($\beta = 0, 1, 2, 3$) are simply the components of the vector V^β expressed in the original x coordinate system. We can therefore say

$$\frac{dy'^0}{d\tau} = \frac{\partial y'^0}{\partial x^0} V^0 + \frac{\partial y'^0}{\partial x^1} V^1 + \frac{\partial y'^0}{\partial x^2} V^2 + \frac{\partial y'^0}{\partial x^3} V^3.$$

Similarly, for the other three components $\left(\frac{dy'^1}{d\tau}, \frac{dy'^2}{d\tau}, \frac{dy'^3}{d\tau} \right)$ of V'^α

$$\frac{dy'^1}{d\tau} = \frac{\partial y'^1}{\partial x^0} V^0 + \frac{\partial y'^1}{\partial x^1} V^1 + \frac{\partial y'^1}{\partial x^2} V^2 + \frac{\partial y'^1}{\partial x^3} V^3,$$

$$\frac{dy'^2}{d\tau} = \frac{\partial y'^2}{\partial x^0} V^0 + \frac{\partial y'^2}{\partial x^1} V^1 + \frac{\partial y'^2}{\partial x^2} V^2 + \frac{\partial y'^2}{\partial x^3} V^3,$$

$$\frac{dy'^3}{d\tau} = \frac{\partial y'^3}{\partial x^0} V^0 + \frac{\partial y'^3}{\partial x^1} V^1 + \frac{\partial y'^3}{\partial x^2} V^2 + \frac{\partial y'^3}{\partial x^3} V^3.$$

We can write all this more succinctly in index notation as

$$\frac{dy'^\alpha}{d\tau} = \frac{\partial y'^\alpha}{\partial x^\beta} V^\beta$$

or even better, seeing that

$$\frac{dy'^\alpha}{d\tau} = V'^\alpha,$$

we can say

$$V'^\alpha = \frac{\partial y'^\alpha}{\partial x^\beta} V^\beta,$$

which should be familiar as it's the coordinate transformation rule (5.3.1) for the components of contravariant vector.

The term $\frac{\partial y'^{\alpha}}{\partial x^{\beta}}$ represents the 4×4 transformation matrix

$$\left[\frac{\partial y'^{\alpha}}{\partial x^{\beta}}\right] = \begin{pmatrix} \frac{\partial y'^0}{\partial x^0} & \frac{\partial y'^0}{\partial x^1} & \frac{\partial y'^0}{\partial x^2} & \frac{\partial y'^0}{\partial x^3} \\ \frac{\partial y'^1}{\partial x^0} & \frac{\partial y'^1}{\partial x^1} & \frac{\partial y'^1}{\partial x^2} & \frac{\partial y'^1}{\partial x^3} \\ \frac{\partial y'^2}{\partial x^0} & \frac{\partial y'^2}{\partial x^1} & \frac{\partial y'^2}{\partial x^2} & \frac{\partial y'^2}{\partial x^3} \\ \frac{\partial y'^3}{\partial x^0} & \frac{\partial y'^3}{\partial x^1} & \frac{\partial y'^3}{\partial x^2} & \frac{\partial y'^3}{\partial x^3} \end{pmatrix}. \tag{5.3.5}$$

So, to transform the components of a contravariant vector in one coordinate system to another coordinate system we multiply the original components by the transformation matrix (which will depend on the coordinate systems we are changing from and to). The result will be the components of the vector in the new coordinate system. For example, say our original vector V^{β} has components (x^0, x^1, x^2, x^3) in the old coordinate system, and our transformed vector V'^{α} has components (y'^0, y'^1, y'^2, y'^3) in the new coordinate system., then

$$V'^{\alpha} = \begin{pmatrix} y'^0 \\ y'^1 \\ y'^2 \\ y'^3 \end{pmatrix} = \begin{pmatrix} \frac{\partial y'^0}{\partial x^0} & \frac{\partial y'^0}{\partial x^1} & \frac{\partial y'^0}{\partial x^2} & \frac{\partial y'^0}{\partial x^3} \\ \frac{\partial y'^1}{\partial x^0} & \frac{\partial y'^1}{\partial x^1} & \frac{\partial y'^1}{\partial x^2} & \frac{\partial y'^1}{\partial x^3} \\ \frac{\partial y'^2}{\partial x^0} & \frac{\partial y'^2}{\partial x^1} & \frac{\partial y'^2}{\partial x^2} & \frac{\partial y'^2}{\partial x^3} \\ \frac{\partial y'^3}{\partial x^0} & \frac{\partial y'^3}{\partial x^1} & \frac{\partial y'^3}{\partial x^2} & \frac{\partial y'^3}{\partial x^3} \end{pmatrix} \begin{pmatrix} x^0 \\ x^1 \\ x^2 \\ x^3 \end{pmatrix}.$$

Note the (often ignored) convention that contravariant vectors are written as a column matrix.

Problem 5.1. (From Schutz [28]). Returning to the flat space of special relativity, assume that an inertial frame S′ is moving with velocity v in the x direction relative to another frame S. A contravariant vector V^{β} in frame S has components $(5, 0, 0, 2)$. Find its components in frame S′.

Contravariant vectors transform according to (from 5.3.1)

$$V'^{\alpha} = \frac{\partial y'^{\alpha}}{\partial x^{\beta}} V^{\beta}.$$

Because we are in Minkowski space coordinate transformations are described using the Lorentz transformations (3.4.2)

$$t' = \gamma \left(t - \frac{vx}{c^2} \right),$$

$$x' = \gamma \left(x - vt \right),$$

$$y' = y,$$

$$z' = z,$$

where the Lorentz factor (3.4.1) $\gamma = \frac{1}{\sqrt{1-(v/c)^2}}$.

The transformation matrix for contravariant vectors (5.3.5) is given by

$$\left[\frac{\partial y'^{\alpha}}{\partial x^{\beta}}\right] = \begin{pmatrix} \frac{\partial y'^0}{\partial x^0} & \frac{\partial y'^0}{\partial x^1} & \frac{\partial y'^0}{\partial x^2} & \frac{\partial y'^0}{\partial x^3} \\ \frac{\partial y'^1}{\partial x^0} & \frac{\partial y'^1}{\partial x^1} & \frac{\partial y'^1}{\partial x^2} & \frac{\partial y'^1}{\partial x^3} \\ \frac{\partial y'^2}{\partial x^0} & \frac{\partial y'^2}{\partial x^1} & \frac{\partial y'^2}{\partial x^2} & \frac{\partial y'^2}{\partial x^3} \\ \frac{\partial y'^3}{\partial x^0} & \frac{\partial y'^3}{\partial x^1} & \frac{\partial y'^3}{\partial x^2} & \frac{\partial y'^3}{\partial x^3} \end{pmatrix}.$$

We calculate the first row of the matrix using $t' = \gamma \left(t - \frac{vx}{c^2}\right)$, substituting $y'^0 = t'$, $x^0 = t$ and $x^1 = x$, giving $\frac{\partial y'^0}{\partial x^0} = \gamma$ and $\frac{\partial y'^0}{\partial x^1} = -v\gamma/c^2$.

We calculate the second row of the matrix using $x' = \gamma \left(x - vt\right)$, substituting $y'^1 = x'$, $x^0 = t$ and $x^1 = x$, giving $\frac{\partial y'^1}{\partial x^0} = -v\gamma$ and $\frac{\partial y'^1}{\partial x^1} = \gamma$.

We calculate the third row of the matrix using $y' = y$, substituting $y'^2 = y'$ and $x^2 = y$, giving $\frac{\partial y'^2}{\partial x^2} = 1$.

We calculate the fourth row of the matrix using $z' = z$, substituting $y'^3 = z'$ and $x^3 = z$, giving $\frac{\partial y'^3}{\partial x^3} = 1$.

All the other terms in the matrix being zero, we can write

$$\left[\frac{\partial y'^\alpha}{\partial x^\beta}\right] = \begin{pmatrix} \frac{\partial y'^0}{\partial x^0} & \frac{\partial y'^0}{\partial x^1} & \frac{\partial y'^0}{\partial x^2} & \frac{\partial y'^0}{\partial x^3} \\ \frac{\partial y'^1}{\partial x^0} & \frac{\partial y'^1}{\partial x^1} & \frac{\partial y'^1}{\partial x^2} & \frac{\partial y'^1}{\partial x^3} \\ \frac{\partial y'^2}{\partial x^0} & \frac{\partial y'^2}{\partial x^1} & \frac{\partial y'^2}{\partial x^2} & \frac{\partial y'^2}{\partial x^3} \\ \frac{\partial y'^3}{\partial x^0} & \frac{\partial y'^3}{\partial x^1} & \frac{\partial y'^3}{\partial x^2} & \frac{\partial y'^3}{\partial x^3} \end{pmatrix} = \begin{pmatrix} \gamma & -v\gamma/c^2 & 0 & 0 \\ -v\gamma & \gamma & 0 & 0 \\ 0 & 0 & 1 & 0 \\ 0 & 0 & 0 & 1 \end{pmatrix}.$$

This is another version of the Lorentz transformation matrix (3.4.12) that we met earlier when looking at the Lorentz transformations in special relativity. (The matrices are not identical as here we use t as the time coordinate, whereas in (3.4.12) we use ct.) The components of V^β in frame S' are thus given by

$$V'^\alpha = \begin{pmatrix} \gamma & -v\gamma/c^2 & 0 & 0 \\ -v\gamma & \gamma & 0 & 0 \\ 0 & 0 & 1 & 0 \\ 0 & 0 & 0 & 1 \end{pmatrix} \begin{pmatrix} 5 \\ 0 \\ 0 \\ 2 \end{pmatrix}.$$

Therefore, working out each of the four components of V'^α, we get

$$V'^0 = (\gamma \times 5) + \left(-\frac{v\gamma}{c^2} \times 0\right) + (0 \times 0) + (0 \times 2) = 5\gamma,$$

$$V'^1 = (-v\gamma \times 5) + (\gamma \times 0) + (0 \times 0) + (0 \times 2) = -5v\gamma,$$

$$V'^2 = (0 \times 5) + (0 \times 0) + (1 \times 0) + (0 \times 2) = 0,$$

$$V'^3 = (0 \times 5) + (0 \times 0) + (0 \times 0) + (1 \times 2) = 2.$$

Therefore, the components of V'^α are $(5\gamma, -5v\gamma, 0, 2)$.

5.3.8 And more on the coordinate transformation of one-forms

Just as we did when looking at the transformation properties of contravariant vectors, we want to see what happens when we use a different coordinate system to describe our one-form $\tilde{d}\phi$, with components $\frac{\partial \phi}{\partial x^\beta}$ (which we'll call V_β). As before, the one-form itself won't change, but its components $\left(\frac{\partial \phi}{\partial x^0}, \frac{\partial \phi}{\partial x^1}, \frac{\partial \phi}{\partial x^2}, \frac{\partial \phi}{\partial x^3}\right) = \frac{\partial \phi}{\partial x^\beta}$ will. We therefore need to determine how the components change under a coordinate transformation.

Our original coordinate system is x^β and again we'll call our new coordinate system y'^α. Using index notation, the components $\frac{\partial \phi}{\partial y'^\alpha}$ (which we'll call V'_α) are given by $\frac{\partial \phi}{\partial y'^0}, \frac{\partial \phi}{\partial y'^1}, \frac{\partial \phi}{\partial y'^2}, \frac{\partial \phi}{\partial y'^3} = \frac{\partial \phi}{\partial y'^\alpha}$.

It bears repeating that V_β and V'_α are components of the same one-form $\tilde{d}\phi$ – the same geometrical 'thing' in space, but with different components because they are being described using different coordinate systems.

A version of the chain rule 1.10.4 tells us how to transform partial derivatives

$$\frac{\partial \phi}{\partial y'^\alpha} = \frac{\partial \phi}{\partial x^\beta} \frac{\partial x^\beta}{\partial y'^\alpha}$$

or, because $\frac{\partial \phi}{\partial y'^\alpha} = V'_\alpha$ and $\frac{\partial \phi}{\partial x^\beta} = V_\beta$, we can write

$$V'_\alpha = \frac{\partial x^\beta}{\partial y'^\alpha} V_\beta, \tag{5.3.6}$$

which is the coordinate transformation rule for the components of one-forms.

The term $\frac{\partial x^\beta}{\partial y'^\alpha}$ in the above equation is shorthand notation for the 4×4 transformation matrix

$$\left[\frac{\partial x^\beta}{\partial y'^\alpha}\right] = \begin{pmatrix} \frac{\partial x^0}{\partial y'^0} & \frac{\partial x^0}{\partial y'^1} & \frac{\partial x^0}{\partial y'^2} & \frac{\partial x^0}{\partial y'^3} \\ \frac{\partial x^1}{\partial y'^0} & \frac{\partial x^1}{\partial y'^1} & \frac{\partial x^1}{\partial y'^2} & \frac{\partial x^1}{\partial y'^3} \\ \frac{\partial x^2}{\partial y'^0} & \frac{\partial x^2}{\partial y'^1} & \frac{\partial x^2}{\partial y'^2} & \frac{\partial x^2}{\partial y'^3} \\ \frac{\partial x^3}{\partial y'^0} & \frac{\partial x^3}{\partial y'^1} & \frac{\partial x^3}{\partial y'^2} & \frac{\partial x^3}{\partial y'^3} \end{pmatrix}.$$

So, for example, if our original one-form V_β components were (x^0, x^1, x^2, x^3) in the old coordinate system, and our transformed one-form V'_α components are (y'^0, y'^1, y'^2, y'^3) in the new coordinate system, then

$$V'_\alpha = (y'^0, y'^1, y'^2, y'^3) = (x^0, x^1, x^2, x^3) \begin{pmatrix} \frac{\partial x^0}{\partial y'^0} & \frac{\partial x^0}{\partial y'^1} & \frac{\partial x^0}{\partial y'^2} & \frac{\partial x^0}{\partial y'^3} \\ \frac{\partial x^1}{\partial y'^0} & \frac{\partial x^1}{\partial y'^1} & \frac{\partial x^1}{\partial y'^2} & \frac{\partial x^1}{\partial y'^3} \\ \frac{\partial x^2}{\partial y'^0} & \frac{\partial x^2}{\partial y'^1} & \frac{\partial x^2}{\partial y'^2} & \frac{\partial x^2}{\partial y'^3} \\ \frac{\partial x^3}{\partial y'^0} & \frac{\partial x^3}{\partial y'^1} & \frac{\partial x^3}{\partial y'^2} & \frac{\partial x^3}{\partial y'^3} \end{pmatrix}.$$

Note the convention that one-forms are written as a row matrix.

5.4 Tensors

As long as we are dealing with the sort of differentiable coordinate transformations we've discussed so far, the laws of physics must be invariant – a principle known as 'general covariance'. In plain English, this means the mathematical laws of physics must be the same irrespective of the coordinate system we are using. The fundamental importance of tensors is that:

- Because the transformation properties of tensor components are basis-independent, if a tensor equation is true in one coordinate system it will be true in all coordinate systems.

As there are no preferred coordinate systems in general relativity, it makes sense to use tensor equations to formulate that theory, and that is what we need to do.

We've already met scalars, vectors and one-forms, which in fact are all simple tensors. Later on, when we look at curvature, we'll meet the Riemann curvature tensor $(R^l_{\ ijk})$, a four index monster tensor that completely describes the curvature of any space. We then say hello to, among others, the Ricci tensor, the energy-momentum tensor, the Einstein tensor, and finally, the star of the show, the Einstein field equations. Along the way we need to learn how to manipulate and differentiate tensors. In short, there's no avoiding tensors if you want to understand general relativity.

So, what's a tensor?

For our purposes we can regard a tensor as:

- A mathematical object represented by a collection of components that transform a certain way.

- A linear machine that produces numbers when fed with vectors and one-forms. Each lower index represents a slot that can be fed a vector; each upper index represents a slot that can be fed a one-form. When all the slots are filled, out pops a number.

We have already discussed, at some length, the transformation properties of contravariant vectors and one-forms. The good news is that tensors transform in exactly the same way: using partial derivatives of the coordinate functions. However, tensors can have multiple components because they are built up from vectors, one-forms, scalars and other tensors. Before looking at how tensors transform, we'll look at some basic tensor notation.

Tensors are classified according to their rank or order, which is just another way of saying how many indices they have. They are also classified by type (n, m) where n is the number of upper indices and m the number of lower indices. For example:

- A scalar, T for example, representing temperature, has no indices and is therefore a tensor of rank 0, type $(0,0)$.

- A contravariant vector V^β has one (upper) index and is therefore a tensor of rank 1, type $(1,0)$.

- Similarly, a one-form V_α also has one (lower) index and is also a tensor of rank 1, type $(0,1)$.

- A hypothetical tensor $T^\alpha_{\ \gamma}$ has two indices (one upper, one lower) and is therefore a tensor of rank 2, type $(1,1)$.

- Another hypothetical tensor $T^{\beta\alpha}$ also has two indices (two upper) and is also a tensor of rank 2, type $(2,0)$.

- The Riemann curvature tensor $R^l_{\ ijk}$, which we'll be meeting later on, has four indices (one upper, three lower) and is therefore a tensor of rank 4, type $(1,3)$.

A type $(0,0)$ tensor is therefore just a number, a scalar, 106 for example.

A type $(1,0)$ tensor (aka a contravariant vector) is conventionally written in column form, eg

$$V^\alpha = \begin{pmatrix} a \\ b \\ c \\ d \end{pmatrix}.$$

A type $(0,1)$ tensor (aka a one-form) is conventionally written in row form, eg

$$X_\beta = (e, f, g, h).$$

These two tensors can be multiplied together (known as taking the **tensor product**) to form a rank 2 tensor, which can be written as a 4×4 matrix:

$$\left[T^{\alpha}{}_{\beta}\right] = \left[V^{\alpha}X_{\beta}\right] = \begin{pmatrix} ae & af & ag & ah \\ be & bf & bg & bh \\ ce & cf & cg & ch \\ de & df & dg & dh \end{pmatrix}.$$

So, for example, $T^{3}{}_{0} = V^{3}X_{0} = de$. In a similar fashion, two contravariant vectors or two one-forms may also be multiplied together to give a rank 2 tensor.

Tensors of rank 3 can be thought of as a stack of matrices. Higher rank tensors are harder to visualise, but you can still think of them as products of scalars, vectors and one-forms – the building blocks from which we construct larger tensors.

Seeing tensors expressed as rows, columns and matrices leads to another informal definition of tensors as:

- A multidimensional array of covariant (lower) and contravariant (upper) indices.

We have already met tensors in the shape of various scalars, vectors and one-forms. We have also come across the vitally important rank 2 Minkowski metric tensor (3.5.2)

$$\left[\eta_{\mu\nu}\right] = \begin{pmatrix} \eta_{00} & \eta_{01} & \eta_{02} & \eta_{03} \\ \eta_{10} & \eta_{11} & \eta_{12} & \eta_{13} \\ \eta_{20} & \eta_{21} & \eta_{22} & \eta_{23} \\ \eta_{30} & \eta_{31} & \eta_{32} & \eta_{33} \end{pmatrix} = \begin{pmatrix} 1 & 0 & 0 & 0 \\ 0 & -1 & 0 & 0 \\ 0 & 0 & -1 & 0 \\ 0 & 0 & 0 & -1 \end{pmatrix}$$

and used this to calculate the scalar product of two four-vectors (3.6.3)

$$\mathbf{A} \cdot \mathbf{B} = \eta_{\mu\nu}A^{\mu}B^{\nu} = A^{0}B^{0} - A^{1}B^{1} - A^{2}B^{2} - A^{3}B^{3},$$

which tells us where the plus and minus signs go in front of the terms $A^{\mu}B^{\nu}$. We used this scalar product when we looked at the equations for four-velocity (3.6.4) and the energy-momentum relation (3.6.18), the latter eventually leading us to the famous mass-energy equation $E = mc^{2}$.

It's worth making a small detour here in order to see just why the metric tensor is involved when taking the scalar product of two vectors. We first need to note that the components g_{ij} of the metric are *defined* in terms of the coordinate basis vectors as

$$g_{ij} = e_{i} \cdot e_{j}$$

or, in spacetime, using Greek indices

$$g_{\mu\nu} = e_{\mu} \cdot e_{\nu}.$$

Using this definition, and recalling that an infinitesimal displacement vector in spacetime is given by

$$d\vec{x} = dx^{\mu}e_{\mu},$$

we can see that the scalar product of this vector with itself will be

$$\left(d\vec{x}\right)^{2} = \left(dx^{\mu}e_{\mu}\right) \cdot \left(dx^{\nu}e_{\nu}\right) = \left(e_{\mu} \cdot e_{\nu}\right)dx^{\mu}dx^{\nu}$$

$$= g_{\mu\nu}dx^\mu dx^\nu.$$

This equation is the spacetime version of (4.2.1)

$$dl^2 = g_{ij}dx^i dx^j,$$

the line element for an n-dimensional manifold that we met in the previous chapter. In a similar fashion, the scalar product of two different vectors must then be given by

$$\vec{V} \cdot \vec{W} = (V^\mu e_\mu) \cdot (W^\nu e_\nu) = g_{\mu\nu}V^\mu W^\nu.$$

Hence the need for the metric tensor.

Incidentally, when we looked at the scalar product of simple Cartesian vectors (1.13.4)

$$\mathbf{A} \cdot \mathbf{B} = A_x B_x + A_y B_y + A_z B_z = A_1 B_1 + A_2 B_2 + A_3 B_3$$

this should, strictly speaking, be written (in index notation) as

$$\mathbf{A} \cdot \mathbf{B} = g_{ij}A^i B^j = \left(1 \times A^1 B^1\right) + \left(1 \times A^2 B^2\right) + \left(1 \times A^3 B^3\right),$$

where g_{ij} is Euclidean metric tensor (1.14.6)

$$[g_{ij}] = \begin{pmatrix} 1 & 0 & 0 \\ 0 & 1 & 0 \\ 0 & 0 & 1 \end{pmatrix},$$

which tells us that a plus sign should go in front of the terms $A^i B^j$. We don't normally bother doing this when using Cartesian vectors because we take for granted that we are in Euclidean space.

5.4.1 Tensor transformation

We're metaphorically diving back under the bonnet again. What follows is the behind the scenes mathematics that explains why the various operations of tensor algebra work. Just to repeat: YOU DON'T NEED TO USE THIS STUFF TO ACTUALLY MANIPULATE TENSORS.

We have already said that tensors transform in the same way as contravariant vectors and one-forms. To see why this is, recall that vector transformations are given by (5.3.1)

$$V'^\alpha = \frac{\partial x'^\alpha}{\partial x^\beta}V^\beta.$$

And one-form transformations are given by (5.3.2)

$$V'_\alpha = \frac{\partial x^\beta}{\partial x'^\alpha}V_\beta.$$

One method of constructing a new tensor is by multiplying the components of vectors and one-forms. For example, we could make a tensor by multiplying the components of two vectors F^β and W^λ and a one-form A_η to give a tensor $T^{\beta\lambda}{}_\eta$

$$T^{\beta\lambda}{}_\eta = F^\beta W^\lambda A_\eta.$$

We know from (5.3.1) that F^β transforms as

$$F'^\alpha = \frac{\partial x'^\alpha}{\partial x^\beta} F^\beta,$$

and W^λ transforms as

$$W'^\epsilon = \frac{\partial x'^\epsilon}{\partial x^\lambda} W^\lambda.$$

And we know from (5.3.2) that A_η transforms as

$$A'_\tau = \frac{\partial x^\eta}{\partial x'^\tau} A_\eta.$$

Therefore, because our new tensor $T^{\beta\lambda}{}_\eta$ has been constructed by multiplying the components of vectors and one-forms, its transformation equation will be the product of the transformation equations of those constituent vectors and one-forms, ie

$$T'^{\alpha\epsilon}{}_\tau = F'^\alpha W'^\epsilon A'_\tau = \frac{\partial x'^\alpha}{\partial x^\beta} \frac{\partial x'^\epsilon}{\partial x^\lambda} \frac{\partial x^\eta}{\partial x'^\tau} F^\beta W^\lambda A_\eta = \frac{\partial x'^\alpha}{\partial x^\beta} \frac{\partial x'^\epsilon}{\partial x^\lambda} \frac{\partial x^\eta}{\partial x'^\tau} T^{\beta\lambda}{}_\eta.$$

In general terms this means that a tensor of contravariant rank m and covariant rank n has components given by the tensor transformation law:

$$T'^{\mu_1\mu_2\ldots\mu_m}{}_{\alpha_1\alpha_2\ldots\alpha_n} = \frac{\partial x'^{\mu_1}}{\partial x^{\nu_1}} \frac{\partial x'^{\mu_2}}{\partial x^{\nu_2}} \cdots \frac{\partial x'^{\mu_m}}{\partial x^{\nu_m}} \times \frac{\partial x^{\beta_1}}{\partial x'^{\alpha_1}} \frac{\partial x^{\beta_2}}{\partial x'^{\alpha_2}} \cdots \frac{\partial x^{\beta_n}}{\partial x'^{\alpha_n}} \times T^{\nu_1\nu_2\ldots\nu_m}{}_{\beta_1\beta_2\ldots\beta_n}. \tag{5.4.1}$$

Of course, this looks horribly complicated. But all it actually represents is a series of vector transformations (the partial derivatives with the μ and ν indices) and one-form transformation (the partial derivatives with the β and α indices) all strung together to give the total tensor transformation.

Crucially, 5.4.1 means the rules of tensor algebra actually work! We'll now look at those rules.

5.4.2 The rules of tensor algebra

In order to manipulate tensors we make use of the following rules:

Scaling   A tensor may be multiplied by a scalar a to produce a new tensor

$$T_{\alpha\beta} = aX_{\alpha\beta}.$$

Addition and subtraction   Tensors can be added or subtracted to form a new tensor

$$T^\alpha{}_\beta = A^\alpha{}_\beta + B^\alpha{}_\beta.$$

Multiplication   We've already seen this one. Tensors can be multiplied together to form a new tensor. The rank of the new tensor will be the sum of the ranks of the constituent tensors

$$T_\lambda{}^{\alpha\beta\eta} = A_\lambda{}^\alpha B^\beta C^\eta.$$

Contraction A tensor can have its rank lowered by multiplying it by another tensor with an equal index in the opposite position, ie by summing over the two indices. In this example the upper and lower α indices are summed over

$$T_\gamma = A_{\alpha\gamma}B^\alpha.$$

A single tensor can also be contracted by summing over identical upper and lower indices. For example, we can contract a tensor $T^{\beta\epsilon}{}_\alpha$ to give another tensor $T^{\alpha\epsilon}{}_\alpha = X^\epsilon$, which is a contravariant vector.

Let's see how this contraction works. The tensor $X^\epsilon = T^{\alpha\epsilon}{}_\alpha$ will transform as follows:

$$X'^\epsilon = T'^{\alpha\epsilon}{}_\alpha = \frac{\partial x'^\alpha}{\partial x^\beta}\frac{\partial x'^\epsilon}{\partial x^\lambda}\frac{\partial x^\eta}{\partial x'^\alpha}T^{\beta\lambda}{}_\eta$$

$$= \frac{\partial x^\eta}{\partial x^\beta}\frac{\partial x'^\epsilon}{\partial x^\lambda}T^{\beta\lambda}{}_\eta = \delta^\eta_\beta\frac{\partial x'^\epsilon}{\partial x^\lambda}T^{\beta\lambda}{}_\eta.$$

Remember that the x coordinates (represented by the β, λ, η indices) are independent of each other. The Kronecker delta in the final term tells us that the derivative of any of these coordinates with respect to any other is zero, and the derivative of a coordinate with respect to itself is 1 (if $x = x$ then $\frac{dx}{dx} = 1$). We can therefore write (effectively, the Kronecker delta relabels the η index as β)

$$X'^\epsilon = T'^{\alpha\epsilon}{}_\alpha = \delta^\eta_\beta\frac{\partial x'^\epsilon}{\partial x^\lambda}T^{\beta\lambda}{}_\eta = \frac{\partial x'^\epsilon}{\partial x^\lambda}T^{\beta\lambda}{}_\beta = \frac{\partial x'^\epsilon}{\partial x^\lambda}X^\lambda,$$

meaning the contracted tensor transforms, as it should, as a contravariant vector.

We've already seen that if we contract a vector with a one-form we obtain a scalar

$$S = A_\alpha B^\alpha.$$

Raising and lowering indices An operation of fundamental importance involves contraction with the metric tensor $g_{\mu\nu}$ and its inverse $g^{\mu\nu}$ (sometimes called the dual metric) to manipulate expressions by raising and lowering indices. For example

$$T_\nu = g_{\mu\nu}T^\mu$$

$$T^\nu = g^{\mu\nu}T_\mu.$$

This gives us a way of directly relating one-forms to vectors and vice versa (see example below).

How does this work? First, recall that a tensor is a machine that produces numbers when fed with vectors and one-forms. The metric tensor $g_{\mu\nu}$ has two lower indices and therefore two slots that need to be filled by two vectors (one for each slot) to output a number

$$g_{\mu\nu}T^\mu W^\nu = e_\mu \cdot e_\nu T^\mu W^\nu = (T^\mu e_\mu) \cdot (W^\nu e_\nu) = \vec{T} \cdot \vec{W}.$$

A metric with only one of its slots filled (with, for example, a vector $\vec{T}$) can effectively be regarded as a one-form – it requires a vector $\vec{W}$ in order to give the number $\vec{T} \cdot \vec{W}$. We'll call this one-form $\tilde{T}$. If we let $\tilde{T}$ act on $\vec{W}$, we can rewrite the above equation as

$$\tilde{T}\left(\vec{W}\right) = \vec{T} \cdot \vec{W}.$$

Because of (5.1.3) $e^\alpha(e_\beta) = \delta^\alpha_\beta$, the components of $\tilde{T}$ are given by $T_\nu = \tilde{T}(e_\nu)$. Substituting e_ν for $\vec{W}$ in $\tilde{T}(\vec{W}) = \vec{T} \cdot \vec{W}$ we can then say

$$T_\nu = \tilde{T}(e_\nu) = \vec{T} \cdot e_\nu = e_\nu \cdot \vec{T} = e_\nu \cdot T^\mu e_\mu = e_\nu \cdot e_\mu T^\mu = g_{\mu\nu} T^\mu.$$

Hence we have shown why

$$T_\nu = g_{\mu\nu} T^\mu.$$

Note that if we multiply the metric tensor $g_{\mu\nu}$ by its inverse $g^{\mu\nu}$ we obtain the identity matrix

$$\begin{pmatrix} 1 & 0 & 0 & 0 \\ 0 & 1 & 0 & 0 \\ 0 & 0 & 1 & 0 \\ 0 & 0 & 0 & 1 \end{pmatrix}.$$

We can write this in terms of the Kronecker delta (1.12.1) as

$$g^{\mu\nu} g_{\nu\lambda} = \delta^\mu_\lambda,$$

where we have again used the Kronecker delta as a tensor with an upper and lower index.

Double contraction to zero   In order to understand this somewhat counter-intuitive result we need to introduce the tensor properties of symmetry/antisymmetry.

A rank 2 tensor is **symmetric** if $S_{\mu\nu} = S_{\nu\mu}$. Such a tensor can be represented by a symmetric matrix

$$\begin{pmatrix} w & a & b & c \\ a & x & d & e \\ b & d & y & f \\ c & e & f & z \end{pmatrix}.$$

So, for example, $S_{02} = S_{20} = b$, $S_{23} = S_{32} = f$, etc. Symmetric tensors we meet in this book include the metric tensor, the energy-momentum tensor, the Ricci tensor and the Einstein tensor.

Conversely, a rank 2 tensor is **antisymmetric** if $A_{\mu\nu} = -A_{\nu\mu}$. If you think about it, this is only possible if the tensor is of the matrix form

$$\begin{pmatrix} 0 & a & b & c \\ -a & 0 & d & e \\ -b & -d & 0 & f \\ -c & -e & -f & 0 \end{pmatrix},$$

where the diagonal terms are all zero. The Riemann curvature tensor, written as R_{hijk}, is antisymmetric on the first and second pair of indices, meaning $R_{hijk} = -R_{hikj} = -R_{ihjk}$.

The double contraction of a symmetric tensor $S^{\mu\nu}$ and an antisymmetric tensor $A_{\mu\nu}$ is zero, ie $A_{\mu\nu} S^{\mu\nu} = 0$. We can show this as follows:

$$A_{\mu\nu} S^{\mu\nu} = (-A_{\nu\mu})(S^{\nu\mu})$$
$$= -(A_{\nu\mu} S^{\nu\mu}).$$

As μ and ν are dummy indices, we can swap them to give

$$A_{\mu\nu}S^{\mu\nu} = -\left(A_{\mu\nu}S^{\mu\nu}\right),$$

which can only be true if

$$A_{\mu\nu}S^{\mu\nu} = -\left(A_{\mu\nu}S^{\mu\nu}\right) = 0.$$

This may be clearer if we write out the double contraction in component form. We'll use two simple tensors, a symmetrical one

$$\left[S^{ij}\right] = \begin{pmatrix} x & a & b \\ a & y & c \\ b & c & z \end{pmatrix},$$

and an antisymmetrical one

$$\left[A_{ij}\right] = \begin{pmatrix} 0 & d & e \\ -d & 0 & f \\ -e & -f & 0 \end{pmatrix},$$

giving

$$A_{ij}S^{ij} = (0 \times x) + da + eb - da + (0 \times y) + fc - eb - fc + (0 \times z) = 0.$$

Problem 5.2. In spherical coordinates a contravariant vector $A^a = (1, r, 0)$ and a one-form $B_a = \left(0, -r^2, \cos^2\theta\right)$. Find A_a and B^a.

We can use the metric to convert a one-form to a contravariant vector:

$$B^a = g^{ab}B_b,$$

and to convert a contravariant vector to a one-form:

$$A_a = g_{ab}A^b.$$

For spherical coordinates (r, θ, ϕ), the metric is (1.14.11)

$$[g_{ab}] = \begin{pmatrix} 1 & 0 & 0 \\ 0 & r^2 & 0 \\ 0 & 0 & r^2\sin^2\theta \end{pmatrix}$$

and the inverse metric (1.14.12) is

$$[g^{ab}] = \begin{pmatrix} 1 & 0 & 0 \\ 0 & \frac{1}{r^2} & 0 \\ 0 & 0 & \frac{1}{r^2\sin^2\theta} \end{pmatrix}.$$

Therefore, $g^{rr} = g_{rr} = 1$, $g^{\theta\theta} = \frac{1}{r^2}$, $g_{\theta\theta} = r^2$, $g^{\phi\phi} = \frac{1}{r^2\sin^2\theta}$, $g_{\phi\phi} = r^2\sin^2\theta$.

Thus $A_r = g_{rr}A^r = (1)(1) = 1$, $A_\theta = g_{\theta\theta}A^\theta = \left(r^2\right)(r) = r^3$, $A_\phi = g_{\phi\phi}A^\phi = \left(r^2\sin^2\theta\right)(0) = 0$.

Therefore, $A_a = \left(1, r^3, 0\right)$.

$B^r = g^{rr}B_r = (1)(0) = 0$, $B^\theta = g^{\theta\theta}B_\theta = \left(\frac{1}{r^2}\right)\left(-r^2\right) = -1$, $B^\phi = g^{\phi\phi}B_\phi = \left(\frac{1}{r^2\sin^2\theta}\right)\left(\cos^2\theta\right) = \frac{\cos^2\theta}{r^2\sin^2\theta} = \frac{\cot^2\theta}{r^2}$ (because $\cot\theta = \frac{\cos\theta}{\sin\theta}$).

Therefore, $B^a = \left(0, -1, \frac{\cot^2\theta}{r^2}\right)$.

Problem 5.3. Show that the tensor equation $R_{\alpha_1 \alpha_2} = 0$ is true in any coordinate system.

The indices α_1 and α_2 are, of course, free indices representing an arbitrary coordinate system and we can change them to whatever we like. Using α_1 and α_2 means we can make direct use of the tensor transformation law as written in (5.4.1)

$$T'^{\mu_1 \mu_2 \ldots \mu_m}_{\quad \alpha_1 \alpha_2 \ldots \alpha_n} = \frac{\partial x'^{\mu_1}}{\partial x^{\nu_1}} \frac{\partial x'^{\mu_2}}{\partial x^{\nu_2}} \cdots \frac{\partial x'^{\mu_m}}{\partial x^{\nu_m}} \times \frac{\partial x^{\beta_1}}{\partial x'^{\alpha_1}} \frac{\partial x^{\beta_2}}{\partial x'^{\alpha_2}} \cdots \frac{\partial x^{\beta_n}}{\partial x'^{\alpha_n}} \times T^{\nu_1 \nu_2 \ldots \nu_m}_{\quad \beta_1 \beta_2 \ldots \beta_n}$$

and assume that our tensor $R_{\alpha_1 \alpha_2}$ is the result of a coordinate transformation of the same tensor $R_{\beta_1 \beta_2}$ expressed using different coordinates β_1 and β_2, meaning we can write

$$R'_{\alpha_1 \alpha_2} = \frac{\partial x^{\beta_1}}{\partial x'^{\alpha_1}} \frac{\partial x^{\beta_2}}{\partial x'^{\alpha_2}} \times R_{\beta_1 \beta_2},$$

where $\frac{\partial x^{\beta_1}}{\partial x'^{\alpha_1}}$ and $\frac{\partial x^{\beta_2}}{\partial x'^{\alpha_2}}$ are the two transformation matrices we need to multiply by $R_{\beta_1 \beta_2}$ in order to get our original tensor $R'_{\alpha_1 \alpha_2}$. As $R_{\alpha_1 \alpha_2} = R'_{\alpha_1 \alpha_2} = 0$ we can say

$$\frac{\partial x^{\beta_1}}{\partial x'^{\alpha_1}} \frac{\partial x^{\beta_2}}{\partial x'^{\alpha_2}} \times R_{\beta_1 \beta_2} = 0.$$

Dividing both sides by $\frac{\partial x^{\beta_1}}{\partial x'^{\alpha_1}} \frac{\partial x^{\beta_2}}{\partial x'^{\alpha_2}}$ gives

$$R_{\beta_1 \beta_2} = 0,$$

meaning that $R_{\alpha_1 \alpha_2} = 0$ is true in any coordinate system. A similar argument shows that *any* tensor equation if true in one coordinate system will be true in all coordinate systems.

Problem 5.4. Show that, in special relativity, if

$$V_\alpha = g_{\alpha\beta} V^\beta$$

then

$$\frac{\partial V_\alpha}{\partial x^\lambda} = g_{\alpha\beta} \frac{\partial V^\beta}{\partial x^\lambda}.$$

The right-hand side of this equation is a product. Using the product rule (1.10.3) gives

$$\frac{\partial V_\alpha}{\partial x^\lambda} = \frac{\partial \left(g_{\alpha\beta} V^\beta \right)}{\partial x^\lambda} = V^\beta \frac{\partial \left(g_{\alpha\beta} \right)}{\partial x^\lambda} + g_{\alpha\beta} \frac{\partial V^\beta}{\partial x^\lambda}.$$

But we can ignore the first term on the right-hand side because, in special relativity $g_{\mu\nu} = \eta_{\mu\nu}$, which is constant, ie

$$\frac{\partial \left(\eta_{\alpha\beta} \right)}{\partial x^\lambda} = 0$$

giving

$$\frac{\partial V_\alpha}{\partial x^\lambda} = g_{\alpha\beta} \frac{\partial V^\beta}{\partial x^\lambda}.$$

5.5 Trying to visualise all this

We've now looked at scalars, vectors, one-forms and tensors, and we know that these objects live on the manifold. This is all quite abstract. Can we picture these things in the real world?

Visualising four-dimensional spacetime is beyond the imaginations of most of us, so we'll simplify the situation and model a manifold using a rough and ready mental picture of a real two-dimensional surface. I'm going to visualise a pebble, but any smooth surface will do – an apple, a bottle, the bodywork of a car etc. The surface has to be smooth because it needs to be differentiable, so use an imaginary file to remove any imaginary sharp edges from your imaginary surface. Picture the surface in your mind (or put a real pebble on the table in front of you) and using our imagination we'll construct our simple model of a manifold.

The surface of the pebble is our manifold, or curved space. It's two-dimensional because if we choose the right coordinate system we only need two coordinates to describe any point on the surface of the pebble. Next, imagine we stretch a red elastic net over the pebble to represent an arbitrary x, y coordinate system, followed by a blue elastic net to represent an arbitrary x', y' coordinate system. Now we can describe any point on the surface of the pebble using both x, y and x', y' coordinates. Let's consider a point P on the surface of the pebble.

First, because our manifold is a Riemannian manifold, we can assign a symmetric metric tensor $g_{\mu\nu} = g_{\nu\mu}$ to P and to every other point on the surface of the pebble. In the jargon, we say we have defined a metric tensor field over the surface of our pebble (a **tensor field** assigns a tensor to each point of a manifold, just as a vector field assigns a vector to each point). The metric defines the separation of infinitesimally adjacent points on the manifold. Because our manifold is two-dimensional, the metric will be a 2×2 matrix. We don't know what the metric actually is, but we do know its coefficients will not be constant (otherwise our pebble would be flat!), and that the metric for the red coordinate system will be different to the one for the blue.

Next, we can use our coordinates to construct a scalar field on the surface of the pebble. I'm going to make one up, call it S, and say

$$S = x^2 + 3y.$$

Now we can assign a scalar value S to the point P. For example if $x = 7, y = 2$, then $S = 55$.

By finding the components of the gradient ∇S we can define a one-form at P, with components (5.3.4)

$$\frac{\partial \phi}{\partial x^\beta} = \frac{\partial S}{\partial x^\beta} = \left(\frac{\partial S}{\partial x}, \frac{\partial S}{\partial y} \right).$$

But we can define an infinite number of scalar fields, and therefore an infinite number of one-forms at P. We therefore have a vector space at P, which is called the cotangent or dual-space.

Now draw a number of imaginary curves on the surface of the pebble, each going through P. These are our parameterised curves. If we were clever enough, each curve could be expressed in terms of a parameter (t for example), so $x = f(t)$ and $y = f(t)$ would define one curve. Now take a flat piece of paper, draw a point on it and carefully draw a number of infinitely short (just do your best) straight lines through that point. Rest the paper on the pebble so that the point on the paper is touching point P on the pebble. By rotating the paper around P, any particular straight line can be made to point along any particular curve and is thus a tangent to that curve at P. Each of these tangents to a parameterised curve represents a contravariant vector.

The vector tangent to (ie pointing along) our t parameterised curve will have components (5.3.3)

$$V^\beta = \frac{dx^\beta}{dt} = \left(\frac{dx}{dt}, \frac{dy}{dt} \right).$$

There are an infinite number of parameterised curves we could draw on the pebble, and therefore an infinite number of tangent vectors passing through P. We therefore have a vector space at P, which is called the tangent space.

At any point on the surface of the pebble we can now, if we wish, take a contravariant vector or one-form and use the transformation equations to flip between the red and blue coordinate systems. But, as stressed earlier, in practice we don't need to worry about physically calculating transformation equations. The prime advantage of defining the geometry of the pebble's surface in terms of general coordinates is that we can then proceed to define tensors that are true in any coordinate system, tensors that describe both the pebble's curvature (the Riemann curvature tensor) and physics (the field equations, for example). Then, if we can somehow ascertain the metric, we can use these tensors to calculate measurable physical quantities.

6 More on curvature

The most beautiful thing we can experience is the mysterious. It is the source of all true art and all science.

ALBERT EINSTEIN

6.1 Introduction

General relativity treats gravity as a property of curved spacetime, with mass and energy as the source of that curvature. We therefore need to be familiar with the mathematical tools necessary for understanding curved spaces/manifolds.

We have now prepared the ground for a more detailed look at the mathematics of curved spaces. Eventually, we want to be able to understand:

- Geodesics – the 'straightest' possible paths between any two points in a Riemannian manifold, and therefore the means to calculate the motion of free particles moving in curved spacetime.

- The Riemann curvature tensor, which measures the curvature of a manifold. A modified version of this tensor provides the left-hand side of the fundamental mathematical object of general relativity, our Holy Grail: the Einstein field equations.

Parallel transport of a vector is the key to both of these concepts. However, before we tackle parallel transport we need to be able to differentiate tensors. To do this we introduce things called connection coefficients (symbol Γ), which allow us to compare vectors at different points on the manifold. Once we know a manifold's metric, we can calculate the connection coefficients, which in turn opens the door to everything else. That is why the metric is crucial for understanding Riemannian manifolds, and hence spacetime.

6.2 Connection coefficients

Our aim is to express general relativity in the form of valid tensor equations. To do this we need to be able to differentiate tensors. Specifically, we need a means of differentiating a tensor so that we end up with something that reduces to an ordinary partial derivative in (flat) Minkowski space using Cartesian coordinates, but transforms as a tensor when used in curved space. We'll start by showing why finding the derivative of a tensor – we'll use the example of a simple vector – is not as straightforward as it might look.

In Section 1.13.1 we considered a vector field $\mathbf{V}(x, y, z)$ representing air moving in a room. We conjured up an imaginary function describing our vector field

$$\mathbf{V} = (xy)\,\hat{e}_x + (2y + x + 3z)\,\hat{e}_y + (2y)\,\hat{e}_z.$$

If we wanted to, we could easily take partial derivatives of $\mathbf{V}$ to find its rate of change with respect to the (x, y, z) coordinates

$$\frac{\partial \mathbf{V}}{\partial x} = y\hat{e}_x + \hat{e}_y,$$

$$\frac{\partial \mathbf{V}}{\partial y} = x\hat{e}_x + 2\hat{e}_y + 2\hat{e}_z,$$

$$\frac{\partial \mathbf{V}}{\partial z} = 3\hat{e}_y.$$

Crucially, we don't need to worry about differentiating the basis vectors $\hat{e}_x, \hat{e}_y, \hat{e}_z$ because they are constant, each is one unit long and pointing along the x, y, z axes.

This is the crux of our problem: the basis vectors we use in contravariant vectors, one-forms and other tensors are not constant (the exception being when we are using Cartesian coordinates in the Minkowski space of special relativity) because they are coordinate basis vectors, defined in terms of derivatives of the coordinates.

When differentiating a tensor we therefore have to take into account the derivatives of the basis vectors. In order to do this we need to introduce things called **connection coefficients** (also known as **Christoffel symbols**), which are given the symbol Γ (the Greek letter Gamma).

Consider a vector $\vec{V} = V^\alpha e_\alpha$ (ie the vector has components V^α and coordinate basis vectors e_α). Using the product rule $\left(\frac{d(uv)}{dx} = \frac{udv}{dx} + \frac{vdu}{dx}\right)$ gives

$$\frac{\partial \vec{V}}{\partial x^\beta} = \frac{\partial V^\alpha}{\partial x^\beta} e_\alpha + V^\alpha \frac{\partial e_\alpha}{\partial x^\beta}. \tag{6.2.1}$$

$\frac{\partial V^\alpha}{\partial x^\beta}$ represents the rate of change of the components V^α (of the vector $\vec{V}$) with respect to x^β.

$\frac{\partial e_\alpha}{\partial x^\beta}$ represents the rate of change of e_α with respect to x^β, is itself a vector, and can be expressed (in terms of basis vectors e_γ) as

$$\frac{\partial e_\alpha}{\partial x^\beta} = \Gamma^\gamma_{\alpha\beta} e_\gamma. \tag{6.2.2}$$

The interpretation of the connection coefficients $\Gamma^\gamma_{\alpha\beta}$ is that the downstairs indices α and β refer to the rate of change of the components of e_α with respect to x^β in the direction of the basis vector e_γ (γ being the upstairs index). To use a polar coordinate example, the connection coefficient $\Gamma^\theta_{r\theta}$ denotes the rate of change of the component of e_r with respect to θ (r and θ being the downstairs indices) in the direction of the basis vector e_θ (θ being the upstairs index). In general, the connection coefficients $\Gamma^\gamma_{\alpha\beta}$ will vary depending on both the particular n-dimensional space where vector $\vec{V}$ 'lives', and the coordinates used to describe vector $\vec{V}$.

The examples that follow should help make this clearer. First, here are a few key facts about connection coefficients:

- Connection coefficients allow us to differentiate tensors.

- They are called connection coefficients because they provide a connection between different points on the manifold (as we'll see when looking at the parallel transport of a vector along a curve).

- If we know the metric, we can calculate the connection coefficients – see (6.2.5).

- In an n-dimensional Riemannian manifold there are theoretically n^3 different connection coefficients. However, in the standard formulations of general relativity this number is reduced because the lower indices are always assumed to be symmetric, ie $\Gamma^\gamma_{\alpha\beta} = \Gamma^\gamma_{\beta\alpha}$.

- Although they have indices, connection coefficients are not the components of a tensor and cannot be treated like tensors.

Problem 6.1. Calculate the connection coefficients Γ^i_{jk} for two-dimensional Euclidean space using polar coordinates.

Equation (1.9.1) tells us how Cartesian coordinates are related to polar coordinates

$$x = r\cos\theta,$$

$$y = r\sin\theta.$$

For the rules of tensor algebra to work, the basis vectors e_β must transform in the same way as the one-form components V_β. So we can find the transformation rule for basis vectors using (5.3.2)

$$V'_\alpha = \frac{\partial x^\beta}{\partial x'^\alpha} V_\beta.$$

Substituting e for V gives

$$e'_\alpha = \frac{\partial x^\beta}{\partial x'^\alpha} e_\beta,$$

giving (noting that β is a dummy or summation index)

$$e_r = \frac{\partial x}{\partial r} e_x + \frac{\partial y}{\partial r} e_y$$

$$e_r = \cos\theta e_x + \sin\theta e_y. \tag{6.2.3}$$

Similarly

$$e_\theta = \frac{\partial x}{\partial \theta} e_x + \frac{\partial y}{\partial \theta} e_y$$

$$e_\theta = -r\sin\theta e_x + r\cos\theta e_y. \tag{6.2.4}$$

Next we take partial derivatives, $\frac{\partial e_r}{\partial r}$ and $\frac{\partial e_r}{\partial \theta}$ using (6.2.3)

$$\frac{\partial e_r}{\partial r} = 0,$$

$$\frac{\partial e_r}{\partial \theta} = -\sin\theta e_x + \cos\theta e_y.$$

Using (6.2.4), we can rewrite this as

$$\frac{\partial e_r}{\partial \theta} = \frac{1}{r} e_\theta.$$

We now take partial derivatives $\frac{\partial e_\theta}{\partial r}$ and $\frac{\partial e_\theta}{\partial \theta}$ using (6.2.4)

$$\frac{\partial e_\theta}{\partial r} = -\sin\theta e_x + \cos\theta e_y.$$

Using (6.2.4) we can rewrite this as

$$\frac{\partial e_\theta}{\partial r} = \frac{1}{r} e_\theta.$$

$$\frac{\partial e_\theta}{\partial \theta} = -r\cos\theta e_x - r\sin\theta e_y.$$

Using (6.2.3) we can rewrite this as

$$\frac{\partial e_\theta}{\partial \theta} = -r e_r.$$

The connection coefficients are therefore:

$$\Gamma^\theta_{r\theta} = \Gamma^\theta_{\theta r} = \frac{1}{r},$$

$$\Gamma^r_{\theta\theta} = -r,$$

$$\Gamma^r_{r\theta} = \Gamma^r_{\theta r} = 0,$$

$$\Gamma^\theta_{rr} = \Gamma^r_{rr} = 0,$$

$$\Gamma^\theta_{\theta\theta} = 0.$$

In this example we relied on knowing the relationship between Cartesian coordinates (x and y) and polar coordinates (r and θ). There is another, much more useful, method of calculating the Christoffel symbols from just the metric that doesn't require any knowledge of Cartesian coordinates. We do this using the important equation (stated without proof)

$$\Gamma^i_{jk} = \frac{1}{2} g^{il} \left(\frac{\partial g_{lk}}{\partial x^j} + \frac{\partial g_{jl}}{\partial x^k} - \frac{\partial g_{jk}}{\partial x^l} \right). \tag{6.2.5}$$

As you can see, this involves partial derivatives of the metric (the $\frac{\partial g_{lk}}{\partial x^j}$ etc terms) and the inverse metric (the g^{il} terms).

Staying with the convention of using Greek indices for spacetime coordinates we can rewrite (6.2.5) as

$$\Gamma^\sigma_{\mu\nu} = \frac{1}{2} g^{\sigma\rho} \left(\frac{\partial g_{\rho\nu}}{\partial x^\mu} + \frac{\partial g_{\mu\rho}}{\partial x^\nu} - \frac{\partial g_{\mu\nu}}{\partial x^\rho} \right). \tag{6.2.6}$$

Problem 6.2. Calculate the connection coefficients Γ^i_{jk} for two-dimensional Euclidean space in Cartesian coordinates.

Equation (1.14.5) gives us the line element for three-dimensional Euclidean space using Cartesian coordinates

$$dl^2 = dx^2 + dy^2 + dz^2.$$

For two dimension this becomes

$$dl^2 = dx^2 + dy^2.$$

The metric is therefore

$$[g_{ij}] = \begin{pmatrix} 1 & 0 \\ 0 & 1 \end{pmatrix}$$

and the inverse metric would of course be identical

$$[g^{ij}] = \begin{pmatrix} 1 & 0 \\ 0 & 1 \end{pmatrix}.$$

But in this case we don't need the inverse metric because we can see from the equation to calculate the connection coefficients (6.2.5)

$$\Gamma^i_{jk} = \frac{1}{2} g^{il} \left(\frac{\partial g_{lk}}{\partial x^j} + \frac{\partial g_{jl}}{\partial x^k} - \frac{\partial g_{jk}}{\partial x^l} \right)$$

that, as all the components of g_{ij} are constant, all the partial derivatives of g_{ij} (the $\frac{\partial g_{lk}}{\partial x^j}$ etc terms) must therefore equal zero. Therefore, $\Gamma^i_{jk} = 0$ for all values of i, j, k.

Problem 6.3. Calculate the connection coefficients Γ^i_{jk} for the surface of a unit radius sphere using polar coordinates.

Equation (4.3.2) gives the metric for the surface of a sphere radius R

$$[g_{ij}] = \begin{pmatrix} R^2 & 0 \\ 0 & R^2 \sin^2 \theta \end{pmatrix},$$

and the inverse metric (4.3.3) is

$$[g^{ij}] = \begin{pmatrix} \frac{1}{R^2} & 0 \\ 0 & \frac{1}{R^2 \sin^2 \theta} \end{pmatrix}.$$

For the unit radius sphere ($R = 1$) these become (4.3.4)

$$[g_{ij}] = \begin{pmatrix} 1 & 0 \\ 0 & \sin^2 \theta \end{pmatrix}$$

and (4.3.5)

$$[g^{ij}] = \begin{pmatrix} 1 & 0 \\ 0 & \frac{1}{\sin^2 \theta} \end{pmatrix}.$$

To calculate the connection coefficients we use (6.2.5)

$$\Gamma^i_{jk} = \frac{1}{2} g^{il} \left(\frac{\partial g_{lk}}{\partial x^j} + \frac{\partial g_{jl}}{\partial x^k} - \frac{\partial g_{jk}}{\partial x^l} \right).$$

Don't forget that the indices (i, j, k, l) represent the spherical polar coordinates θ and ϕ in various permutations.

The only $\frac{\partial g_{lk}}{\partial x^j}$ terms which do not equal zero are

$$\frac{\partial g_{\phi\phi}}{\partial x^\theta} = \frac{\partial \left(\sin^2 \theta \right)}{\partial \theta} = 2 \sin \theta \cos \theta$$

(found by using the product rule).

Inspection of $\left[g^{ij} \right]$ tells us that this equals zero except when $g^{il} = g^{\theta\theta} = 1$ and $g^{il} = g^{\phi\phi} = \frac{1}{\sin^2 \theta}$.

Substituting these into (6.2.5)

$$\Gamma^i_{jk} = \frac{1}{2} g^{il} \left(\frac{\partial g_{lk}}{\partial x^j} + \frac{\partial g_{jl}}{\partial x^k} - \frac{\partial g_{jk}}{\partial x^l} \right).$$

We see that

$$\Gamma^\theta_{\theta\theta} = \frac{1}{2} g^{\theta\theta} \left(\frac{\partial g_{\theta\theta}}{\partial x^\theta} + \frac{\partial g_{\theta\theta}}{\partial x^\theta} - \frac{\partial g_{\theta\theta}}{\partial x^\theta} \right) = 0,$$

$$\Gamma^\theta_{\theta\phi} = \frac{1}{2} g^{\theta\theta} \left(\frac{\partial g_{\theta\phi}}{\partial x^\theta} + \frac{\partial g_{\theta\theta}}{\partial x^\phi} - \frac{\partial g_{\theta\phi}}{\partial x^\theta} \right) = 0,$$

$$\Gamma^\theta_{\phi\phi} = \frac{1}{2} g^{\theta\theta} \left(\frac{\partial g_{\theta\phi}}{\partial x^\phi} + \frac{\partial g_{\phi\theta}}{\partial x^\phi} - \frac{\partial g_{\phi\phi}}{\partial x^\theta} \right) = -\frac{1}{2} g^{\theta\theta} \frac{\partial g_{\phi\phi}}{\partial x^\theta}$$

$$= -\frac{1}{2} \times 1 \times 2 \sin \theta \cos \theta = -\sin \theta \cos \theta,$$

$$\Gamma^\phi_{\theta\theta} = \frac{1}{2} g^{\phi\phi} \left(\frac{\partial g_{\phi\theta}}{\partial x^\theta} + \frac{\partial g_{\theta\phi}}{\partial x^\theta} - \frac{\partial g_{\theta\theta}}{\partial x^\phi} \right) = 0,$$

$$\Gamma^\phi_{\theta\phi} = \frac{1}{2} g^{\phi\phi} \left(\frac{\partial g_{\phi\phi}}{\partial x^\theta} + \frac{\partial g_{\theta\phi}}{\partial x^\phi} - \frac{\partial g_{\theta\phi}}{\partial x^\phi} \right) = \frac{1}{2} g^{\phi\phi} \frac{\partial g_{\phi\phi}}{\partial x^\theta}$$

$$= \frac{1}{2} \times \frac{1}{\sin^2 \theta} \times 2 \sin \theta \cos \theta = \frac{\cos \theta}{\sin \theta} = \cot \theta,$$

$$\Gamma^\phi_{\phi\phi} = \frac{1}{2} g^{\phi\phi} \left(\frac{\partial g_{\phi\phi}}{\partial x^\phi} + \frac{\partial g_{\phi\phi}}{\partial x^\phi} - \frac{\partial g_{\phi\phi}}{\partial x^\phi} \right) = 0.$$

At last, we can say the only non-zero connection coefficients for the surface of the unit radius sphere in polar coordinates are

$$\Gamma^\theta_{\phi\phi} = -\sin \theta \cos \theta$$

and

$$\Gamma^\phi_{\theta\phi} = \Gamma^\phi_{\phi\theta} = \cot \theta.$$

We've seen that flat Euclidean space only has zero valued connection coefficients when using Cartesian coordinates. In the same space, but using polar coordinates, the connection coefficients are not all zero because in polar coordinates the basis vectors are not constant. In non-Euclidean space (the surface of a sphere for example) the connection coefficients cannot all be zero no matter what coordinate system we use, for the same reason – the basis vectors are not constant.

6.3 Covariant differentiation

After spending some time looking at connection coefficients we can now return to our original problem of how to differentiate a tensor. Equation (6.2.1), telling us how to differentiate a vector $\vec{V} = V^\alpha e_\alpha$, takes us part of the way

$$\frac{\partial \vec{V}}{\partial x^\beta} = \frac{\partial V^\alpha}{\partial x^\beta} e_\alpha + V^\alpha \frac{\partial e_\alpha}{\partial x^\beta}. \tag{6.3.1}$$

The connection coefficients are defined by (6.2.2)

$$\frac{\partial e_\alpha}{\partial x^\beta} = \Gamma^\gamma_{\alpha\beta} e_\gamma. \tag{6.3.2}$$

Substituting (6.3.2) into (6.3.1) gives

$$\frac{\partial \vec{V}}{\partial x^\beta} = \frac{\partial V^\alpha}{\partial x^\beta} e_\alpha + V^\alpha \Gamma^\gamma_{\alpha\beta} e_\gamma. \tag{6.3.3}$$

The right-hand term $V^\alpha \Gamma^\gamma_{\alpha\beta} e_\gamma$ has two dummy indices (ie indices to be summed over): α and γ. We can improve (6.3.3) by changing α to γ and γ to α to give

$$\frac{\partial \vec{V}}{\partial x^\beta} = \frac{\partial V^\alpha}{\partial x^\beta} e_\alpha + V^\gamma \Gamma^\alpha_{\gamma\beta} e_\alpha,$$

and factoring out e_α gives

$$\frac{\partial \vec{V}}{\partial x^\beta} = \left(\frac{\partial V^\alpha}{\partial x^\beta} + V^\gamma \Gamma^\alpha_{\gamma\beta} \right) e_\alpha. \tag{6.3.4}$$

Now we have separated out the basis vectors e_α, we can say the components of $\frac{\partial \vec{V}}{\partial x^\beta}$ are

$$\frac{\partial V^\alpha}{\partial x^\beta} + V^\gamma \Gamma^\alpha_{\gamma\beta}. \tag{6.3.5}$$

This expression is known as the **covariant derivative** of the contravariant vector $\vec{V}$, ie the rate of change of V^α in each of the directions β of the coordinate system x^β. The notation $\nabla_\beta V^\alpha$ is often used to denote the covariant derivative, so (6.3.5) is written

$$\nabla_\beta V^\alpha = \frac{\partial V^\alpha}{\partial x^\beta} + V^\gamma \Gamma^\alpha_{\gamma\beta}. \tag{6.3.6}$$

It can be shown that the covariant derivative of a one-form is given by

$$\nabla_\beta V_\alpha = \frac{\partial V_\alpha}{\partial x^\beta} - V_\gamma \Gamma^\gamma_{\alpha\beta}. \tag{6.3.7}$$

In flat space, using Cartesian coordinates the covariant derivative of a vector or one-form reduces to the partial derivative

$$\nabla_\beta V^\alpha = \frac{\partial V^\alpha}{\partial x^\beta} \tag{6.3.8}$$

and

$$\nabla_\beta V_\alpha = \frac{\partial V_\alpha}{\partial x^\beta},$$

which is what we would expect. Otherwise we need the partial derivative plus a correction provided by the connection coefficient, ie we need the covariant derivative.

The covariant derivatives of higher rank tensors are constructed from the building blocks of (6.3.6) and (6.3.7) as follows:

- take the partial derivative of the tensor,
- add a $\Gamma^{\alpha}_{\gamma\beta}$ term for each upper index,
- subtract a $\Gamma^{\gamma}_{\alpha\beta}$ term for each lower index.

Problem 6.4. Find the covariant derivatives of the tensors $X^{\mu}{}_{\nu}$, $X_{\mu\nu}$ and $X^{\mu\nu}$.

Following the three rules given above, we obtain

$$\nabla_{\beta} X^{\mu}{}_{\nu} = \frac{\partial X^{\mu}{}_{\nu}}{\partial x^{\beta}} + X^{\alpha}{}_{\nu}\Gamma^{\mu}_{\alpha\beta} - X^{\mu}{}_{\alpha}\Gamma^{\alpha}_{\nu\beta},$$

$$\nabla_{\beta} X_{\mu\nu} = \frac{\partial X_{\mu\nu}}{\partial x^{\beta}} - X_{\alpha\nu}\Gamma^{\alpha}_{\mu\beta} - X_{\mu\alpha}\Gamma^{\alpha}_{\nu\beta}, \qquad (6.3.9)$$

$$\nabla_{\beta} X^{\mu\nu} = \frac{\partial X^{\mu\nu}}{\partial x^{\beta}} + X^{\alpha\nu}\Gamma^{\mu}_{\alpha\beta} + X^{\mu\alpha}\Gamma^{\nu}_{\alpha\beta}.$$

This isn't of course the first time we've differentiated vectors. In Section 1.13.3 we looked at the divergence of a Cartesian vector field (remember sources and sinks)

$$\mathbf{V} = (V_x)\,\hat{e}_x + (V_y)\,\hat{e}_y + (V_z)\,\hat{e}_z,$$

where the divergence is given by (1.13.8)

$$\nabla \cdot \mathbf{V} = \frac{\partial V_x}{\partial x} + \frac{\partial V_y}{\partial y} + \frac{\partial V_z}{\partial z}.$$

Using an identical upper and lower index (ie a dummy index), the covariant derivative of a contravariant vector or tensor field is equivalent to the field's divergence. Then, in flat space using Cartesian coordinate, the covariant derivative (6.3.6)

$$\nabla_{\alpha} V^{\alpha} = \frac{\partial V^{\alpha}}{\partial x^{\alpha}} + V^{\gamma}\Gamma^{\alpha}_{\gamma\alpha}$$

reduces to

$$\nabla_{\alpha} V^{\alpha} = \frac{\partial V^{\alpha}}{\partial x^{\alpha}},$$

which is the same as (1.13.8)

$$\nabla \cdot \mathbf{V} = \frac{\partial V_x}{\partial x} + \frac{\partial V_y}{\partial y} + \frac{\partial V_z}{\partial z}.$$

We'll return to the divergence of a tensor field when we look at the energy-momentum tensor.

Sometimes a comma is used to show a partial derivative, and a semi-colon is used instead of the nabla symbol ∇ to denote the covariant derivative. Equation (6.3.6)

$$\nabla_\beta V^\alpha = \frac{\partial V^\alpha}{\partial x^\beta} + V^\gamma \Gamma^\alpha_{\gamma\beta}$$

would then be written as

$$V^\alpha_{;\beta} = V^\alpha_{,\beta} + V^\gamma \Gamma^\alpha_{\gamma\beta}, \tag{6.3.10}$$

and (6.3.7)

$$\nabla_\beta V_\alpha = \frac{\partial V_\alpha}{\partial x^\beta} - V_\gamma \Gamma^\gamma_{\alpha\beta}$$

would be written as

$$V_{\alpha;\beta} = V_{\alpha,\beta} - V_\gamma \Gamma^\gamma_{\alpha\beta}. \tag{6.3.11}$$

Problem 6.5. Show that the covariant derivative of the Euclidean metric tensor g_{ij} for any coordinate system is zero.

Substituting the Euclidean metric tensor for Cartesian coordinates (1.14.6)

$$[g_{ij}] = \begin{pmatrix} 1 & 0 & 0 \\ 0 & 1 & 0 \\ 0 & 0 & 1 \end{pmatrix}$$

into (6.3.9)

$$\nabla_\beta T_{\mu\nu} = \frac{\partial T_{\mu\nu}}{\partial x^\beta} - T_{\alpha\nu} \Gamma^\alpha_{\mu\beta} - T_{\mu\alpha} \Gamma^\alpha_{\nu\beta}$$

gives

$$\nabla_\beta g_{ij} = \frac{\partial g_{ij}}{\partial x^\beta} - g_{\alpha j} \Gamma^\alpha_{i\beta} - g_{i\alpha} \Gamma^\alpha_{j\beta}.$$

In Cartesian coordinates the right-hand side equals zero (because the Euclidean metric is constant, and the connection coefficients are zero), therefore

$$\nabla_\beta g_{ij} = 0.$$

But (and this is the clever thing about tensors) if this equation is true in Cartesian coordinates, it must be true for the Euclidean metric in all coordinate systems.

Actually, it is straightforward to show, by substituting (6.2.6)

$$\Gamma^\sigma_{\mu\nu} = \frac{1}{2} g^{\sigma\rho} \left(\frac{\partial g_{\rho\nu}}{\partial x^\mu} + \frac{\partial g_{\mu\rho}}{\partial x^\nu} - \frac{\partial g_{\mu\nu}}{\partial x^\rho} \right)$$

into (6.3.9)

$$\nabla_\beta T_{\mu\nu} = \frac{\partial T_{\mu\nu}}{\partial x^\beta} - T_{\alpha\nu} \Gamma^\alpha_{\mu\beta} - T_{\mu\alpha} \Gamma^\alpha_{\nu\beta},$$

(and, of course, substituting $g_{\mu\nu}$ for $T_{\mu\nu}$) that the covariant derivative of the general metric tensor is zero for any coordinate system, ie

$$\nabla_\beta g_{\mu\nu} = 0. \tag{6.3.12}$$

Intuitively, we can also show this by knowing that the covariant derivative of the metric tensor must vanish in the flat spacetime of special relativity (where the metric is constant). Being a tensor equation, $\nabla_\beta g_{\mu\nu} = 0$ must also then be true in any coordinate system.

6.4 Parallel transport of vectors

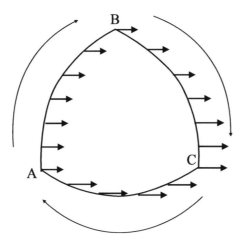

Figure 6.1: A vector parallel transported along a closed loop in flat space.

We continue to home in on the key problem of how to define the curvature of a particular space or manifold. We've seen that curves or surfaces embedded in our familiar Euclidean space can be understood by reference to that space – we can draw a circle on a piece of paper or put a sphere on a table, and measure their extrinsic curvature directly. But we can't do that with intrinsically curved manifolds such as spacetime.

The solution starts with the notion of moving a vector through a space whilst keeping it as constant as possible. The question is: if we move a vector in this way does it remain parallel to itself?

Let's first look at two-dimensional flat space. The answer is straightforward. If we move a vector along a closed loop A, B, C in flat space (see Figure 6.1) the vector doesn't change direction, in other words, the vector that arrives back at point A is parallel to the vector that started from point A.

Notice that this works for a cylinder as well because, as we've seen, its surface is intrinsically flat.

Now let's see what happens if we try to parallel transport a vector around a closed loop A, B, C on the surface of a sphere (see Figure 6.2).

Although each vector is drawn as parallel as possible to the previous one, by the time it arrives back it A it has changed direction by 90°. Furthermore, the change in direction depends on the path taken – different loops give different changes in direction.

This change in direction is a result of the curvature of the surface of the sphere. For curved spaces in general, vector parallel transport is key to understanding curvature. We now need to define parallel transport mathematically.

To do this we return to our old friend, the parameterised curve, which we met in Sections 1.11, 3.6.3 and 5.3.4.2.

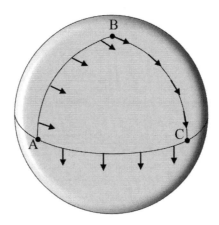

Figure 6.2: Parallel transport around closed loop on a sphere.

One vector we've met so far in relation to a parameterised curve is the tangent vector (5.3.3) with components

$$V^\beta = \frac{dx^\beta}{d\lambda} = \left(\frac{dx^0}{d\lambda}, \frac{dx^1}{d\lambda}, \frac{dx^2}{d\lambda}, \frac{dx^3}{d\lambda} \right),$$

which, to avoid confusion, we relabel as $\vec{U}$ (see Figure 6.3) with components

$$U^\beta = \frac{dx^\beta}{d\lambda} = \left(\frac{dx^0}{d\lambda}, \frac{dx^1}{d\lambda}, \frac{dx^2}{d\lambda}, \frac{dx^3}{d\lambda} \right). \tag{6.4.1}$$

We put $\vec{U}$, our tangent vector, to one side for the moment and consider a different vector $\vec{V}$, which isn't a tangent vector, but is also on the same parameterised curve.

In order to parallel transport vector $\vec{V}$ along the curve the covariant derivative of $\vec{V}$ along the curve must be zero, ie

$$\frac{d\vec{V}}{d\lambda} = 0. \tag{6.4.2}$$

First, we need to find $\frac{d\vec{V}}{d\lambda}$.

Equation (6.2.1) tells us how to find the derivative of a vector

$$\frac{\partial \vec{V}}{\partial x^\beta} = \frac{\partial V^\alpha}{\partial x^\beta} e_\alpha + V^\alpha \frac{\partial e_\alpha}{\partial x^\beta}.$$

Swapping our parameter λ for x^β (so we are now taking ordinary and not partial derivatives) we obtain

$$\frac{d\vec{V}}{d\lambda} = \frac{dV^\alpha}{d\lambda} e_\alpha + V^\alpha \frac{de_\alpha}{d\lambda}.$$

We can rewrite $\frac{de_\alpha}{d\lambda}$ as

$$\frac{de_\alpha}{d\lambda} = \frac{\partial e_\alpha}{\partial x^\beta} \frac{dx^\beta}{d\lambda}.$$

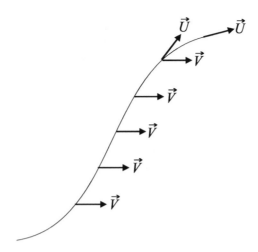

Figure 6.3: Parallel transport of $\vec{V}$ along $\vec{U}$.

Using the definition of the connection coefficients (6.2.2)

$$\frac{\partial e_\alpha}{\partial x^\beta} = \Gamma^\gamma_{\alpha\beta} e_\gamma$$

gives

$$\frac{d\vec{V}}{d\lambda} = \frac{dV^\alpha}{d\lambda} e_\alpha + V^\alpha \Gamma^\gamma_{\alpha\beta} \frac{dx^\beta}{d\lambda} e_\gamma.$$

Swapping α and γ in the final term gives

$$\frac{d\vec{V}}{d\lambda} = \frac{dV^\alpha}{d\lambda} e_\alpha + V^\gamma \Gamma^\alpha_{\gamma\beta} \frac{dx^\beta}{d\lambda} e_\alpha,$$

which can be rewritten as

$$\frac{d\vec{V}}{d\lambda} = \left(\frac{dV^\alpha}{d\lambda} + V^\gamma \Gamma^\alpha_{\gamma\beta} \frac{dx^\beta}{d\lambda} \right) e_\alpha,$$

which is itself a vector with components

$$\frac{DV^\alpha}{d\lambda} = \frac{dV^\alpha}{d\lambda} + V^\gamma \Gamma^\alpha_{\gamma\beta} \frac{dx^\beta}{d\lambda}. \tag{6.4.3}$$

This is an example of an absolute derivative (also known as the intrinsic derivative), recognisable by the upper case D in the left-hand term. An absolute derivative is a covariant derivative along a curve, in this case the curve parameterised by λ.

The absolute derivative can also be expressed in terms of the tangent vector U^β (6.4.1) that we earlier mentioned and then put to one side. U^β has components

$$U^\beta = \frac{dx^\beta}{d\lambda} = \left(\frac{dx^0}{d\lambda}, \frac{dx^1}{d\lambda}, \frac{dx^2}{d\lambda}, \frac{dx^3}{d\lambda} \right).$$

So, using the chain rule, we can write

$$\frac{DV^\alpha}{d\lambda} = \frac{dx^\beta}{d\lambda}\frac{dV^\alpha}{dx^\beta} + V^\gamma \Gamma^\alpha_{\gamma\beta}\frac{dx^\beta}{d\lambda} = U^\beta \left(\frac{dV^\alpha}{dx^\beta} + V^\gamma \Gamma^\alpha_{\gamma\beta}\right).$$

Our original condition (6.4.2) for the parallel transport of vector $\vec{V}$ along a parameterised curve was

$$\frac{d\vec{V}}{d\lambda} = 0,$$

which defines the parallel transport of $\vec{V}$ along $\vec{U}$.

Or, in component form

$$\frac{DV^\alpha}{d\lambda} = \frac{dV^\alpha}{d\lambda} + V^\gamma \Gamma^\alpha_{\gamma\beta}\frac{dx^\beta}{d\lambda} = 0. \tag{6.4.4}$$

This equation is important because we'll shortly be using it to find the geodesic equations. As we'll see, freely falling objects in spacetime follow the most direct path possible – curves known as geodesics.

6.5 Geodesics

In Euclidean space the shortest distance between two points is a straight line. In curved space the shortest distance between two points is a curve called a geodesic. These are curves that, given the curvature of the space, are as 'straight' as possible: great circles on the surface of a sphere for example.

We can now make use of the discussion in the previous section and define a geodesic – the 'straightest possible' line – as a path which parallel transports its own tangent vector. In flat space we can think of a straight line as a succession of tangent vectors all going in the same direction (a straight line is the only curve in Euclidean space that does this).

So, for a curve to be a geodesic the tangent vector must equal the parallel transported vector, ie

$$U^\alpha = \frac{dx^\alpha}{d\lambda} = V^\alpha.$$

And, from (6.4.2)

$$\frac{d\vec{U}}{d\lambda} = 0.$$

Using (6.4.4) we can therefore say

$$\frac{DU^\alpha}{d\lambda} = \frac{dU^\alpha}{d\lambda} + U^\gamma \Gamma^\alpha_{\gamma\beta}\frac{dx^\beta}{d\lambda} = 0$$

$$\frac{d}{d\lambda}\left(\frac{dx^\alpha}{d\lambda}\right) + \Gamma^\alpha_{\gamma\beta}\frac{dx^\beta}{d\lambda}\frac{dx^\gamma}{d\lambda} = 0$$

$$\frac{DU^\alpha}{d\lambda} = \frac{d^2x^\alpha}{d\lambda^2} + \Gamma^\alpha_{\gamma\beta}\frac{dx^\beta}{d\lambda}\frac{dx^\gamma}{d\lambda} = 0, \tag{6.5.1}$$

where we have rewritten the vector U^γ as $\frac{dx^\gamma}{d\lambda}$, and (in case you've forgotten) the upper case D signifies the absolute derivative.

Equations (6.5.1) are known as the **geodesic equations** and define the 'straightest possible' line in any curved space. When these equations are written in the form shown in (6.5.1), where the right-hand side equals zero, the parameter λ is known as an **affine parameter**.

Problem 6.6. Show that the geodesics for two-dimensional Euclidean space are straight lines.

As there is no z coordinate, we use a simplified version of (1.14.5)

$$dl^2 = dx^2 + dy^2.$$

Therefore, the metric is

$$[g_{ij}] = \begin{pmatrix} 1 & 0 \\ 0 & 1 \end{pmatrix}.$$

The geodesic equations (6.5.1), using Latin indices as we are not working in spacetime, are

$$\frac{d^2 x^i}{d\lambda^2} + \Gamma^i_{jk} \frac{dx^j}{d\lambda} \frac{dx^k}{d\lambda} = 0. \tag{6.5.2}$$

We are in two-dimensional Euclidean space, so the index i equals 1 or 2, or x and y in Cartesian coordinates. In order to calculate the connection coefficients Γ^i_{jk} we need (6.2.5)

$$\Gamma^i_{jk} = \frac{1}{2} g^{il} \left(\frac{\partial g_{lk}}{\partial x^j} + \frac{\partial g_{jl}}{\partial x^k} - \frac{\partial g_{jk}}{\partial x^l} \right).$$

But as the values of the metric are constant (they all equal 1) the partial derivatives $\frac{\partial g_{ij}}{\partial x^k} = 0$ for all values of i, j, k. Therefore, $\Gamma^i_{jk} = 0$ for all values of i, j, k, and (6.5.2) becomes

$$\frac{d^2 x^i}{d\lambda^2} = 0.$$

We need to find a function that if we differentiate twice will give us $\frac{d^2 x^i}{d\lambda^2} = 0$. This function is

$$x^i = a\lambda + b, \tag{6.5.3}$$

where a and b are constants. Using Cartesian coordinates where x^i equals x and y, (6.5.3) becomes

$$x = a\lambda + b$$

and

$$y = c\lambda + d.$$

Although these two equations look different to the standard equation for a straight line ((1.5.1) $y = ax + b$) they are actually the same thing expressed in parametric form, using λ as the parameter. We can see this by solving for λ, giving

$$\lambda = \frac{x - b}{a}$$

and

$$\lambda = \frac{y-d}{c}.$$

Therefore

$$\frac{x-b}{a} = \frac{y-d}{c}$$

$$y = c\left(\frac{x-b}{a}\right) + d$$

$$y = \frac{c}{a}x + \left(\frac{ad-bc}{a}\right),$$

which is a straight line with gradient $\frac{c}{a}$ and constant $\left(\frac{ad-bc}{a}\right)$.

Problem 6.7. Show that for the surface of a unit radius sphere (a) part of a meridian joining the equator to the north pole is a geodesic, and (b) a great circle on the surface of the sphere is a geodesic.

(a) In spherical polar coordinates our meridian is part of a great circle with ends at the equator (at points $\theta = \pi/2$, $\phi = 0$) and at the north pole (at points $\theta = 0$, $\phi = 0$).

The geodesic equation (6.5.2) is

$$\frac{d^2 x^i}{d\lambda^2} + \Gamma^i_{jk}\frac{dx^j}{d\lambda}\frac{dx^k}{d\lambda} = 0.$$

In spherical polar coordinates $x^1 = \theta$ and $x^2 = \phi$. We also need the connection coefficients for a unit radius sphere from Problem (6.3), which we found to be

$$\Gamma^\theta_{\phi\phi} = -\sin\theta\cos\theta$$

and

$$\Gamma^\phi_{\theta\phi} = \Gamma^\phi_{\phi\theta} = \frac{\cos\theta}{\sin\theta} = \cot\theta.$$

Equation (6.5.2) thus becomes ($i = \theta$, $j = \phi$, $k = \phi$)

$$\frac{d^2\theta}{d\lambda^2} + \Gamma^\theta_{\phi\phi}\frac{d\phi}{d\lambda}\frac{d\phi}{d\lambda} = 0$$

giving

$$\frac{d^2\theta}{d\lambda^2} - \sin\theta\cos\theta\left(\frac{d\phi}{d\lambda}\right)^2 = 0. \tag{6.5.4}$$

And ($i = \phi$, $j = \theta$, $k = \phi$)

$$\frac{d^2\phi}{d\lambda^2} + \Gamma^\phi_{\theta\phi}\frac{d\theta}{d\lambda}\frac{d\phi}{d\lambda} + \Gamma^\phi_{\phi\theta}\frac{d\phi}{d\lambda}\frac{d\theta}{d\lambda} = 0$$

giving

$$\frac{d^2\phi}{d\lambda^2} + 2\frac{\cos\theta}{\sin\theta}\frac{d\theta}{d\lambda}\frac{d\phi}{d\lambda} = 0. \tag{6.5.5}$$

(We have two Γ terms ($\Gamma^\phi_{\theta\phi}$ and $\Gamma^\phi_{\phi\theta}$) because we are summing over $\Gamma^i_{jk}\frac{dx^j}{d\lambda}\frac{dx^k}{d\lambda}$ where j and k take the value of θ and ϕ.)

We parameterise our meridian by saying $\theta = \lambda$ where $0 \leq \lambda \leq \pi/2$, and $\phi = 0$.

Therefore, $\frac{d\theta}{d\lambda} = 1$, and $\frac{d^2\theta}{d\lambda^2} = \frac{d^2\phi}{d\lambda^2} = \frac{d\phi}{d\lambda} = 0$.

(6.5.4) becomes $0 - \sin\theta\cos\theta \times 0 = 0$

and (6.5.5) becomes $0 + 2\frac{\cos\theta}{\sin\theta} \times 0 = 0$.

Both these equations are true (the left-hand side equals the right-hand side), therefore they satisfy the geodesic equations (6.5.2) meaning we have shown that the meridian is a geodesic. This shouldn't be a surprise as we define a meridian to be a great circle passing through the two poles.

(b) We can use the equator as a simple example of a great circle. In spherical polar coordinates the equator is defined by $\theta = \pi/2$ and $0 \leq \phi \leq 2\pi$.

So, we can parameterise our equator by saying $\theta = \pi/2$ and $\phi = \lambda$, where $0 \leq \phi \leq 2\pi$.

Therefore, $\frac{d\phi}{d\lambda} = 1$, and $\frac{d^2\theta}{d\lambda^2} = \frac{d^2\phi}{d\lambda^2} = \frac{d\theta}{d\lambda} = 0$.

And $\sin\theta = \sin(\pi/2) = 1$ and $\cos\theta = \cos(\pi/2) = 0$.

We use (6.5.4) and (6.5.5) as we did in (a). Equation (6.5.4) becomes $0 - 1 \times 0 \times 1^2 = 0$

and (6.5.5) becomes $0 + 2 \times 0 \times 0 \times 1 = 0$.

Both these equations are true (the left-hand side equals the right-hand side), therefore they satisfy the geodesic equations (6.5.2) meaning we have shown that the equator is a geodesic. Again, this shouldn't be a surprise as we know the equator is a great circle.

Problem 6.8. Show that a circle on the surface of a unit radius sphere defined by $\theta = \pi/4$ and $0 \leq \phi \leq 2\pi$ is not a geodesic.

Such a circle describes a line of latitude on the 'top' part of the sphere and can be parameterised by saying $\theta = \pi/4$ and $\phi = \lambda$ where $0 \leq \phi \leq 2\pi$.

Therefore, $\frac{d\phi}{d\lambda} = 1$, and $\frac{d^2\theta}{d\lambda^2} = \frac{d^2\phi}{d\lambda^2} = \frac{d\theta}{d\lambda} = 0$.

And $\sin\theta = \cos\theta = \sin(\pi/4) = \cos(\pi/4) = 1/\sqrt{2}$.

(6.5.4) becomes $0 - (1/\sqrt{2}) \times (1/\sqrt{2}) \times 1 = -1/2 \neq 0$.

And (6.5.5) becomes $0 + 2 \times 1 \times 0 \times 1 = 0$.

The second equation equals zero, but the first equation does not. Therefore, they do not satisfy the geodesic equations (6.5.2) meaning this line of latitude is not a geodesic. Which again shouldn't be a surprise as the equator is the only line of latitude that is a geodesic.

6.6 The Riemann curvature tensor

If we parallel transport a vector around an infinitesimal loop on a manifold, the vector we end up with will only be equal to the vector we started with if the manifold is flat. Otherwise, we can define the intrinsic curvature of the manifold in terms of the amount the vector has been transformed at the end of its tiny journey. Saying the manifold is curved is the same as saying the vector is changed after being parallel transported around a closed loop. By measuring that change, the **Riemann**

curvature tensor therefore measures the curvature of the manifold. In fact, the Riemann curvature tensor contains *all* the information about the curvature of a manifold.

By considering the process of moving a vector around a closed loop we can get a non-rigorous but intuitive idea of what this tensor is going to look like. This description is taken from Carroll [4].

First, we imagine the loop as an infinitesimal parallelogram with one pair of parallel sides defined by a vector A^ν and the other pair by a vector B^μ. The lengths of the sides of the loop are δa and δb. We wish to parallel transport a third vector V^σ around our parallelogram. After we have done this, V^σ will be transformed (unless the space is flat) by an amount δV^ρ So, we are looking for a mathematical machine that when we feed in the initial conditions (vectors A^ν, B^μ and V^σ) will then tell us the answer δV^ρ, ie how much V^σ has changed. Because the parallel transport of a vector is independent of coordinates we can say that this mathematical machine is a tensor. Using the rules of tensor contraction we can then say that δV^ρ should be given by

$$\delta V^\rho = (\delta a)\,(\delta b)\, A^\nu B^\mu R^\rho{}_{\sigma\mu\nu} V^\sigma,$$

where $R^\rho{}_{\sigma\mu\nu}$ is our long awaited rank 4, type $(1,3)$ Riemann curvature tensor. Notice that this is a balanced tensor equation where all the dummy indices on the right-hand side are summed over to leave a single index ρ on both sides.

That's an intuitive look at what the curvature tensor is about. The Appendix gives a full derivation of the Riemann tensor based on a related operation: the commutator of two covariant derivatives. (The calculations in the Appendix, though somewhat lengthy, not only provide insight into the meaning of the curvature tensor but are also a useful demonstration of covariant differentiation and index manipulation in action.)

The actual form of the Riemann curvature tensor, a glorious mixture of derivatives and products of connection coefficients, is

$$R^l{}_{ijk} = \frac{\partial \Gamma^l_{ik}}{\partial x^j} - \frac{\partial \Gamma^l_{ij}}{\partial x^k} + \Gamma^m_{ik}\Gamma^l_{mj} - \Gamma^m_{ij}\Gamma^l_{mk}. \tag{6.6.1}$$

This looks (and often is) complicated. The following remarks might make things a little less daunting:

- Because the connection coefficients are derived from the metric (6.2.5), the Riemann curvature tensor is also fundamentally derived from the metric and its derivatives.

- As with the connection coefficients, the indices (i, j, k, l) represent the coordinates of the particular n-dimensional space we are using. For example, if we were trying to find the Riemann curvature tensor for the two-dimensional surface of a sphere in polar coordinates, the (i, j, k, l) indices would represent the coordinates θ and ϕ. In three-dimensional (x, y, z) Cartesian coordinates the (i, j, k, l) indices would represent the coordinates (x, y, z).

- Using the metric, we can lower the upstairs index to give

$$R_{hijk} = g_{hl} R^l{}_{ijk}.$$

- Having four indices, in n-dimensions the Riemann curvature tensor has n^4 components, ie $2^4 = 16$ in two-dimensional space, $3^4 = 81$ in three dimensions and $4^4 = 256$ in four dimensions (as in spacetime). However, the good news is that the tensor has various symmetries that reduce the number of independent components to only one in two-dimensional space, 6 in three dimensions, and 20 in four dimensions. These symmetries include:

$$R^l{}_{ijk} = -R^l{}_{ikj},$$

$$R_{hijk} = R_{jkhi} = -R_{hikj} = -R_{ihjk}.$$

(As mentioned when looking at tensor symmetry/antisymmetry in Section 5.4.2, the Riemann curvature tensor R_{hijk} is antisymmetric on the first and second pair of indices.) The Appendix gives the derivation of some of these symmetries.

- The third and fourth terms both contain a dummy or summation index m. These terms therefore need to be summed over all their possible values.

Problem 6.9. Calculate the components of the Riemann curvature tensor for the surface of a unit radius sphere.

As noted above, for a two-dimensional surface of a sphere in polar coordinates, the (i, j, k, l) indices represent the coordinates θ and ϕ. We also saw that for a two-dimensional space (such as the surface of a sphere) the Riemann curvature tensor has only one independent component. We'll take one of these, $R^\theta{}_{\phi\theta\phi}$ for example, and calculate it.

The Riemann curvature tensor is given by (6.6.1)

$$R^l{}_{ijk} = \frac{\partial \Gamma^l_{ik}}{\partial x^j} - \frac{\partial \Gamma^l_{ij}}{\partial x^k} + \Gamma^m_{ik}\Gamma^l_{mj} - \Gamma^m_{ij}\Gamma^l_{mk}.$$

After substituting the indices for $R^\theta{}_{\phi\theta\phi}$, (6.6.1) becomes

$$R^\theta{}_{\phi\theta\phi} = \frac{\partial \Gamma^\theta_{\phi\phi}}{\partial x^\theta} - \frac{\partial \Gamma^\theta_{\phi\theta}}{\partial x^\phi} + \Gamma^m_{\phi\phi}\Gamma^\theta_{m\theta} - \Gamma^m_{\phi\theta}\Gamma^\theta_{m\phi}.$$

We now sum over m to give

$$R^\theta{}_{\phi\theta\phi} = \frac{\partial \Gamma^\theta_{\phi\phi}}{\partial x^\theta} - \frac{\partial \Gamma^\theta_{\phi\theta}}{\partial x^\phi} + \Gamma^\theta_{\phi\phi}\Gamma^\theta_{\theta\theta} + \Gamma^\phi_{\phi\phi}\Gamma^\theta_{\phi\theta} - \Gamma^\theta_{\phi\theta}\Gamma^\theta_{\theta\phi} - \Gamma^\phi_{\phi\theta}\Gamma^\theta_{\phi\phi}.$$

How complicated is that? But from Problem (6.3), where we calculated the connection coefficients for the surface of a unit radius sphere using polar coordinates, we found that

$$\Gamma^\theta_{\theta\theta} = \Gamma^\theta_{\theta\phi} = \Gamma^\theta_{\phi\theta} = \Gamma^\phi_{\theta\theta} = \Gamma^\phi_{\phi\phi} = 0.$$

And we can therefore simplify to

$$R^\theta{}_{\phi\theta\phi} = \frac{\partial \Gamma^\theta_{\phi\phi}}{\partial x^\theta} - \Gamma^\phi_{\phi\theta}\Gamma^\theta_{\phi\phi}.$$

From Problem (6.3) we also know

$$\Gamma^\theta_{\phi\phi} = -\sin\theta\cos\theta$$

and

$$\Gamma^\phi_{\theta\phi} = \Gamma^\phi_{\phi\theta} = \frac{\cos\theta}{\sin\theta} = \cot\theta.$$

So we can simplify

$$R^\theta{}_{\phi\theta\phi} = \frac{\partial \Gamma^\theta_{\phi\phi}}{\partial x^\theta} - \Gamma^\phi_{\phi\theta}\Gamma^\theta_{\phi\phi}$$

to

$$R^\theta{}_{\phi\theta\phi} = \frac{\partial\left(-\sin\theta\cos\theta\right)}{\partial x^\theta} - \frac{\cos\theta}{\sin\theta}\left(-\sin\theta\cos\theta\right)$$

$$= -\cos^2\theta + \sin^2\theta + \cos^2\theta$$

$$R^\theta{}_{\phi\theta\phi} = \sin^2\theta.$$

At last!

6.7 The Ricci tensor and Ricci scalar

The complete, unadulterated form of the Riemann curvature tensor doesn't appear in the Einstein field equations. Instead, it is contracted to give two other important measures of curvature known as the **Ricci tensor** and the **Ricci scalar**. Together, these are used to define the Einstein tensor, the left-hand side of the Einstein field equations.

6.7.1 The Ricci tensor

We contract the first and last indices of the Riemann curvature tensor to give

$$R^l{}_{ijl} = \frac{\partial \Gamma^l_{il}}{\partial x^j} - \frac{\partial \Gamma^l_{ij}}{\partial x^l} + \Gamma^m_{il}\Gamma^l_{mj} - \Gamma^m_{ij}\Gamma^l_{ml} = R_{ij}, \tag{6.7.1}$$

where R_{ij} is the Ricci tensor. Applying the convention of using Greek indices for spacetime coordinates we can rewrite this as

$$R_{\mu\nu} = \frac{\partial \Gamma^\alpha_{\mu\alpha}}{\partial x^\nu} - \frac{\partial \Gamma^\alpha_{\mu\nu}}{\partial x^\alpha} + \Gamma^\beta_{\mu\alpha}\Gamma^\alpha_{\beta\nu} - \Gamma^\beta_{\mu\nu}\Gamma^\alpha_{\beta\alpha}. \tag{6.7.2}$$

The Ricci tensor can also be formed from other contractions of the Riemann curvature tensor (many authors use one and three). However, because of the symmetries of the Riemann tensor, they all give either zero or $\pm R_{\mu\nu}$.

Problem 6.10. Calculate the components of the Ricci tensor for the surface of a unit radius sphere.

In Problem 6.9 we found that, for the surface of a unit sphere,

$$R^\theta{}_{\phi\theta\phi} = \sin^2\theta.$$

We can lower an index of $R^\theta{}_{\phi\theta\phi}$ using the metric (4.3.4)

$$[g_{ij}] = \begin{pmatrix} 1 & 0 \\ 0 & \sin^2\theta \end{pmatrix}.$$

And then

$$R_{\theta\phi\theta\phi} = g_{\theta\theta}R^\theta{}_{\phi\theta\phi} = 1 \times \sin^2\theta = \sin^2\theta.$$

We can then use the metric (4.3.5)

$$[g^{ij}] = \begin{pmatrix} 1 & 0 \\ 0 & \frac{1}{\sin^2\theta} \end{pmatrix}$$

to calculate the Ricci tensor components

$$R_{hk} = g^{ij} R_{hijk}$$

(for example) and obtain

$$R_{\theta\theta} = g^{\phi\phi} R_{\theta\phi\theta\phi} = \frac{1}{\sin^2\theta} \times \sin^2\theta = 1,$$

$$R_{\theta\phi} = R_{\phi\theta} = g^{\theta\theta} R_{\theta\phi\theta\phi} = 0 \times \sin^2\theta = 0,$$

$$R_{\phi\phi} = g^{\theta\theta} R_{\theta\phi\theta\phi} = 1 \times \sin^2\theta = \sin^2\theta.$$

But note that not all products of the metric and Riemann tensor give the Ricci tensor. Recall "Double contraction to zero", that we discussed in Section 5.4.2. The metric tensor is symmetric, the Riemann tensor, written as R_{hijk}, is antisymmetric on the first and second pair of indices. If the metric tensor indices equal the first or second pair of the Riemann tensor R_{hijk} indices then the result will be zero:

$$g^{hi} R_{hijk} = g^{jk} R_{hijk} = 0.$$

6.7.2 The Ricci scalar

By contracting the Ricci tensor we derive the Ricci scalar R,

$$R = g^{ij} R_{ij}. \tag{6.7.3}$$

Problem 6.11. Show that the Ricci scalar $R = 2$ for the surface of a unit radius sphere.

We previously found $R_{\theta\theta} = 1$ and $R_{\phi\phi} = \sin^2\theta$. Using (6.7.3)

$$R = g^{ij} R_{ij}.$$

We sum over the i and j indices to give

$$R = g^{\theta\theta} R_{\theta\theta} + g^{\phi\phi} R_{\phi\phi}$$

$$R = (1 \times 1) + \left(\frac{1}{\sin^2\theta} \times \sin^2\theta \right) = 2.$$

Incidentally, it's straightforward to show that if we had used the metric for the surface of a sphere radius a instead of unit radius, the Ricci scalar would be given by

$$R = \frac{2}{a^2},$$

meaning the Ricci scalar decreases as the radius increase and becomes almost zero for large radii.

7 General relativity

I must observe that the theory of relativity resembles a building consisting of two separate stories, the special theory and the general theory. The special theory, on which the general theory rests, applies to all physical phenomena with the exception of gravitation; the general theory provides the law of gravitation and its relations to the other forces of nature.

ALBERT EINSTEIN

7.1 Introduction

We start this section by introducing the three key principles that guided Einstein in formulating his theory of general relativity. We then turn our attention to:

- the spacetime of general relativity,
- how objects move in spacetime,
- how the curvature of spacetime is generated,

before finally introducing the Einstein field equations.

7.2 The three key principles

In his search for a gravitational theory of relativity Einstein formulated three principles:

- The principle of equivalence
- The principle of general covariance
- The principle of consistency

7.2.1 The principle of equivalence

In terms of understanding general relativity, this is the BIG one. There are two closely related principles of equivalence, the first being a weaker version of the second. From now on we'll be using the second formulation, which we'll refer to as *the* **equivalence principle**:

- The weak equivalence principle states that, in a sufficiently small frame of reference, freely falling objects in a gravitational field cannot be distinguished by any experiment from uniformly accelerating objects.

- The strong equivalence principle states that, in a sufficiently small frame of reference, the physical behaviour of objects in a gravitational field cannot be distinguished by any experiment from uniformly accelerating objects.

To understand these principles, it's helpful to consider what Einstein described as the 'happiest thought of my life.' This was his realisation that, 'for an observer falling freely from the roof of a house, the gravitational field does not exist.' Why doesn't it exist? Because, as we saw when we looked at Newtonian gravitational acceleration (Section 2.5.3), inertial mass equals (to an accuracy of at least one part in 10^{11}) gravitational mass, meaning that the acceleration of a body due to gravity does not depend on that body's mass. If acceleration did depend on mass, you could tell if you were in a freely falling frame (a plunging elevator for example) because objects of different masses released in that frame would move at different accelerations.

Therefore, the movement of objects under the influence of gravity does not depend on the composition or other properties of the objects. This is totally different to other forces in nature. The acceleration of electrically charged particles, for example, depends on the size and polarity (whether negative or positive) of the charge.

Einstein was a great one for thought experiments, asking himself what would happen in various imaginary scenarios. His 'happiest thought' was a thought experiment about falling from the roof of a house. He also famously imagined what would happen to an observer in an elevator in different situations. These observations are only valid locally, in regions of spacetime small enough to ignore tidal forces (the difference in the gravitational forces acting on two neighbouring particles due to variations in the gravitational field). Bear in mind that deep space, away from any gravitational field, is the same as the flat, Minkowski space of special relativity. Consider an elevator:

1. The elevator is resting on the surface of the Earth. The observer would experience the everyday force that we call gravity pulling him down. If he drops a small ball it, of course, falls straight to the floor.

2. The elevator is in deep space, away from the effects of any gravitational force, but is undergoing constant acceleration. The observer again feels a force pulling him down. If he drops a ball, it accelerates to the floor just as if it was experiencing a gravitational force.

3. The elevator is now plunging down a mineshaft in free fall. The observer feels weightless as there is no force pulling him down. If he gently lets go of a ball, it will hover in mid air.

4. The elevator is again in deep space, away from the effects of any gravitational force, but this time it is moving with a constant velocity. Again the observer feels weightless, and again, if he gently releases a ball, it will appear to hover in mid air.

Two notions of equivalence arise from these thought experiments. These are:

- Constant acceleration in deep/Minkowski space is equivalent to being in a gravitational field.

- Uniform motion in deep/Minkowski space is equivalent to being in free fall in a gravitational field.

From these insights, Einstein was able to draw the following conclusions:

- Gravity is an aspect of the geometry of spacetime, not of the composition of the object moving through spacetime.

- Small regions of spacetime are locally flat, and special relativity applies.

- The laws of physics in a uniformly accelerating frame should be the same as in a frame in a gravitational field (this is the statement of the strong equivalence principle).

Mathematically, all this begins to make sense if spacetime is regarded as a Riemannian manifold (maths-speak for a curved space) that is locally flat and equipped with a metric tensor. The actual form of the metric (reflecting the degree of curvature of spacetime) will vary depending on the proximity of mass-energy. Freely falling test bodies follow paths that are geodesics of that metric. This is what Einstein proposed in his theory of general relativity.

Also, because the laws of physics should be the same in a uniformly accelerating frame and a gravitational field, Einstein was able to make testable predictions that a gravitational field would:

- Slow down clocks (a phenomenon known as gravitational time dilation) and cause the wavelength of light to be lengthened (an effect known as gravitational redshift). Later, in Section 9.4, we'll use the equivalence principle to derive locally-valid equations for gravitational time dilation and gravitational redshift. We'll also use the Schwarzschild metric to derive general equations for these two phenomena that are valid in widely separated local frames.

- Cause light to be deflected by the sun. Imagine standing in an 'upwardly' accelerating elevator in deep space, away from the effects of a gravitational field. Now shine a torch horizontally across the elevator onto the opposite wall. Because the elevator is accelerating, the beam of light will strike the wall at a lower point than if the elevator were in uniform motion. To an observer in the elevator, the light beam would appear to curve downwards. The equivalence principle requires that the same phenomenon should also occur in a gravitational field. In Section 9.4.8 we'll take a closer look at how the Schwarzschild metric can be used to quantify the gravitational deflection of light.

One of the key concepts of special relativity was the existence of Lorentz frames of reference in (flat) Minkowski space, where objects obey Newton's first law, ie remain at rest or in uniform motion in a straight line unless acted upon by an external force.

Now let's examine Lorentz frames using the perspective of the equivalence principle, which tells us that uniform accelerated motion is indistinguishable from being at rest in a gravitational field. This means that an observer at rest in a gravitational field is equivalent to one in a uniformly accelerating frame. A frame that is uniformly accelerating cannot be inertial, therefore an observer in a gravitational field cannot be in an inertial frame.

Imagine a uniformly accelerating spaceship in deep space unaffected by gravity. Just ahead of the spaceship is a small rock, happily minding its own business and obeying Newton's first law – ie in a state of uniform motion – and therefore constituting an inertial frame of reference. Someone has forgotten to close the skylight in the spaceship (there's always someone – luckily all the occupants are wearing spacesuits) and the rock plunges through the opening and crashes onto the upper deck. If the people on board the spaceship didn't know any better they could reasonably assume that they were stationary in a gravitational field, and the rock had fallen due to the effects of gravity.

Now take an observer sitting in a real gravitational field, on a chair on the Earth's surface, for example. An apple falls from a nearby tree and hits the ground. 'Pesky gravity', is the observer's explanation of the apple's fall. But the equivalence principle tells us that this observer, at rest in a gravitational field, cannot carry out any experiment that will distinguish her reference frame from one that is constantly accelerating. Therefore, she is effectively in a non-inertial accelerating frame, identical to the occupants of the spaceship.

But if that's the case, the freely falling apple must be equivalent to the uniformly moving rock, and they *both* must constitute inertial frames. It must follow that if the apple is in a state of uniform motion then it isn't the apple that's accelerating down, it's the Earth's surface accelerating up – because, locally:

- *a gravitational field is equivalent to a uniformly accelerating frame.*

The equivalence principle means that locally gravity is indistinguishable from the inertial forces we met earlier when looking at Newtonian mechanics (Section 2.3). Recall that these inertial or fictitious forces (think of the force pushing you back into your seat in an accelerating car) are forces caused by an observer in a non-inertial reference frame. Newton's second law (2.4.1) tells us that if two objects of mass m_1 and m_2 are to fall identically (ie with equal acceleration a, as demanded by the equivalence principle) then

$$F_1 = m_1a \text{ and } F_2 = m_2a,$$

which, for constant a, means the force acting on each object must be proportional to the object's mass. Because inertial forces result from the acceleration of a non-inertial reference frame, they must be (in accordance with Newton's second law (2.4.1)) proportional to the mass of the object acted upon. We have just shown that the gravitational force acting on two objects is proportional to their masses, meaning gravity is an inertial force.

From all of this we can conclude that free fall is the natural state of motion, and physics is most straightforward when everything is freely falling, as summarised by Misner et al [23]:

> 'Forego talk of acceleration! That, paradoxically, is the lesson of the circumstance that "all objects fall with the same acceleration." Whose fault were those accelerations after all? They came from allowing a ground-based observer into the act. The push of the ground under his feet was driving him away from a natural world line. Through that flaw in his arrangements, he became responsible for all those accelerations. Put him in space and strap rockets to his legs. No difference! Again the responsibility for what he sees is his. Once more he notes that "all objects fall with the same acceleration". Physics looks as complicated to the jet-driven observer as it does to the man on the ground. Rule out both observers to make physics look simple. Instead, travel aboard the freely moving spaceship. Nothing could be more natural than what one sees: every object moves in a straight line with uniform velocity. This is the way to do physics!'

And all this hinges on the equivalence principle. Testability is the essence of a good scientific theory. To use a rather brutal metaphor, think of a scientific theory as a champion prizefighter offering to take on all comers. The prizefighter has to suffer only one defeat to lose his crown. Likewise, find just one object that accelerates at a different rate in a gravitational field to other objects, and the equivalence principle and hence general relativity would, at the very least, be in serious trouble.

So now we know how to construct an inertial frame in a gravitational field: in a sufficiently small frame of reference, gravity is equivalent to acceleration and a freely falling frame will be an inertial frame. Imagine a laboratory on Earth full of scientists busily measuring things with their clocks and rulers. As we'll see later, when looking at the Schwarzschild metric, these measurements are affected (admittedly to a tiny, tiny extent) by the Earth's gravitational field. Drop that laboratory off a cliff or from an aeroplane and, for a few seconds at least, the scientists would be working in an excellent approximation to an inertial frame in the Minkowski space of special relativity.

But why, in the previous paragraph, did we need to stipulate a 'sufficiently small frame of reference'? One reason is the tidal forces that we met in Section 2.5.6, the consequence of the fact that any gravitational field is non-uniform in space because:

- it varies horizontally in space because objects in free fall don't move parallel to each other but radially towards the centre of the planet, star or other massive body that is 'attracting' them;

- it varies vertically in space because the acceleration due to gravity varies with height.

The other reason is that gravitational fields also vary with time, because the longer an object is falling the more pronounced will be the above horizontal and vertical effects.

As an illustration of the need for strictly local freely falling frames, consider two such frames, one in the UK and one in Australia. Observers in the southern hemisphere would measure the acceleration of the British freely falling coordinate system to be almost opposite to their own.

Notwithstanding these non-uniformities in the gravitational field, the equivalence principle allows us to state that small enough freely falling frames will be inertial. This is true for any gravitational field. If we choose a sufficiently small (in terms of space and time) freely falling frame, it will be a **locally inertial frame** (LIF). At the origin of a LIF it is possible to choose coordinates where the first derivatives of the metric are zero, meaning that the connection coefficients are also zero. However, the second derivatives of the metric (and therefore the derivatives of the connection coefficients) are not necessarily zero.

7.2.2 The principle of general covariance

One of the postulates of special relativity is the principle of relativity, ie that the laws of physics are the same in any inertial frame of reference. The **principle of general covariance** extends that requirement to say that the *form* of the laws of physics should be the same in all – inertial and accelerating – frames. The rationale behind this principle is that physical phenomena shouldn't depend on the choice of coordinate systems used to describe them, and that all frames therefore are equally valid. Foster and Nightingale [9] summarise the principle of general covariance as follows:

> 'A physical equation of general relativity is generally true in all coordinate systems if (a) the equation is a tensor equation (ie it preserves its form under general coordinate transformations), and (b) the equation is true in special relativity.'

Taken together, these two conditions mean that if we have a valid tensor equation that is true in special relativity we can, with a little bit of twiddling, magic up an equation that is true in general relativity. Shortly, we'll look at what that 'twiddling' actually involves.

There appears to be an ongoing debate about the meaning and significance of this principle, and Einstein appeared to have had a bit of a love-hate relationship with it. But he did believe it was a sort of useful guide for him, and 'of great heuristic value' (a heuristic being a mental short cut, a rule of thumb or an educated guess) in helping him formulate his theory of general relativity. At our introductory level, we can understand the principle of general covariance as simply demanding that in general relativity physical laws must be expressed as valid tensor equations.

7.2.3 The principle of consistency

The **principle of consistency** requires that a new scientific theory must be able to account for the successful predictions of the old theories it seeks to replace. What this means is that, given the appropriate conditions, general relativity should reduce to both the laws of Newtonian mechanics and, in the absence of gravity, to the formulations of special relativity.

In the next chapter we'll look in detail how general relativity approximates Newtonian gravity in what is called the Newtonian limit, which assumes slowly moving objects, and a weak and static gravitational field.

7.3 Spacetime

Mathematically, we can model spacetime using a four dimensional Riemannian manifold (that therefore locally approximates the flat spacetime of special relativity). More precisely, we can say that in a small enough region of curved spacetime we can construct a Cartesian coordinate system where the metric tensor approximates to (3.5.2)

$$[\eta_{\mu\nu}] = \begin{pmatrix} 1 & 0 & 0 & 0 \\ 0 & -1 & 0 & 0 \\ 0 & 0 & -1 & 0 \\ 0 & 0 & 0 & -1 \end{pmatrix}.$$

Another way of saying this is that we can find a coordinate transformation matrix that allows us to transform the metric tensor $g_{\mu\nu}$ so that

$$g_{\mu\nu} \approx \eta_{\mu\nu}.$$

In free fall, of course, the spacetime inside our plunging laboratory hasn't changed: you can't change spacetime by jumping over a cliff. What has happened though is that, because of the equivalence principle, we have in effect constructed a coordinate system where the metric tensor in our falling laboratory, $g_{\mu\nu}$ – the metric for curved spacetime, approximates $\eta_{\mu\nu}$ – the Minkowski metric of flat space. None of this would work of course if the equivalence principle was not true, if objects of different mass moved at different accelerations in a gravitational field. Fortunately, for us and for general relativity, no experiment to date has contradicted this principle.

We can use Cartesian coordinates (in the form of spacetime diagrams) in the flat space of special relativity. In general relativity this usually isn't possible, and we use general coordinates instead – the things that allow vectors, one-forms and tensors to work properly.

The principle of general covariance allows us to take valid tensor equations from special relativity and extend them to general relativity. We do this by the 'twiddling' referred to in Section 7.2.2, specifically by:

1. replacing the Minkowski metric $\eta_{\mu\nu}$ with the general metric tensor $g_{\mu\nu}$,

2. and, if they exist, replacing partial derivatives with covariant derivatives.

The second rule is often called the 'comma goes to semi-colon' rule because, as we saw when looking at covariant differentiation in Section 6.3, commas and semi-colons can be used as shorthand notation for partial and covariant derivatives. This rule means that where equations valid in Cartesian coordinates in special relativity contain derivatives ('commas'), these can be changed to covariant derivatives ('semi-colons') to give an equation valid in any coordinate system in general relativity.

An example using the first rule would be that the equation (3.5.5) for proper τ time in special relativity

$$c^2 d\tau^2 = \eta_{\mu\nu} dx^\mu dx^\nu$$

can be changed to

$$c^2 d\tau^2 = g_{\mu\nu} dx^\mu dx^\nu, \tag{7.3.1}$$

giving proper time in curved spacetime in general relativity, by replacing $\eta_{\mu\nu}$ with the general metric tensor $g_{\mu\nu}$.

Two examples using the second rule are:

- The law of conservation of particles in special relativity is given by

$$n U^\alpha_{,\alpha} = n \frac{\partial U^\alpha}{\partial x^\alpha} = 0,$$

where n is the density of the particles in their rest frame, and U^α is their four-velocity. In general relativity, this law takes the form, valid in all frames, of

$$n U^\alpha_{;\alpha} = n \nabla_\alpha U^\alpha = 0.$$

- The law of conservation of four-momentum changes from

$$T^{\mu\nu}_{,\nu} = \frac{\partial T^{\mu\nu}}{\partial x^\nu} = 0$$

in special relativity to

$$T^{\mu\nu}_{;\nu} = \nabla_\nu T^{\mu\nu} = 0$$

in general relativity, where $T^{\mu\nu}$ is the energy-momentum tensor (see Section 7.5) defined as (7.5.8)

$$T^{\mu\nu} = \left(\rho + p/c^2\right) U^\mu U^\nu - p g^{\mu\nu}.$$

7.4 Geodesics in spacetime

A fundamentally important result derived from the 'comma goes to semi-colon rule' allows us to describe how free particles move in spacetime. In Section 6.5 we looked at geodesics on manifolds, including examples finding shortest distances on the surface of a sphere. We've already claimed that spacetime can be modelled using a Riemannian manifold, so it shouldn't be a huge surprise to find that the geodesic equations given for curved spaces in general should also apply to spacetime. Now we can show this.

Newton's first law (2.4) states that an object will remain at rest or in uniform motion in a straight line unless acted upon by an external force. As noted in Section 7.2.1, if an object obeys this law it must be in an inertial frame, ie is not subject to a gravitational field. Therefore, we can say that the four-force (3.6.17) acting on the object must be zero

$$F^\mu = \frac{dP^\mu}{d\tau} = 0,$$

where proper time τ is given by (3.5.5) and P^μ is the four-momentum (3.6.14)

$$P^\mu = mU^\mu = m\frac{dx^\mu}{d\tau}.$$

If the object is at rest or in uniform motion in a straight line it must have zero acceleration, ie

$$\frac{dU^\mu}{d\tau} = 0. \tag{7.4.1}$$

We can now use the 'comma goes to semi-colon' rule and move from special to general relativity by replacing the derivative $\frac{dU^\mu}{d\tau}$ with the absolute derivative $\frac{DU^\mu}{d\tau}$ that we first met in Section 6.4.

But we've already met something very similar to $\frac{DU^\mu}{d\tau}$ in the form of (6.5.1)

$$\frac{DU^\alpha}{d\lambda} = \frac{d^2x^\alpha}{d\lambda^2} + \Gamma^\alpha_{\gamma\beta}\frac{dx^\beta}{d\lambda}\frac{dx^\gamma}{d\lambda} = 0,$$

which are the geodesic equations of a Riemannian manifold.

All we need do is replace the general parameter λ with proper time τ, the time measured by a clock attached to a particle moving along a world-line through spacetime (given by (7.3.1)) and we've found the geodesic equations for a free particle moving in curved spacetime:

$$\frac{d^2x^\alpha}{d\tau^2} + \Gamma^\alpha_{\gamma\beta}\frac{dx^\beta}{d\tau}\frac{dx^\gamma}{d\tau} = 0. \tag{7.4.2}$$

We'll return to this equation later when we look at the 'Newtonian limit' of general relativity.

There are three types of geodesic in spacetime:

- time-like geodesics
- null-geodesics
- space-like geodesics.

To understand the difference, recall that an n-dimensional Riemannian manifold has a line element (4.2.1) given by

$$dl^2 = g_{ij}dx^idx^j,$$

where g_{ij} is the metric. In spacetime this can be expressed using the spacetime metric tensor $g_{\mu\nu}$ as

$$ds^2 = g_{\mu\nu}dx^\mu dx^\nu,$$

where ds^2 gives the interval between two infinitesimally close events in spacetime.

Also, proper time τ is given by (7.3.1)

$$c^2d\tau^2 = g_{\mu\nu}dx^\mu dx^\nu,$$

which links proper time to the spacetime line element by

$$c^2d\tau^2 = ds^2. \tag{7.4.3}$$

Recalling that spacetime is modelled using a pseudo-Riemannian manifold where ds^2 may be positive, zero or negative, the three types of spacetime geodesic are defined as:

- Time-like geodesics (where $ds^2 > 0$ and proper time is real, ie $d\tau \neq 0$), which describe the paths of freely falling massive particles. By 'massive' we just mean they have a mass of some sort.

- Null geodesics (where $ds^2 = 0$ and proper time is unchanging, ie $d\tau = 0$), which describe the paths of light rays (composed, of course, of massless photons).

- Space-like geodesics (where $ds^2 < 0$), which have no physical significance.

Time-like geodesics can be described using (7.4.2)

$$\frac{d^2 x^\alpha}{d\tau^2} + \Gamma^\alpha_{\gamma\beta} \frac{dx^\beta}{d\tau} \frac{dx^\gamma}{d\tau} = 0.$$

For null geodesics, we cannot use (7.4.2) because light rays travel along world-lines of unchanging proper time. Instead we use another parameter, such as in the geodesic equations 6.5.1

$$\frac{d^2 x^\alpha}{d\lambda^2} + \Gamma^\alpha_{\gamma\beta} \frac{dx^\beta}{d\lambda} \frac{dx^\gamma}{d\lambda} = 0.$$

7.4.1 Geodesics on Earth

Our description of geodesics as defining the 'straightest possible' line in a curved space may seem at odds with what we observe when we throw an object, a ball for example, here on Earth. Irrespective of whether the ball is thrown straight up or up at an angle, a graph of the ball's height plotted against time gives an obvious parabolic curve, which doesn't look at all like a straight line. The problem is that the ball's trajectory through three-dimensional space is not the same as its trajectory (or world-line) through four-dimensional spacetime. Let's take the example of a ball thrown to a height of 5 m, as shown in Figure 7.1, a three-dimensional spacetime diagram.

The parabolic curve OA shows the spatial trajectory of the ball on the xy plane where it reaches a maximum height of 5 m. Using the relevant equation of uniform motion (1.10.15)

$$s = ut + \frac{at^2}{2},$$

we can easily calculate the total flight time of the ball to be about 2 seconds. Using our normal ct time units means 2 seconds is represented by $ct = 6 \times 10^8$ metres on the time axis. The parabolic curve OB represents the world-line of the ball, with a spatial height of 5 m and a ct time axis length of 6×10^8 m. If we were to plot this curve using a scale of 1 cm equals 1 m, it would have a height of 5 cm and a length (measured along the ct axis) of 6,000,000 m = 6000 km, the approximate distance between London and Washington, D.C. The fact that, to all intents and purposes, this curve is a straight line illustrates the weakness of the terrestrial gravitational field and how little the Earth's mass curves spacetime.

7.4.2 Geodesic deviation

It's worth mentioning that the behaviour of two neighbouring parallel geodesics provides a means of measuring spacetime curvature. Two such geodesics in flat space will remain parallel but will

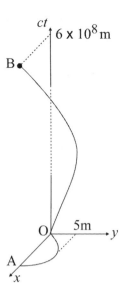

Figure 7.1: World-line of a ball thrown to a height of $5\,\mathrm{m}$ on Earth.

eventually meet in curved spacetime. This is the underlying mechanism behind the phenomenon of tidal forces that we mentioned earlier. Mathematically, the relative acceleration between two such geodesics is defined by something called the **equation of geodesic deviation**, which describes the behaviour of a small 'connecting vector' joining the two geodesics. Without going into complicated details, the two geodesics (A and B, for example) are defined in terms of a pair of parameterised curves. So A is given by $x_A^\mu(\lambda)$ and B is given by $x_B^\mu(\lambda)$, where λ is an affine parameter. Each value of λ then corresponds to a point on A and a neighbouring point on B. These two points are joined by a connecting vector $\xi^\mu(\lambda)$ (where ξ is the Greek letter xi), and we can say

$$\xi^\mu(\lambda) = x_B^\mu(\lambda) - x_A^\mu(\lambda).$$

It can then be shown that changes in the connecting vector are described by the equation of geodesic deviation

$$\frac{D^2\xi^\mu}{D\lambda^2} + R^\mu{}_{\beta\alpha\gamma}\xi^\alpha\frac{dx^\beta}{d\lambda}\frac{dx^\gamma}{d\lambda} = 0.$$

The upper case D signifies the absolute derivative that we met earlier in (6.4.3)

$$\frac{DV^\alpha}{d\lambda} = \frac{dV^\alpha}{d\lambda} + V^\gamma\Gamma^\alpha_{\gamma\beta}\frac{dx^\beta}{d\lambda},$$

and $R^\mu{}_{\beta\alpha\gamma}$ is the Riemann curvature tensor.

7.5 The energy-momentum tensor

Newtonian theory gives mass as the 'source' of the gravitational field. However, we saw in Section 3.6.9 that, in relativistic theory, mass, energy and momentum are all related, as expressed in (3.6.18), the energy momentum relation.

$$E^2 = p^2 c^2 + m^2 c^4.$$

It therefore seems reasonable to assume that the source of the gravitational field in general relativity should include momentum and energy as well as mass. Recall that four-momentum (3.6.16) is given by

$$P^\mu = (E/c, \mathbf{p}) = (E/c, p_x, p_y, p_z)$$

and provides a complete description of the total relativistic energy E (its time component) and relativistic momentum $\mathbf{p}$ (its spatial components) of a particle. Misner et al [23] speak of a 'river' of four-momentum flowing through spacetime, 'Each particle carries its four-momentum vector with itself along its world line. Many particles, on many world lines... produce a continuum flow – a river of four-momentum.'

The density and flow of this river of four-momentum is the source of the gravitational field in general relativity. We need a machine to describe the momentum and energy of not one particle but many. That machine is a rank 2 tensor known as the **energy-momentum tensor** $T^{\mu\nu}$, also known as the **stress-energy tensor**. This tensor is the thing that describes the total energy of a particular physical system, curves spacetime and sits proudly on the right-hand side of the Einstein field equations. The energy-momentum tensor is symmetric, so $T^{\mu\nu} = T^{\nu\mu}$, and therefore has ten independent components. $T^{\mu\nu}$ is defined as:

- the rate of flow of the μth component of four-momentum across a surface of constant ν.

What does this mean? The components of four-momentum are $(E/c, p_x, p_y, p_z)$. If we use spacetime coordinates of ct, x, y, z, the components of the energy-momentum tensor are the rate of flow of each four-momentum component across a surface of each constant spacetime coordinate. It's easy enough to imagine the rate of flow of something or other through a spatial x, y, z surface. Think of water shooting through a metre square hole in a wall (which could represent one of the spatial surfaces, x for example). The volume of water pouring through the hole in 1 second is the water's rate of flow. But what's the meaning of rate of flow of a four-momentum component across a surface of constant time ct? Let's look at this question in a little more detail.

Consider a volume of non-interacting particles at rest with respect to each other in a Lorentz frame. Particles of this sort are collectively known as **dust** and are the simplest type of fluid. In fact, as we'll see shortly, dust is the special case of a pressure-free perfect fluid.

A tiny cube containing a number of these particles has volume $dx \times dy \times dz$. If there are n_0 particles per unit volume then n_0 (a scalar) is the particles' number density. In a different Lorentz frame, where the particles are moving in the x direction, Lorentz contraction will cause the volume of the cube to change to $(dx/\gamma) \times dy \times dz$ (where γ is the Lorentz factor), and the new number density n will be

$$n = \gamma n_0. \tag{7.5.1}$$

Now consider the four-vector

$$\vec{N} = n_0 U^\mu,$$

where $U^\mu = \gamma(c, \mathbf{v})$ is the particles' four-velocity (3.6.2), where $\mathbf{v} = (v_x, v_y, v_z)$. The time component N^0 of the four-vector $\vec{N}$ is given by

$$N^0 = \gamma n_0 c,$$

which, from (7.5.1) gives

$$N^0 = nc,$$

which is proportional to the number density (and would be the number density if we let $c = 1$). The spatial components N^i of $\vec{N}$ (where $i = 1, 2, 3$) are the rates of flow (particles per unit time per unit area) across a surface of constant x, y, z and are given by

$$N^i = \gamma n_0 v^i = nv^i. \tag{7.5.2}$$

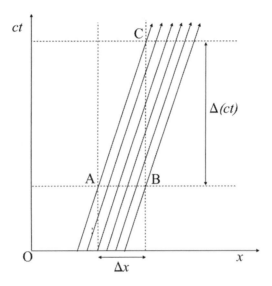

Figure 7.2: Spacetime diagram showing dust particles crossing surfaces of constant x and constant ct.

Because rate of flow per unit area is known as **flux**, $\vec{N}$ is called the **number-flux four-vector**. Figure 7.2 (from Marsh [19]) shows a spacetime diagram with the world-lines of particles travelling at constant velocity v along the x axis. These world-lines cross two surfaces: BC of constant x and AB of constant time ct. The world-lines crossing BC represent the flux across constant x, (ie nv from (7.5.2)). The *same* world-lines also cross AB, a surface of constant time. We can obtain a new flux by scaling N^i as follows

$$N^i \frac{BC}{AB} = N^i \frac{\Delta(ct)}{\Delta x} =$$

(as $v = x/t$)

$$N^i \frac{c}{v} = \frac{nvc}{v} = nc = N^0,$$

and we have shown that N^0, the number density, is also a flux or rate of flow across a surface, a surface of constant time. There are four $T^{\mu\nu}$ components where ν equals time ct. If we use the usual spacetime coordinate labels $\mu, \nu = 0, 1, 2, 3$ these 'density-type' components are T^{00}, T^{10}, T^{20} and T^{30}.

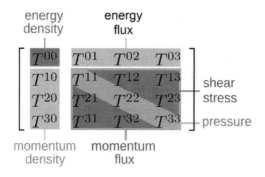

Figure 7.3: The components of the energy-momentum tensor.

Figure 7.3 shows all sixteen components of $T^{\mu\nu}$, which can have various energy density units, such as Jm^{-3} or kgm^{-3}.

These components can be described as follows:

- T^{00} is the flux (rate of flow) of the 0th component of P^{μ} across a surface of $t = $ constant, and equals energy density.

- T^{i0} is the flux of the ith component of P^{μ} across a surface of $t = $ constant, and equals momentum density.

- $T^{0i} \left(= T^{i0}\right)$ is the flux of energy across a surface of $x^{i} = $ constant, and equals energy flux.

- T^{ij} is the j flux of i momentum, and equals pressure when $i = j$ and shear stress when $i \neq j$.

To further illustrate the physical meaning of these various fluxes and densities Figure 7.4 (from Lambourne[17]) shows a simple system of non-interacting particles, each of mass m, moving with common velocity $\mathbf{v} = (v_x, v_y, 0)$, ie there is no velocity component in the z direction. (A parallelepiped is a sort of squashed rectangular box where the opposite faces remain parallel. The volume of a parallelepiped equals the area of its base multiplied by its height.)

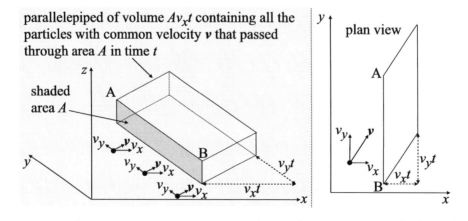

Figure 7.4: Particles moving with common velocity $\mathbf{v}$ passing through area A.

Equation (3.6.12) tells us that in an inertial frame a particle's total relativistic energy is given by

$$E = \gamma m c^2.$$

If there are n particles per unit volume, the energy density, ie the T^{00} component of the energy-momentum tensor, will be

$$T^{00} = n\gamma m c^2.$$

Note that if the particles are at rest in some frame, then $\gamma = 1$ and $T^{00} = nmc^2$.

After time t, all the particles that have passed through an area A perpendicular to the x axis will be contained in a parallelepiped of volume $Av_x t$, and the number of those particles will equal $nv_x At$. As each particle has energy $\gamma m c^2$, the total energy of all these particles equals $nv_x At\gamma m c^2$. The energy-momentum component T^{01} is the rate of flow of energy across a surface perpendicular to the x axis. To find this quantity we need to divide the total energy $nv_x At\gamma m c^2$ by Atc (the c is necessary to keep the units correct) to give

$$T^{01} = \frac{nv_x At\gamma m c^2}{Atc} = nmv_x\gamma c.$$

Now imagine rotating the area A so it is perpendicular to the y axis. We could then carry out a similar calculation to find the rate of flow of energy across a surface perpendicular to the y axis, which would be given by

$$T^{02} = \frac{nv_y At\gamma m c^2}{Atc} = nmv_y\gamma c.$$

We've assumed there is no v_z velocity component, but if there were we would have $T^{03} = nmv_z\gamma c$.

From the equation for four-momentum (3.6.15)

$$P^\mu = (mc\gamma, m\gamma\mathbf{v})$$

we can see that each particle has an x component of four-momentum given by $m\gamma v_x$. We earlier saw (when looking at N^0) that number density is a rate of flow across a surface of constant time. The density of the x component of four-momentum is therefore given by $nm\gamma v_x c$ (again, the c is necessary to keep the units correct), ie

$$T^{10} = nm\gamma v_x c,$$

and we can see that $T^{10} = T^{01}$. Similarly,

$$T^{20} = nm\gamma v_y c,$$

and, if there were a v_z velocity component, we would have $T^{30} = nm\gamma v_z c$.

Finally, the particles crossing area A perpendicular to the x axis will have a y component of four-momentum equal to $m\gamma v_y$. The rate of flow of these particles across A equals $nv_x At$ divided by At $= nv_x$. T^{21}, the rate of flow of the y component of four-momentum across A, a surface perpendicular to the x axis, is therefore given by

$$T^{21} = m\gamma v_y \times nv_x = nm\gamma v_y v_x,$$

and describes a shear stress, a force (force being rate of change of momentum) that *isn't* perpendicular to A.

A similar calculation gives $T^{11} = nm\gamma v_x^2$, which describes the rate of flow of the x component of four-momentum across a surface perpendicular to the x axis. This force, being perpendicular to A, defines pressure in the x direction.

Eventually, we end up with all the components of the energy-momentum tensor:

$$
[T^{\mu\nu}] =
\begin{pmatrix}
T^{00} & T^{01} & T^{02} & T^{03} \\
T^{10} & T^{11} & T^{12} & T^{13} \\
T^{20} & T^{21} & T^{22} & T^{23} \\
T^{30} & T^{31} & T^{32} & T^{33}
\end{pmatrix}
$$

$$
=
\begin{pmatrix}
n\gamma mc^2 & nm\gamma v_x c & nm\gamma v_y c & nm\gamma v_z c \\
nm\gamma v_x c & nm\gamma v_x^2 & nm\gamma v_x v_y & nm\gamma v_x v_z \\
nm\gamma v_y c & nm\gamma v_y v_x & nm\gamma v_y^2 & nm\gamma v_y v_z \\
nm\gamma v_z c & nm\gamma v_z v_x & nm\gamma v_z v_y & nm\gamma v_z^2
\end{pmatrix}.
\tag{7.5.3}
$$

This is the general form of the energy-momentum tensor, which describes the density and flow of four-momentum at a point in spacetime. Now let's look at the two most common specific examples of an energy-momentum tensor: dust and a perfect fluid.

7.5.1 Dust

As previously mentioned, dust is a collection of non-interacting particles at rest with respect to each other in a Lorentz frame. In cosmology, dust is a good approximation to the matter-dominated later universe. Although a hypothetical cloud of dust may be swirling around in a complicated fashion we can consider a small region known as a **fluid element**, where the particles have an approximately equal average velocity. Schutz [28] refers to the Lorentz frame containing this homogeneous region as the **momentarily comoving rest frame** (MCRF) of the fluid element. Because the dust particles are at rest in the MCRF, they have no velocity component, no momentum, $\gamma = 1$, and (7.5.3) nicely simplifies to

$$
[T^{\mu\nu}] =
\begin{pmatrix}
\rho c^2 & 0 & 0 & 0 \\
0 & 0 & 0 & 0 \\
0 & 0 & 0 & 0 \\
0 & 0 & 0 & 0
\end{pmatrix},
\tag{7.5.4}
$$

where $\rho = nm$ is the mass density. For any Lorentz frame (not just the MCRF) the energy-momentum tensor for dust is given by

$$
T^{\mu\nu} = \rho U^\mu U^\nu.
\tag{7.5.5}
$$

In the MCRF, $U^0 = c$, $U^1 = U^2 = U^3 = 0$ and we get back to (7.5.4).

7.5.2 Perfect fluid

In physics, **perfect fluids** are used to model many different systems including, in cosmology, the universe. A perfect fluid can be thought of as dust plus pressure p. Random motion of the fluid's particles is what causes pressure, which acts with equal magnitude in all directions. Unlike a real fluid, perfect fluids are assumed to have zero viscosity (ie they aren't 'sticky') and zero heat conductance in the MCRF. We can use these restrictions to deduce the energy-momentum tensor of a perfect fluid.

Viscosity is essentially liquid friction between two adjacent particles. That implies a shear stress, a force that is not perpendicular to the interface between particles. We know that shear stresses are described by the T^{ij} components of the energy-momentum tensor, where $i \neq j$. Therefore, for a perfect fluid, $T^{ij} = 0$ for $i \neq j$.

Although particles aren't moving in the MCRF, energy may still be transferred by the mechanism of heat conduction. Moving energy will have associated momentum described by the T^{i0} components, and the heat conduction itself is described by the T^{0i} components. No heat conduction, as in a perfect fluid, therefore implies $T^{i0} = T^{0i} = 0$.

We earlier saw that pressure p is the result of a force perpendicular to the interface between particles and is described by the components T^{ij}, where $i = j$.

These restrictions mean that T^{ij} must be a diagonal matrix. Because T^{ij} must be diagonal in all frames it must be a scalar multiple of the identity matrix

$$\begin{pmatrix} 1 & 0 & 0 \\ 0 & 1 & 0 \\ 0 & 0 & 1 \end{pmatrix},$$

which means $T^{11} = T^{22} = T^{33}$. Putting all this together, we find for a perfect fluid in the MCRF

$$[T^{\mu\nu}] = \begin{pmatrix} \rho c^2 & 0 & 0 & 0 \\ 0 & p & 0 & 0 \\ 0 & 0 & p & 0 \\ 0 & 0 & 0 & p \end{pmatrix}. \tag{7.5.6}$$

The energy-momentum tensor for a perfect fluid in any Lorentz frame (not just the MCRF) is given by

$$T^{\mu\nu} = \left(\rho + p/c^2 \right) U^\mu U^\nu - p\eta^{\mu\nu}. \tag{7.5.7}$$

Note that if (as is the case in many 'ordinary' relativistic circumstances not involving exotic, superdense objects such as neutron stars) pressure is very small compared to density ($p \ll \rho$), then we can assume that $p \approx 0$ and (7.5.7) reduces to the energy-momentum tensor for dust (7.5.5) $T^{\mu\nu} = \rho U^\mu U^\nu$.

It's instructive to use (7.5.7) to get back to the particular MCRF case of (7.5.6). From (3.5.2) we know the Minkowski metric is

$$[\eta_{\mu\nu}] = \begin{pmatrix} 1 & 0 & 0 & 0 \\ 0 & -1 & 0 & 0 \\ 0 & 0 & -1 & 0 \\ 0 & 0 & 0 & -1 \end{pmatrix},$$

and from (3.6.2) we know the components of the four-velocity U^μ are given by

$$U^\mu = \left(U^0, U^1, U^2, U^3 \right) = (c\gamma, \gamma\mathbf{v}) = \gamma \left(c, \mathbf{v} \right).$$

So $U^0 = c\gamma$, but in the MCRF $\gamma = 1$, and therefore $U^0 = c$, $U^0 U^0 = c^2$, and thus

$$T^{00} = \rho c^2 + \frac{pc^2}{c^2} - p = \rho c^2.$$

Also, in the MCRF the spatial components of the four-velocity U^i (where $i = 1, 2, 3$) will $= 0$. And also $\eta^{0i} = 0$ and $\eta^{i0} = 0$, therefore

$$T^{0i} = T^{i0} = 0.$$

But $\eta^{ii} = -1$, therefore

$$T^{ii} = p.$$

Finally, when $i \neq j$, in the MCRF again $U^i = 0$, and $\eta^{ij} = 0$, therefore

$$T^{ij} = 0.$$

Putting all this together we can reaffirm that $T^{\mu\nu}$ in matrix form is indeed

$$[T^{\mu\nu}] = \begin{pmatrix} \rho c^2 & 0 & 0 & 0 \\ 0 & p & 0 & 0 \\ 0 & 0 & p & 0 \\ 0 & 0 & 0 & p \end{pmatrix}.$$

The principle of general covariance allows us to replace the Minkowski metric and rewrite (7.5.7)

$$T^{\mu\nu} = \left(\rho + p/c^2 \right) U^\mu U^\nu - p\eta^{\mu\nu}$$

so that it applies at any point in curved spacetime, where the metric is $g^{\mu\nu}$, giving

$$T^{\mu\nu} = \left(\rho + p/c^2 \right) U^\mu U^\nu - pg^{\mu\nu}. \tag{7.5.8}$$

7.5.3 Covariant divergence of $T^{\mu\nu}$

We mentioned earlier that, using a dummy index, the covariant derivative of a tensor field is equivalent to the divergence of that field. In an isolated system, energy and momentum should be conserved (ie there are flows but no sources or sinks of energy-momentum), and we would expect the divergence of the energy-momentum tensor to equal zero. Although zero divergence of $T^{\mu\nu}$ is true for both flat and curved spacetime, its meaning differs in the two cases. In fact, as noted at the end of this section, the concept of energy conservation is not an unambiguous one in general relativity.

In the flat spacetime of special relativity, the fundamental laws of conservation of energy and momentum can indeed be expressed by saying

$$\frac{\partial T^{\mu\nu}}{\partial x^\mu} = 0$$

or, using comma notation,

$$T^{\mu\nu}_{,\mu} = 0. \tag{7.5.9}$$

These equations describe conservation of energy when $\nu = 0$, and conservation of the ith component of momentum when

$$\frac{\partial T^{\mu i}}{\partial x^\mu} = 0.$$

In other words, as there are no sources or sinks of energy-momentum, the increase or decrease of total energy within the system equals the rate at which total energy is entering or leaving that system.

In curved spacetime things are not quite so straightforward. Using the 'comma goes to semi-colon' rule (7.5.9) becomes

$$\nabla_\mu T^{\mu\nu} = 0. \tag{7.5.10}$$

Or, using a semi-colon to represent the covariant derivative,

$$T^{\mu\nu}_{;\mu} = 0,$$

which is sometimes known as the **covariant divergence** of $T^{\mu\nu}$. Although the energy-momentum tensor has zero divergence in curved spacetime, this does not imply a true conservation law as it does in special relativity. That is because in curved spacetime there is an additional source of energy that isn't included in the energy-momentum tensor, and that is gravitational energy.

Lambourne [17] says, 'In the presence of gravitation (ie curvature), the conservation of energy is not expected to apply to matter and radiation alone – we also have to take the gravitational energy into account, and that is not included in the energy-momentum tensor.'

Carroll [4] comments that, 'The gravitational field ... does not have an energy-momentum tensor. In fact it is very hard to come up with a sensible local expression for the energy of a gravitational field; a number of suggestions have been made, but they all have their drawbacks. Although there is no "correct" answer, it is an important issue from the point of view of asking seemingly reasonable questions such as, "What is the energy emitted per second from a binary pulsar as the result of gravitational radiation?"'

The authoritative answer to the sixty-four-thousand-dollar question – 'Is energy conserved in general relativity?' – appears to be a definite maybe. For example, Weiss and Baez's [31] response is, 'In special cases, yes. In general – it depends on what you mean by "energy", and what you mean by "conserved".'

Carroll [3] says, 'Einstein tells us that space and time are dynamical, and in particular that they can evolve with time. When the space through which particles move is changing, the total energy of those particles is not conserved.'

Interested readers may wish to consult these sources.

7.6 The Einstein field equations

From 1907 to 1915 Einstein worked to develop his theory of general relativity. It was a long and tortuous struggle, summed up by Earman and Glymour in *Lost in the Tensors: Einstein's Struggles with Covariance Principles 1912-1916* [6]:

> 'The magnificence of Einstein's intellectual odyssey lies not only in the grandeur of its conclusion, but also in its chaos, in the indirectness of the paths that led to home. One cannot read this history without amazement at Einstein's intellect; for much of the period between 1912 and 1916 he was truly lost in the tensors, quite completely on the wrong path, accompanied by erroneous reasons he claimed to be fundamental. And yet, quite singularly, in the course of a month he abandoned his errors and their justifications. The moral, perhaps, is that a certain fickleness is more conducive to theoretical progress than is any abundance of conceptual clarity – at least if one is Einstein.'

So – and we all breathe a sigh of relief – even Einstein had problems with tensors!

The heart of Einstein's problem was finding the correct relationship

$$A = B,$$

where A on the left-hand side of what are known as the **field equations** describes the curvature of spacetime, and B on the right-hand side gives the mass-energy source of that curvature. By 'correct' relationship we mean an expression that:

- is a valid tensor equation;
- doesn't contradict any of the fundamental laws of physics;
- approximates, given the appropriate conditions, to the Newtonian description of gravity as demanded by the principle of consistency.

We saw in Section 7.5 that the energy-momentum tensor $T^{\mu\nu}$ describes, in theory at least, the total energy of a particular physical system, and is therefore an excellent candidate for the source of the gravitational field.

But, we have also seen that the metric tensor $g_{\mu\nu}$ is of fundamental importance in describing the curvature of a Riemannian manifold, including four-dimensional spacetime. So maybe

$$g^{\mu\nu} = \kappa T^{\mu\nu} \qquad (7.6.1)$$

is a solution, with κ (the Greek letter kappa) some unknown constant. The good news is that this is a valid tensor equation, ie the μ and ν indices are upstairs on both sides. Also, we saw from (6.3.12) that

$$\nabla_\mu g^{\mu\nu} = 0,$$

and from (7.5.10) that

$$\nabla_\mu T^{\mu\nu} = 0.$$

Taking the covariant derivative of both sides of (7.6.1) gives zero on both sides, which is encouraging. Unfortunately, (7.6.1) does not reduce to Poisson's equation, the defining field equation of Newtonian gravitation, so we must reject it.

Although the Riemann curvature tensor $R^l{}_{ijk}$ (6.6.1) contains all the information about the curvature of a manifold we cannot use it uncontracted in the field equations as it is of a higher rank than $T^{\mu\nu}$, and therefore wouldn't give a valid tensor equation. However, in Section 6.7.2 we referred to the Ricci tensor, an object formed from the Riemann curvature tensor. In 1915 Einstein had high hopes of the Ricci tensor and suggested

$$R^{\mu\nu} = \kappa T^{\mu\nu} \qquad (7,6.2)$$

as a candidate for the field equations. Using (7.6.2) he was able to resolve a long standing problem concerning a slight variation in Mercury's orbit that had puzzled astronomers for many years, causing him to write in November of that year that, 'For a few days I was beside myself with joyous excitement.' He also corrected a previous tiny error he'd made when calculating the deflection of light as it grazed the sun – a prediction famously confirmed by Eddington in 1919, using photographs of a total solar eclipse taken on the island of Príncipe off the west coast of Africa.

However, there was a problem with (7.6.2) in that

$$\nabla_\mu R^{\mu\nu} \neq 0,$$

which it has to because of (7.5.10).

Eventually, Einstein had to reject (7.6.2) (his Mercury and light deflection triumphs were unaffected), and later in November 1915 he published his final field equations

$$R^{\mu\nu} - \frac{1}{2}Rg^{\mu\nu} = -\kappa T^{\mu\nu} \qquad (7.6.3)$$

or, in covariant form,

$$R_{\mu\nu} - \frac{1}{2}Rg_{\mu\nu} = -\kappa T_{\mu\nu}, \qquad (7.6.4)$$

where $\kappa = 8\pi G/c^4$ is the **Einstein constant**. By defining the **Einstein tensor** $G_{\mu\nu}$ (incorporating both the Ricci tensor (6.7.1) and the Ricci scalar (6.7.3)) as

$$G_{\mu\nu} = R_{\mu\nu} - \frac{1}{2}Rg_{\mu\nu}. \qquad (7.6.5)$$

(7.6.4) can be written more succinctly as

$$G_{\mu\nu} = -\kappa T_{\mu\nu}.$$

You may also see the field equations written in terms of what are known as **geometrised units**, where $G = c = 1$. Equation (7.6.4) then becomes

$$G_{\mu\nu} = -8\pi T_{\mu\nu}.$$

We'll see later (the end of Section 8.5, when looking at the Newtonian limit of Einstein's field equations) why κ has to equal $8\pi G/c^4$ in order to be consistent with Newtonian gravitation.

Sometimes, the geometrised units $8\pi G = c = 1$ are used, giving the ultra-succinct

$$G_{\mu\nu} = -T_{\mu\nu},$$

keeping most of its charms discreetly hidden.

Different authors use different sign (+ or -) conventions for the field equations. Foster and Nightingale [9], for example, give the form

$$R^{\mu\nu} - \frac{1}{2}Rg^{\mu\nu} = \kappa T^{\mu\nu}.$$

It all depends on the signs chosen to define the metric tensor, the Riemann curvature tensor and the Ricci tensor.

Problem 7.1. Show that (7.6.4) can also be written in the form

$$R_{\mu\nu} = -\kappa\left(T_{\mu\nu} - \frac{1}{2}g_{\mu\nu}T\right). \qquad (7.6.6)$$

First, multiply both sides of (7.6.4) by $g^{\mu\nu}$ and sum over both indices

$$g^{\mu\nu}\left(R_{\mu\nu} - \frac{1}{2}Rg_{\mu\nu} = -\kappa T_{\mu\nu}\right)$$

$$R - \frac{1}{2}Rg_{\mu\nu}g^{\mu\nu} = -\kappa T.$$

Using the Kronecker delta (with dummy indices), we can then sum over all possible values of ν

$$g^{\mu\nu}g_{\mu\nu} = \delta^\nu_\nu = \delta^0_0 + \delta^1_1 + \delta^2_2 + \delta^3_3 = 1 + 1 + 1 + 1 = 4.$$

Therefore

$$R - \frac{1}{2}R \times 4 = -\kappa T$$

$$R - 2R = -\kappa T$$

$$R = \kappa T,$$

which we can substitute into (7.6.4) to give

$$R_{\mu\nu} - \frac{1}{2}\kappa T g_{\mu\nu} = -\kappa T_{\mu\nu}$$

$$R_{\mu\nu} = -\kappa T_{\mu\nu} + \frac{1}{2}\kappa T g_{\mu\nu}$$

$$R_{\mu\nu} = -\kappa \left(T_{\mu\nu} - \frac{1}{2}g_{\mu\nu}T \right).$$

For the remainder of this book, we'll be exploring the meaning and consequences of the Einstein field equations.

It's worth at this point mentioning a plain English, single sentence summary of these equations, in terms of the motion of freely falling test particles, given by Baez and Bunn [1]. Although this simple formulation is of limited practical use in exploring the full implications of general relativity, the authors do use it to, 'Derive a few of its consequences concerning tidal forces, gravitational waves, gravitational collapse, and the Big Bang cosmology.'

First, recall that the energy-momentum tensor $T^{\mu\nu}$ is defined as the rate of flow of the μth component of four-momentum across a surface of constant ν. Next, we assume that we are working with a perfect fluid, which can be used to describe a range of phenomena including, on a sufficiently large scale, the universe. We start with a round ball of test particles that are all initially at rest relative to each other. Those particles are then allowed to fall freely through spacetime. If $V(t)$ is the volume of the ball after a proper time t has passed, then it can be shown that

$$\left[\frac{1}{V}\frac{d^2V}{dt^2} \right]_{t=0} = -\frac{1}{2} \left(\begin{array}{l} \text{flow of } t\text{momentum in } t\text{direction } + \\ \text{flow of } x\text{momentum in } x\text{direction } + \\ \text{flow of } y\text{momentum in } y\text{direction } + \\ \text{flow of } z\text{momentum in } z\text{direction} \end{array} \right),$$

where these flows are measured at the centre of the ball at $t = 0$ using local inertial coordinates. We met these flows in the previous section when discussing the energy-momentum tensor. In this context, in plain English, Baez and Bunn summarise the field equations as follows:

'Given a small ball of freely falling test particles initially at rest with respect to each other, the rate at which it begins to shrink is proportional to its volume times: the energy density at the center of the ball, plus the pressure in the x direction at that point, plus the pressure in the y direction, plus the pressure in the z direction.'

An even simpler formulation applies if the pressure of the perfect fluid is the same in all directions. As we'll see when we come to look at relativistic cosmology, this is a working assumption that is thought to apply, on a sufficiently large scale, to the universe. The field equations can then be summarised as:

> 'Given a small ball of freely falling test particles initially at rest with respect to each other, the rate at which it begins to shrink is proportional to its volume times: the energy density at the center of the ball plus three times the pressure at that point.'

Now the energy-momentum tensor is that of a perfect fluid, which in the MCRF is given by (7.5.6)

$$[T^{\mu\nu}] = \begin{pmatrix} \rho c^2 & 0 & 0 & 0 \\ 0 & p & 0 & 0 \\ 0 & 0 & p & 0 \\ 0 & 0 & 0 & p \end{pmatrix}.$$

The authors use an even simpler version of this formulation (with pressure equal to zero) to derive Newton's law of universal gravitation.

7.7 The cosmological constant

In 1917 Einstein proposed a modification to his field equations in the form of an additional term $\Lambda g_{\mu\nu}$, where Λ (the Greek letter Lambda) is called the **cosmological constant**. His reason for doing this was his conviction, at that time, that the universe is static. The obvious problem with assuming a static universe is that gravity, being a force of attraction, would eventually act to make the universe collapse onto itself. In order to 'fix' the field equations, Einstein needed a mechanism to prevent this collapse, hence the cosmological constant, which describes a sort of 'anti-gravity' repulsive force or negative pressure. The modified field equations with the cosmological constant are

$$R_{\mu\nu} - \frac{1}{2}Rg_{\mu\nu} + \Lambda g_{\mu\nu} = -\kappa T_{\mu\nu}. \tag{7.7.1}$$

Shortly after Einstein's modification it was found that 'static universe' solutions to the field equations were unstable. Then came the momentous discovery in 1929 by the American astronomer Edwin Hubble (1889–1953) that the universe is expanding. Furthermore, it was found that this expansion is consistent with cosmological solutions derived from the original, unmodified field equations. This led Einstein to remark that the cosmological constant was 'his greatest blunder'.

In 1998 observations from the Hubble Space Telescope showed that the universe is not only expanding, but expanding at an accelerating rate. One way to explain this expansion is in terms of a hypothetical **dark energy** that permeates all space and which may possibly be described by the cosmological constant. Ironically, Einstein's 'greatest blunder' has therefore now regained a degree of scientific credibility. Dark energy and the cosmological constant may be theoretically necessary to account for the accelerated expansion of the universe, but their physical meaning is to date a complete mystery (a leading current explanation is in terms of another mysterious quantity known as **vacuum energy**).

We later discuss dark energy in greater detail when looking at relativistic cosmology. Both dark energy and the cosmological constant can be ignored when using the field equations to understand phenomena on a sub-cosmic scale.

8 The Newtonian limit

There could be no fairer destiny for any physical theory than that it should point the way to a more comprehensive theory in which it lives on as a limiting case.

ALBERT EINSTEIN

8.1 Introduction

Newtonian mechanics provides an extremely accurate description of many gravitational phenomena. The principle of consistency demands that general relativity, which is a theory of gravity, must make the same predictions as Newtonian gravitation given appropriate non-relativistic conditions. Assuming we are using a test particle of negligible mass, we take these conditions to be that:

- The particle is moving relatively slowly (compared to the speed of light).
- The gravitational field is weak.
- The field does not change with time, ie it is static.

These conditions are known as the **Newtonian limit** of general relativity. We'll now look at what happens when we apply these conditions to the equations of general relativity. Specifically, we'll try to find relativistic approximations to:

1. Newton's three laws of motion (Section 2.4).
2. The relationship between the gravitational field and gravitational potential (2.5.27) $\mathbf{g} = -\nabla\phi$.
3. Newton's law of universal gravitation (2.5.2) $\mathbf{F} = -\frac{Gm_1m_2}{\mathbf{r}^2}\hat{\mathbf{r}}$.
4. Poisson's equation (2.5.28) $\nabla^2\phi = 4\pi G\rho$.

8.2 Newton's three laws of motion

8.2.1 Newton's first law of motion

This states that a particle will remain at rest or in uniform motion in a straight line unless acted upon by an external force. We saw in Section 7.4 that in spacetime freely moving or falling particles move along geodesics described by the geodesic equations (7.4.2)

$$\frac{d^2x^\alpha}{d\tau^2} + \Gamma^\alpha_{\gamma\beta}\frac{dx^\beta}{d\tau}\frac{dx^\gamma}{d\tau} = 0.$$

We can retrieve Newton's first law from the geodesic equations by considering a particle in a Lorentz frame, ie in a Newtonian inertial frame in the absence of gravity. Minkowski space, in other words, where we can use ordinary Cartesian coordinates and the geodesic equations can then be written as

$$\frac{d^2x^i}{d\tau^2} + \Gamma^i_{jk}\frac{dx^j}{d\tau}\frac{dx^k}{d\tau} = 0,$$

where i, j, k are the three spatial coordinates x, y, z. The connection coefficients Γ^i_{jk} will then equal zero, and the geodesic equations become

$$\frac{d^2x^i}{d\tau^2} = 0.$$

For non-relativistic speeds proper time τ approximates to coordinate time t, and we can rewrite this equation as

$$\frac{d^2x^i}{dt^2} = 0,$$

which is just another way of saying the acceleration of the particle equals zero, ie the particle is either stationary or moving with constant velocity, which is equivalent to Newton's first law.

8.2.2 Newton's second law of motion

This states that a net force acting on a particle causes an acceleration of that particle described by (2.4.1)

$$\mathbf{F} = m\mathbf{a}.$$

Or, equivalently, the force on the particle equals the rate of change of its momentum $\mathbf{p}$

$$\mathbf{F} = \frac{d\mathbf{p}}{dt} = \frac{d(m\mathbf{v})}{dt} = m\frac{d\mathbf{v}}{dt}.$$

In special relativity (Section 3.6.8) we defined the four-force (3.6.17) as the rate of change of four-momentum

$$F^\mu = \frac{dP^\mu}{d\tau},$$

using (3.5.5) for proper τ time in special relativity

$$c^2d\tau^2 = \eta_{\mu\nu}dx^\mu dx^\nu.$$

The principle of general covariance allows us to replace the Minkowski metric $\eta_{\mu\nu}$ with the general metric tensor $g_{\mu\nu}$, giving proper time in curved spacetime (7.3.1)

$$c^2d\tau^2 = g_{\mu\nu}dx^\mu dx^\nu,$$

which we can substitute into the special relativistic equation (3.6.17) for four-force to obtain

$$F^\mu = \frac{DP^\mu}{d\tau}, \tag{8.2.1}$$

the upper case D signifying the absolute derivative. Equation (8.2.1) is the general relativity version of Newton's second law (2.4.1) and is true for any coordinate system.

8.2.3 Newton's third law of motion

Foster and Nightingale [9] state that this law – for every action there is an equal and opposite reaction -

> 'Is true in general relativity also. However, we must be careful, because Newton's gravitational force is now replaced by Einstein's idea that a massive body causes curvature of the spacetime around it, and a free particle responds by moving along a geodesic in that spacetime. It should be noted that this viewpoint ignores any curvature produced by the particle following the geodesic. That is, the particle is a *test particle*, and there is no question of its having any affect on the body producing the gravitational field.
>
> The gravitational interaction of two large bodies is not directly addressed by Einstein's theory, although it is of importance in astronomy, as for example in the famous pair of orbiting neutron stars PSR 1913+16. Approximation methods for such cases ... are beyond the scope of our book.'

If it's beyond the scope of their book, we can happily assume that it's beyond the scope of this one.

8.3 Newton's field equation for how matter responds to gravity

Equation (2.5.27) $\mathbf{g} = -\nabla\phi$ describes the relationship between the Newtonian gravitational field $\mathbf{g}$ and the Newtonian gravitational potential ϕ. Because $\mathbf{g}$ equals acceleration, the equation $\mathbf{g} = -\nabla\phi$ is therefore the Newtonian equation of motion for a particle moving in a gravitational field of potential ϕ.

We'll now try to derive an approximation of the Newtonian gravitational equation $\mathbf{g} = -\nabla\phi$ from the mathematics of general relativity.

Remember that in the Newtonian limit we make three assumptions:

1. The particle is moving relatively slowly (compared to the speed of light).

2. The gravitational field is weak.

3. The field does not change with time, ie it is static.

We've seen that proper time is measured by a clock travelling on a time-like world-line, so we can take the proper time τ as the parameter of the world-line. The first assumption, that the particle is moving slowly, implies that the time-component (ie the $0th$ component of the particle's four-velocity) dominates the other (spatial) components, ie

$$\frac{dx^i}{d\tau} \ll \frac{dt}{d\tau},$$

and therefore the geodesic equations (6.5.2)

$$\frac{d^2x^\mu}{d\lambda^2} + \Gamma^\mu_{jk}\frac{dx^j}{d\lambda}\frac{dx^k}{d\lambda} = 0$$

become

$$\frac{d^2x^\mu}{d\tau^2} + \Gamma^\mu_{00}\left(\frac{dt}{d\tau}\right)^2 = 0. \qquad (8.3.1)$$

To calculate the connection coefficients Γ^μ_{00} from the metric we use (a form of – we've changed the indices) (6.2.6)

$$\Gamma^\mu_{\alpha\nu} = \frac{1}{2}g^{\mu\lambda}\left(\frac{\partial g_{\lambda\nu}}{\partial x^\alpha} + \frac{\partial g_{\alpha\lambda}}{\partial x^\nu} - \frac{\partial g_{\alpha\nu}}{\partial x^\lambda}\right)$$

to give

$$\Gamma^\mu_{00} = \frac{1}{2}g^{\mu\lambda}\left(\frac{\partial g_{\lambda 0}}{\partial x^0} + \frac{\partial g_{0\lambda}}{\partial x^0} - \frac{\partial g_{00}}{\partial x^\lambda}\right).$$

However, we can simplify this because of our third assumption, that the gravitational field is static, meaning we can ignore the $\frac{\partial g_{\lambda 0}}{\partial x^0}$ and $\frac{\partial g_{0\lambda}}{\partial x^0}$ terms (as these are with respect to time). This gives

$$\Gamma^\mu_{00} = -\frac{1}{2}g^{\mu\lambda}\left(\frac{\partial g_{00}}{\partial x^\lambda}\right). \tag{8.3.2}$$

Our second assumption, that the gravitational field is weak, allows us to write

$$g_{\mu\nu} = \eta_{\mu\nu} + h_{\mu\nu}, \tag{8.3.3}$$

where we assume that spacetime is only slightly changed compared to the zero gravity Minkowski space of special relativity. The metric we are looking for $g_{\mu\nu}$ therefore equals the Minkowski metric $\eta_{\mu\nu}$ plus a small perturbation (an extra little bit of metric) $h_{\mu\nu}$, which is due to the weak gravitational field. This equation is the jump-off to what is known as the **linearized theory of gravity** – an extremely useful approximation used for many practical calculations in general relativity. The components of $h_{\mu\nu}$ are small compared to $\eta_{\mu\nu}$, meaning $|h_{\mu\nu}| \ll 1$.

We next need to find an approximation for $g^{\mu\lambda}$ in (8.3.2).

Recall that multiplying a metric by its inverse gives the identity matrix I, which we can write as

$$g^{\mu\nu}g_{\nu\alpha} = \delta^\mu_\alpha = I.$$

Also, if we ignore the A^2 term (which we can do if $|A| \ll 1$), then $(I - A)(I + A) \approx I$.

Therefore, if we let $(I + A) = (\eta_{\mu\nu} + h_{\mu\nu})$, we can say

$$(\eta^{\mu\nu} - h^{\mu\nu})(\eta_{\mu\nu} + h_{\mu\nu}) = I = g^{\mu\nu}g_{\nu\alpha},$$

meaning

$$g^{\mu\nu} = \eta^{\mu\nu} - h^{\mu\nu},$$

where (using the metric to juggle the indices) $h^{\mu\nu} = \eta^{\mu\rho}\eta^{\nu\sigma}h_{\rho\sigma}$.

Also, as $g_{\mu\nu} = \eta_{\mu\nu} + h_{\mu\nu}$ then $g_{00} = 1 + h_{00}$ and $\frac{\partial g_{00}}{\partial x^\lambda} = \frac{\partial h_{00}}{\partial x^\lambda}$ (because $\frac{\partial(\eta_{00}=1)}{\partial x^\lambda} = 0$).

We can now rewrite (8.3.2) as

$$\Gamma^\mu_{00} = -\frac{1}{2}\eta^{\mu\lambda}\left(\frac{\partial h_{00}}{\partial x^\lambda}\right), \tag{8.3.4}$$

and the geodesic equations (8.3.1) become

$$\frac{d^2 x^\mu}{d\tau^2} = \frac{1}{2}\eta^{\mu\lambda}\left(\frac{\partial h_{00}}{\partial x^\lambda}\right)\left(\frac{dt}{d\tau}\right)^2. \tag{8.3.5}$$

As the gravitational field is static we can assume that $\frac{\partial h_{00}}{\partial x^0} = 0$ and (8.3.5) becomes (where $x^\mu = x^0 = t$)

$$\frac{d^2 t}{d\tau^2} = 0,$$

meaning $\frac{dt}{d\tau}$ is a constant.

Next we look at the spatial ($\mu \neq 0$) components of (8.3.5), which are given by

$$\frac{d^2 x^i}{d\tau^2} = -\frac{1}{2}\left(\frac{\partial h_{00}}{\partial x^i}\right)\left(\frac{cdt}{d\tau}\right)^2. \tag{8.3.6}$$

We've introduced a minus sign because the spatial components of the Minkowski metric $\eta_{\mu\nu}$ (3.5.2) are negative. We write cdt instead of dt because we are using ct units of time (as a constant, c is written to the left of dt). We now need to change the derivative on the left-hand side from τ to t. To do this we need to play around with $\frac{d^2 x^i}{d\tau^2}$ a little. By definition

$$\frac{d^2 x^i}{d\tau^2} = \frac{d}{d\tau}\frac{dx^i}{d\tau}$$

$$= \frac{d}{d\tau}\left(\frac{dt}{d\tau}\frac{dx^i}{dt}\right).$$

We then use the product rule:

$$= \frac{dt}{d\tau}\left(\frac{d}{d\tau}\frac{dx^i}{dt}\right) + \frac{dx^i}{dt}\left(\frac{d}{d\tau}\frac{dt}{d\tau}\right)$$

$$= \frac{dt}{d\tau}\left(\frac{dt}{d\tau}\frac{d}{dt}\frac{dx^i}{dt}\right) + \frac{dx^i}{dt}\left(\frac{d}{d\tau}\frac{dt}{d\tau}\right)$$

$$= \left(\frac{dt}{d\tau}\right)^2\left(\frac{d^2 x^i}{dt^2}\right) + \frac{dx^i}{dt}\left(\frac{d^2 t}{d\tau^2}\right).$$

But we already know that $\frac{d^2 t}{d\tau^2} = 0$, meaning the right-hand term vanishes and

$$\frac{d^2 x^i}{d\tau^2} = \left(\frac{dt}{d\tau}\right)^2\left(\frac{d^2 x^i}{dt^2}\right).$$

Therefore, if we multiply both sides of (8.3.6) by $\left(\frac{d\tau}{cdt}\right)^2$ we change the left-hand term from $\frac{d^2 x^i}{d\tau^2}$ to $\frac{d^2 x^i}{c^2 dt^2}$ giving

$$\frac{d^2 x^i}{c^2 dt^2} = -\frac{1}{2}\left(\frac{\partial h_{00}}{\partial x^i}\right). \tag{8.3.7}$$

But if we assume that

$$h_{00} = \frac{2\phi}{c^2} \tag{8.3.8}$$

(which also implies $g_{00} = \left(1 + \frac{2\phi}{c^2}\right)$) then (8.3.7) becomes (the c^2s cancel)

$$\frac{d^2 x^i}{dt^2} = -\left(\frac{\partial \phi}{\partial x^i}\right), \tag{8.3.9}$$

which is just another way of writing the Newtonian gravitational Equation (2.5.27) $\mathbf{g} = -\nabla\phi$.

8.4 Newton's law of universal gravitation

In order to derive Newton's law of universal gravitation from general relativity we need to jump ahead of ourselves a little and make use of the Schwarzschild metric, one of the most useful solutions to Einstein's field equations and something we'll be looking at in greater detail in the next chapter.

For now, we note that this metric can be used to describe the gravitational field of a slowly rotating massive body (of mass M) such as the Earth or Sun. The metric can be written as a four-dimensional line element

$$ds^2 = \left(1 - \frac{2GM}{c^2r}\right)c^2dt^2 - \frac{(dr)^2}{1 - \frac{2GM}{c^2r}} - r^2d\theta^2 - r^2sin^2\theta d\phi^2. \tag{8.4.1}$$

By looking at this equation we can see that as the term $\frac{2GM}{c^2r}$ decreases (either by decreasing M or increasing r) the metric approaches the line element of flat spacetime in spherical coordinates (3.5.6)

$$ds^2 = c^2dt^2 - dr^2 - r^2d\theta^2 - r^2\sin^2\theta d\phi^2.$$

We are thus here dealing with the 'almost flat' spacetime of special relativity and can use a 'nearly Cartesian' coordinate system.

So for very small values of $\frac{2GM}{c^2r}$, we can do as we did in Section 8.3 and assume that for a weak gravitational field we can use a metric of the form

$$g_{\mu\nu} = \eta_{\mu\nu} + h_{\mu\nu},$$

where $h_{\mu\nu}$ is a little additional tweak to the Minkowski metric.

The g_{00} term of our 'almost flat' metric $g_{\mu\nu} = \eta_{\mu\nu} + h_{\mu\nu}$ then becomes (from 8.4.1)

$$g_{00} = \left(1 - \frac{2GM}{c^2r}\right),$$

giving

$$h_{00} = -\frac{2GM}{c^2r}. \tag{8.4.2}$$

We know (2.5.21) that Newtonian gravitational potential ϕ at a point in a gravitational field is given by $\phi = \frac{-GM}{r}$. We can therefore rewrite (8.4.2) as

$$h_{00} = \frac{2\phi}{c^2}, \tag{8.4.3}$$

which we met (8.3.8) in the previous Section.

(8.3.9) tells us that

$$\frac{d^2x^i}{dt^2} = -\left(\frac{\partial\phi}{\partial x^i}\right), \tag{8.4.4}$$

where $\frac{d^2x^i}{dt^2}$ is acceleration, and $\left(\frac{\partial\phi}{\partial x^i}\right) = \left(\frac{\partial\phi}{\partial x}, \frac{\partial\phi}{\partial y}, \frac{\partial\phi}{\partial z}\right) = \frac{\partial\phi}{\partial \mathbf{r}}$, where $\mathbf{r}$ is the particle's radius vector in our Cartesian x, y, z coordinates. Substituting $\mathbf{r}$ and using the vector form of $\phi = \frac{-GM}{r} = \frac{-GM}{\mathbf{r}}$ we can rewrite (8.4.4) as

$$\frac{d^2\mathbf{r}}{dt^2} = -\left(\frac{\partial\left(\frac{-GM}{\mathbf{r}}\right)}{\partial\mathbf{r}}\right) = GM\frac{\partial\left(\mathbf{r}^{-1}\right)}{\partial\mathbf{r}} = -\frac{GM}{\mathbf{r}^2}\hat{\mathbf{r}}, \tag{8.4.5}$$

where $\hat{\mathbf{r}}$ is a unit vector in the direction of $\mathbf{r}$. We know from Newton's second law (2.4.1) that $\mathbf{F} = m\mathbf{a}$ and can therefore multiply both sides of (8.4.5) by the particle's mass m to give

$$\frac{md^2\mathbf{r}}{dt^2} = -\frac{GMm}{\mathbf{r}^2}\hat{\mathbf{r}},$$

which we can now identify as the vector form (2.5.2) of Newton's law of universal gravitation

$$\mathbf{F} = -\frac{Gm_1m_2}{\mathbf{r}^2}\hat{\mathbf{r}}.$$

8.5 Poisson's equation

In Section 2.5.7.3 we met Poisson's equation (2.5.28)

$$\nabla^2\phi = 4\pi G\rho,$$

which describes how mass produces the Newtonian gravitational field.

We are now going to try to find a way of deriving an approximation of Poisson's equation from the Einstein field equations.

As in Section 8.3 we'll assume that for a weak gravitational field

$$g_{\mu\nu} = \eta_{\mu\nu} + h_{\mu\nu},$$

where $h_{\mu\nu}$ is a small perturbation (an extra little bit of metric) and the components of $h_{\mu\nu}$ are small compared to $\eta_{\mu\nu}$, ie $|h_{\mu\nu}| \ll 1$.

We'll also assume that a weak gravitational field approximates the energy momentum tensor for dust (7.5.5) so $T^{\mu\nu} = \rho U^\mu U^\nu$ and $T = \rho c^2$.

(Note: $T^{\mu\nu}$ is the overall equation for the energy-momentum tensor, T is the actual value of it.)

Substituting $T^{\mu\nu}$ and T into the Einstein field equations (7.6.6)

$$R_{\mu\nu} = -\kappa\left(T_{\mu\nu} - \frac{1}{2}g_{\mu\nu}T\right)$$

we get

$$R_{\mu\nu} = -\kappa\left(\rho U^\mu U^\nu - \frac{1}{2}g_{\mu\nu}\rho c^2\right).$$

Substituting into this equation the weak gravitational field metric $g_{\mu\nu} = \eta_{\mu\nu} + h_{\mu\nu}$ gives

$$R_{\mu\nu} = -\kappa\left(\rho U^\mu U^\nu - \frac{1}{2}\left(\eta_{\mu\nu} + h_{\mu\nu}\right)\rho c^2\right). \tag{8.5.1}$$

As the only non-zero component of $T^{\mu\nu}$ is $T^{00} = \rho c^2$, it seems reasonable to focus on the $\mu = \nu = 0$ components of $R_{\mu\nu}$, ie the R_{00} term.

Remember that when we looked at the definition of four-velocity in Section 3.6.3 we saw from (3.6.2) that

$$U^\mu = \left(U^0, U^1, U^2, U^3\right) = \frac{dx^\mu}{d\tau} = (c\gamma, \gamma\mathbf{v}) = \gamma\left(c, \mathbf{v}\right).$$

In the Newtonian limit we are assuming that speeds are low compared to the speed of light, and therefore $\gamma \approx 1$, $U^0 \approx c$ and $U^0 U^0 = c^2$.

Also, as $\eta_{00} = 1$ and $|h_{\mu\nu}| \ll 1$ we can say $(\eta_{00} + h_{\mu\nu}) \approx 1$.

We can therefore write (8.5.1) as

$$R_{00} = -\kappa \left(\rho c^2 - \frac{1}{2}\rho c^2 \right) = -\kappa \frac{1}{2}\rho c^2. \tag{8.5.2}$$

Now we need to refer back to Section 6.7 where we saw that the Ricci tensor (6.7.2) is

$$R_{\mu\nu} = \frac{\partial \Gamma^\alpha_{\mu\alpha}}{\partial x^\nu} - \frac{\partial \Gamma^\alpha_{\mu\nu}}{\partial x^\alpha} + \Gamma^\beta_{\mu\alpha}\Gamma^\alpha_{\beta\nu} - \Gamma^\beta_{\mu\nu}\Gamma^\alpha_{\beta\alpha}.$$

For the R_{00} term this becomes

$$R_{00} = \frac{\partial \Gamma^\alpha_{0\alpha}}{\partial x^0} - \frac{\partial \Gamma^\alpha_{00}}{\partial x^\alpha} + \Gamma^\beta_{0\alpha}\Gamma^\alpha_{\beta 0} - \Gamma^\beta_{00}\Gamma^\alpha_{\beta\alpha}. \tag{8.5.3}$$

A weak gravitational field assumes we are in 'almost' Minkowski space and that our coordinate system is 'nearly Cartesian'. Therefore, the connection coefficients $\Gamma^\gamma_{\delta\rho}$ are small and we can ignore the last two terms in (8.5.3), which then becomes

$$R_{00} = \frac{\partial \Gamma^\alpha_{0\alpha}}{\partial x^0} - \frac{\partial \Gamma^\alpha_{00}}{\partial x^\alpha}. \tag{8.5.4}$$

Because our field is assumed to be static we can say $\frac{\partial \Gamma^\alpha_{0\alpha}}{\partial x^0} \approx 0$ and (8.5.4) becomes

$$R_{00} = -\frac{\partial \Gamma^\alpha_{00}}{\partial x^\alpha}. \tag{8.5.5}$$

But, we've already seen from (8.3.4) in Section 8.3 that

$$\Gamma^\mu_{00} = -\frac{1}{2}\eta^{\mu\lambda}\left(\frac{\partial h_{00}}{\partial x^\lambda} \right),$$

and so (changing the indices slightly) we can rewrite

$$\Gamma^\alpha_{00} = -\frac{1}{2}\eta^{\alpha\lambda}\left(\frac{\partial h_{00}}{\partial x^\lambda} \right). \tag{8.5.6}$$

Substituting (8.5.6) into (8.5.5) gives

$$R_{00} = -\frac{\partial \left(-\frac{1}{2}\eta^{\alpha\lambda}\left(\frac{\partial h_{00}}{\partial x^\lambda} \right) \right)}{\partial x^\alpha}$$

$$R_{00} = \frac{1}{2}\eta^{\alpha\lambda}\frac{\partial^2 h_{00}}{\partial x^\alpha \partial x^\lambda}. \tag{8.5.7}$$

But, from the definition (1.13.9) of the Laplacian operator ∇^2 we can rewrite the right-hand side as

$$\eta^{\alpha\lambda}\frac{\partial^2 h_{00}}{\partial x^\alpha \partial x^\lambda} = -\nabla^2 h_{00},$$

giving

$$R_{00} = -\frac{1}{2}\nabla^2 h_{00}. \tag{8.5.8}$$

The reason for the minus sign is because we are only using the spatial components of $\eta^{\alpha\lambda}$ and as

$$\eta^{\alpha\lambda} = \begin{pmatrix} 1 & 0 & 0 & 0 \\ 0 & -1 & 0 & 0 \\ 0 & 0 & -1 & 0 \\ 0 & 0 & 0 & -1 \end{pmatrix},$$

these all $= -1$.

We now have two expressions for R_{00}: (8.5.2) and (8.5.8) and can therefore write

$$-\kappa\frac{1}{2}\rho c^2 = -\frac{1}{2}\nabla^2 h_{00}$$

$$\kappa\frac{1}{2}\rho c^2 = \frac{1}{2}\nabla^2 h_{00}. \tag{8.5.9}$$

But we found in Section 8.3 that $h_{00} = \frac{2\phi}{c^2}$, meaning (8.5.9) becomes

$$\kappa\rho c^2 = \nabla^2\left(\frac{2\phi}{c^2}\right)$$

$$\nabla^2\phi = \frac{1}{2}\kappa\rho c^4. \tag{8.5.10}$$

And, if we let the Einstein constant $\kappa = 8\pi G/c^4$, (8.5.10) approximates to Poisson's equation

$$\nabla^2\phi = 4\pi G\rho.$$

9 The Schwarzschild metric

It is always pleasant to have exact solutions in simple form at your disposal.

KARL SCHWARZSCHILD

9.1 Introduction

Figure 9.1: Karl Schwarzschild (1873–1916).

Minkowski space is known – somewhat disconcertingly to the student who has exerted much effort trying to understand special relativity – as the 'trivial' solution to the Einstein field equations. Triviality may be in the eye of the beholder, but if both the energy momentum tensor and the Riemann curvature tensor equal zero, the result must be the flat spacetime of special relativity.

Ignoring the flat option, the Einstein field equations – effectively a set of non-linear differential equations – are notoriously difficult to solve exactly. Einstein himself used approximation methods when working out the predictions of general relativity, such as accounting for the small discrepancy in

Mercury's orbit. We used an approximation in the previous chapter when looking at the Newtonian limit (weak, static gravitational fields and light, slow-moving particles) and assumed the metric (8.3.3) to be $g_{\mu\nu} = \eta_{\mu\nu} + h_{\mu\nu}$.

There's nothing wrong of course with approximate solutions as long as they produce reasonably accurate results. But in 1916, shortly after he had proposed his general theory of relativity, Einstein was pleasantly surprised when a German astrophysicist Karl Schwarzschild (Figure 9.1) published an exact solution to the field equations. Schwarzschild posted his results to Einstein, who wrote back in January of that year:

> 'I have read your paper with the utmost interest. I had not expected that one could formulate the exact solution of the problem in such a simple way. I liked very much your mathematical treatment of the subject. Next Thursday I shall present the work to the Academy with a few words of explanation.'

Sadly, Karl Schwarzschild never had much time to savour his triumph. He died in May 1916 as a result of a disease he contracted while serving in the German army during World War I.

The **Schwarzschild metric** describes a static, spherically symmetric gravitational field in the empty region of spacetime near a massive spherical object. Strictly speaking, the solution only applies to non-rotating spherical masses. However, the Schwarzschild metric also provides a good approximation to the gravitational field of slowly rotating bodies such as the Sun or Earth. It can also describe the simplest type of black hole.

Lambourne [17] calls the Schwarzschild solution, 'The first and arguably the most important non-trivial solution of the Einstein field equations.'

9.2 Forms of the Schwarzschild metric

It's probably a good idea at this point to simply state the Schwarzschild metric so at least we know what it looks like. Or rather, what 'they' look like, as the same metric can be expressed in several different forms. Here is the full version in line element form:

$$ds^2 = \left(1 - \frac{2GM}{c^2 r}\right) c^2 dt^2 - \frac{dr^2}{1 - \frac{2GM}{c^2 r}} - r^2 d\theta^2 - r^2 \sin^2\theta d\phi^2. \tag{9.2.1}$$

Where M is the mass of our spherically symmetric gravitational source, G is the gravitational constant, and t, r, θ, ϕ are known as **Schwarzschild coordinates**. Keep an open mind about these coordinates for the time being as they are not quite as straightforward as they appear.

Some textbooks simplify (9.2.1) by using the substitution $m = GM/c^2$ giving

$$ds^2 = \left(1 - \frac{2m}{r}\right) c^2 dt^2 - \frac{dr^2}{1 - \frac{2m}{r}} - r^2 d\theta^2 - r^2 \sin^2\theta d\phi^2. \tag{9.2.2}$$

Finally, the metric may be expressed in terms of a quantity known as the Schwarzschild radius ($R_S = \frac{2GM}{c^2}$ – see Section 9.4.1) as

$$ds^2 = \left(1 - \frac{R_S}{r}\right) c^2 dt^2 - \frac{dr^2}{1 - \frac{R_S}{r}} - r^2 d\theta^2 - r^2 \sin^2\theta d\phi^2.$$

To add to the confusion, don't forget that some authorities assume that $c = 1$, so don't even show the blessed thing in their equations!

Finally, to give one example of the metric in matrix form, we can write (9.2.1) as

$$[g_{\mu\nu}] = \begin{pmatrix} 1 - \frac{2GM}{c^2 r} & 0 & 0 & 0 \\ 0 & -\frac{1}{1 - \frac{2GM}{c^2 r}} & 0 & 0 \\ 0 & 0 & -r^2 & 0 \\ 0 & 0 & 0 & -r^2 \sin^2 \theta \end{pmatrix}. \tag{9.2.3}$$

Now, let's list the assumptions that guided Schwarzschild in his search for this metric. They are:

- The metric describes the geometry of empty spacetime surrounding a spherically symmetric body, such as a star or planet.

- The gravitational field is static – we don't want the complication of a metric that changes with time.

- The gravitational field is spherically symmetric – the field will be the same at equal distances from the source mass.

- The metric is **asymptotically flat**, which simply means that the metric must be constructed in such a way that it ends up describing the flat spacetime of special relativity at a sufficiently large distance from the source mass.

Because we are considering a spherically symmetric body, it makes sense to try to use spherical coordinates t, r, θ, ϕ. As we'll see, using such coordinates is not quite as straightforward as might first appear, but we'll make a start with them for now.

It can be shown that in order for the above four conditions to be met the metric must take the following general form:

$$ds^2 = Uc^2 dt^2 - V dr^2 - W r^2 \left(d\theta^2 + \sin^2 \theta d\phi^2 \right), \tag{9.2.4}$$

where U, V and W are some unknown functions of r. In this metric the coordinate r is the usual radial distance from the origin (the centre of the source mass). If we assume that $W = 1$, we have changed r and can no longer assume that it is a simple radial distance from the origin. This step makes sense, however, as we can now rewrite (9.2.4) as

$$ds^2 = Ac^2 dt^2 - B dr^2 - r^2 \left(d\theta^2 + \sin^2 \theta d\phi^2 \right), \tag{9.2.5}$$

where A and B are some new unknown functions of r, ie $A = A(r)$ and $B = B(r)$. Now we can take a magnifying glass to (9.2.5) and see how it conforms to the above conditions. We note that this proposed metric has the following properties:

- None of the metric components depend on time – meaning the metric is static.

- If we let r and t be constant, then $dt^2 = 0$ and $dr^2 = 0$ and the metric becomes

$$ds^2 = r^2 \left(d\theta^2 + \sin^2 \theta d\phi^2 \right),$$

which is the line element for the surface of a sphere (4.3.1) – meaning the metric is spherically symmetric.

- The functions $A(r)$ and $B(r)$ must be consistent with $T_{\mu\nu} = 0$. This is because the metric describes the properties of empty spacetime in the vicinity of a source mass. If the spacetime is empty at a particular point, then the energy-momentum tensor at that point must equal zero. As we'll shortly see, saying $T_{\mu\nu} = 0$ implies both the Ricci tensor and Ricci scalar are also zero in the Einstein field equations (7.6.4):

$$R_{\mu\nu} - \frac{1}{2}Rg_{\mu\nu} = -\kappa T_{\mu\nu}.$$

- Both $A(r)$ and $B(r)$ must approach 1 as r approaches infinity. If this happens, the metric becomes (3.5.6), the Minkowski metric in spherical coordinates

$$ds^2 = c^2dt^2 - dr^2 - r^2d\theta^2 - r^2\sin^2\theta d\phi^2,$$

and the fourth condition is satisfied – the metric is asymptotically flat.

We'll now derive the Schwarzschild metric, before examining some of its most important properties.

9.3 Derivation of the Schwarzschild metric

9.3.1 Vacuum field equations

The Schwarzschild metric describes the gravitational field of empty spacetime surrounding a spherically symmetric body. In empty space, the energy-momentum tensor $T_{\mu\nu}$ on the right-hand term of the Einstein field equations (7.6.4) is equal to zero, and we have

$$R_{\mu\nu} - \frac{1}{2}Rg_{\mu\nu} = 0,$$

which are known as the **vacuum field equations**. If we multiply both sides by $g^{\mu\nu}$ we get

$$g^{\mu\nu}\left(R_{\mu\nu} - \frac{1}{2}Rg_{\mu\nu}\right) = 0$$

$$R^\nu_\nu - \frac{1}{2}\delta^\nu_\nu R = 0.$$

If we sum R^ν_ν over all values of ν we obtain the Ricci scalar (6.7.3) R, and

$$R - \frac{1}{2}\delta^\nu_\nu R = 0$$

and if we sum the Kronecker delta δ^ν_ν over all possible values of ν

$$\delta^\nu_\nu = \delta^0_0 + \delta^1_1 + \delta^2_2 + \delta^3_3 = 1 + 1 + 1 + 1 = 4.$$

Therefore

$$R - \frac{1}{2}4R = 0,$$

meaning R must equal zero. This means that for vacuum solutions of the Einstein field equations both the Ricci tensor and the Ricci scalar vanish. However, this does not necessarily mean that spacetime is flat.

Recall that the Riemann curvature tensor (6.6.1) is the absolute acid test for determining whether a manifold is flat or curved. Only if the Riemann curvature tensor is zero for all points in a particular space, is that space flat. The Ricci tensor and the Ricci scalar are both derived from the Riemann curvature tensor. Curved space is a tricky beast, however, and a zero valued Ricci tensor and Ricci scalar does not necessarily imply that the underlying Riemann curvature tensor must also equal zero. That is why the field equations can have non-trivial solutions even when the energy-momentum tensor equals zero.

In passing, we can namedrop **Birkhoff's theorem** – named after the American mathematician G D Birkhoff (1844–1944) – which states that the Schwarzschild solution we are deriving is the *only possible* spherically symmetric vacuum solution to Einstein's field equations.

9.3.2 Using our proposed metric

We start with (9.2.4) – the form our proposed metric must take

$$ds^2 = Uc^2dt^2 - Vdr^2 - Wr^2\left(d\theta^2 + \sin^2\theta d\phi^2\right),$$

where U, V, W are functions of r.

If, as we did above, we assume that $W = 1$, we have changed the coordinate r and can no longer assume that it is the usual radial distance from the origin. We can rewrite (9.2.4) as

$$ds^2 = e^{2X}c^2dt^2 - e^{2Y}dr^2 - r^2d\theta^2 - r^2\sin^2\theta d\phi^2. \tag{9.3.1}$$

Where X and Y are functions of r. We use exponential functions e^{2X} and e^{2Y} because these are always positive, meaning we can ensure the $e^{2X}dt^2$ term is always positive and the $-e^{2Y}dr^2$ term is always negative. We do this because the line element (9.3.1) must always have the consistent $+---$ signature, ie the one we have used throughout this book.

From the line element (9.3.1) we can read off the values of $g_{\mu\nu}$. These are $g_{00} = e^{2X}$, $g_{11} = -e^{2Y}$, $g_{22} = -r^2$, $g_{33} = -r^2\sin^2\theta$. All other values of $g_{\mu\nu} = 0$.

Because the metric is a nice, simple diagonal matrix, the components $g^{\mu\nu}$ of the inverse metric are the reciprocals of $g_{\mu\nu}$, ie $g^{00} = 1/e^{2X} = e^{-2X}$, $g^{11} = 1/-e^{2Y} = -e^{-2Y}$, $g^{22} = -1/r^2$, $g^{33} = -1/r^2\sin^2\theta$.

9.3.3 Finding the connection coefficients

We now need to find the connection coefficients by plugging the values of $g^{\mu\nu}$ and $g_{\mu\nu}$ into (6.2.6)

$$\Gamma^\sigma_{\mu\nu} = \frac{1}{2}g^{\sigma\rho}\left(\frac{\partial g_{\rho\nu}}{\partial x^\mu} + \frac{\partial g_{\mu\rho}}{\partial x^\nu} - \frac{\partial g_{\mu\nu}}{\partial x^\rho}\right).$$

There are actually forty independent connection coefficients, but only nine are non-zero. We'll calculate one as an example and state the others. We'll use a prime $'$ to indicate differentiation with respect to r, so Y' means $\frac{dY}{dr}$.

Problem 9.1. Calculate Γ^0_{01} from the line element (9.3.1).

Γ^0_{01} means $\sigma = 0$, $\mu = 0$, $\nu = 1$, so we can say

$$\Gamma^0_{01} = \frac{1}{2}g^{0\rho}\left(\frac{\partial g_{\rho 1}}{\partial x^0} + \frac{\partial g_{0\rho}}{\partial x^1} - \frac{\partial g_{01}}{\partial x^\rho}\right).$$

ρ must equal 0 (otherwise $g^{\sigma\rho} = 0$), therefore we can write

$$\Gamma^0_{01} = \frac{1}{2}g^{00}\left(\frac{\partial g_{01}}{\partial x^0} + \frac{\partial g_{00}}{\partial x^1} - \frac{\partial g_{01}}{\partial x^0}\right).$$

We now plug in the values of $g^{\mu\nu}$ and $g_{\mu\nu}$ to give

$$\Gamma^0_{01} = \frac{1}{2}e^{-2X}\left(\frac{\partial(0)}{\partial t} + \frac{\partial e^{2X}}{\partial r} - \frac{\partial(0)}{\partial t}\right)$$

$$\Gamma^0_{01} = \frac{1}{2}e^{-2X}\frac{\partial e^{2X}}{\partial r}. \tag{9.3.2}$$

We can use a version of the chain rule (1.10.4) to solve $\frac{\partial e^{2X}}{\partial r}$. Let $u = 2X$, giving $\frac{\partial u}{\partial r} = 2X'$ (where $X' = \frac{\partial X}{\partial r}$). Then $\frac{\partial e^u}{\partial u} = e^u$. The chain rule then says

$$\frac{\partial e^{2X}}{\partial r} = \frac{\partial e^u}{\partial r} = \frac{\partial u}{\partial r} \times \frac{\partial e^u}{\partial u}$$

$$= 2X'e^{2X},$$

which we can plug into (9.3.2) to give

$$\Gamma^0_{01} = \frac{1}{2}e^{-2X}2X'e^{2X} = \frac{2X'e^{2X}}{2e^{2X}} = X'$$

$$\Gamma^0_{01} = X'.$$

The other connection coefficients can by similarly calculated. The nine non-zero connection coefficients are:

$$\Gamma^0_{01} = X' = \Gamma^0_{10},$$
$$\Gamma^1_{00} = X'e^{2(X-Y)},$$
$$\Gamma^1_{11} = Y',$$
$$\Gamma^1_{22} = -re^{-2Y},$$
$$\Gamma^1_{33} = -e^{-2Y}r\sin^2\theta,$$
$$\Gamma^2_{12} = \frac{1}{r} = \Gamma^2_{21},$$
$$\Gamma^2_{33} = -\sin\theta\cos\theta,$$
$$\Gamma^3_{13} = \frac{1}{r} = \Gamma^3_{31},$$
$$\Gamma^3_{23} = \cot\theta = \Gamma^3_{32}.$$

9.3.4 The Ricci tensor components

Next we need the relevant Ricci tensor components using (6.7.2)

$$R_{\mu\nu} = \frac{\partial \Gamma^{\alpha}_{\mu\alpha}}{\partial x^{\nu}} - \frac{\partial \Gamma^{\alpha}_{\mu\nu}}{\partial x^{\alpha}} + \Gamma^{\beta}_{\mu\alpha}\Gamma^{\alpha}_{\beta\nu} - \Gamma^{\beta}_{\mu\nu}\Gamma^{\alpha}_{\beta\alpha}.$$

Only four components are non-zero. After substituting the connection coefficients into this equation, plus lots of algebra, these are found to be:

$$R_{00} = -e^{2(X-Y)}\left(X'' + (X')^2 - X'Y' + \frac{2X'}{r}\right),$$

$$R_{11} = X'' + (X')^2 - X'Y' - \frac{2Y'}{r},$$

$$R_{22} = e^{-2Y}\left(1 + r\left(X' - Y'\right)\right) - 1,$$

$$R_{33} = \sin^2\theta\left(e^{-2Y}\left(1 + r\left(X' - Y'\right)\right) - 1\right).$$

9.3.5 Assembling the bits

These equations look horribly complicated, but thankfully things now start to get easier. For a vacuum solution all these components must equal zero: $R_{00} = 0$, $R_{11} = 0$, $R_{22} = 0$, $R_{33} = 0$. We can therefore rearrange the R_{11} equation to give

$$X'' + (X')^2 - X'Y' = \frac{2Y'}{r}.$$

Next, we substitute this into the R_{00} equation to give

$$-e^{2(X-Y)}\left(\frac{2Y'}{r} + \frac{2X'}{r}\right) = 0.$$

Now, we know that $-e^{2(X-Y)}$ cannot equal zero (e to the power of anything cannot equal zero), therefore it must be that

$$\left(\frac{2Y'}{r} + \frac{2X'}{r}\right) = 0$$

or

$$Y' + X' = 0, \qquad (9.3.3)$$

which we can integrate to

$$Y + X = \text{a constant.}$$

The proposed metric (9.3.1) that we started off with

$$ds^2 = e^{2X}c^2 dt^2 - e^{2Y}dr^2 - r^2 d\theta^2 - r^2\sin^2\theta d\phi^2$$

must approximate to the Minkowski flat space metric as r approaches infinity, ie $e^{2X} \to 1$ and $e^{2Y} \to 1$ as $r \to \infty$. For this to happen $2X$ and $2Y$ must equal zero, because $e^0 = 1$. Therefore, it must be that

$$Y + X = 0. \qquad (9.3.4)$$

We can substitute (9.3.3) into the R_{22} equation to eliminate X':

$$e^{-2Y} \left(1 + r \left(X' - Y'\right)\right) - 1 = 0$$

$$e^{-2Y} \left(1 + r \left(-Y' - Y'\right)\right) - 1 = 0$$

$$e^{-2Y} \left(1 - 2rY'\right) = 1. \tag{9.3.5}$$

We can rearrange (9.3.5) so that it becomes

$$e^{-2Y} \left(1 - 2rY'\right) = 1 = \frac{d \left(re^{-2Y}\right)}{dr}. \tag{9.3.6}$$

To show this, we first use the product rule on the right-hand term of (9.3.6) and say

$$\frac{d \left(re^{-2Y}\right)}{dr} = e^{-2Y} + r \frac{de^{-2Y}}{dr}. \tag{9.3.7}$$

We then use the chain rule and let $u = -2Y$, giving $\frac{du}{dr} = -2\frac{dY}{dr}$, and $\frac{de^{-2Y}}{du} = \frac{de^u}{du} = e^u$. The chain rule then says

$$\frac{de^{-2Y}}{dr} = \frac{du}{dr} \times \frac{de^u}{du} = -2e^{-2Y} \frac{dY}{dr},$$

which we substitute into (9.3.7) to give

$$\frac{d \left(re^{-2Y}\right)}{dr} = e^{-2Y} + r \left(-2e^{-2Y} \frac{dY}{dr}\right) = e^{-2Y} \left(1 - 2rY'\right),$$

which is what we were trying to show. So now we can say

$$\frac{d \left(re^{-2Y}\right)}{dr} = 1,$$

which integrates to

$$re^{-2Y} = r + b, \tag{9.3.8}$$

where b is an integration constant. We can rearrange this to get

$$e^{2Y} = \left(1 + \frac{b}{r}\right)^{-1}. \tag{9.3.9}$$

Equation (9.3.4) tells us that $Y = -X$, which we can substitute into (9.3.9) to give

$$e^{-2X} = \left(1 + \frac{b}{r}\right)^{-1},$$

and turn this on its head to give

$$e^{2X} = \left(1 + \frac{b}{r}\right). \tag{9.3.10}$$

9.3.6 Solving for b/r

To find b/r we use the principle of consistency, ie that given the appropriate conditions, general relativity should reduce to the laws of Newtonian gravitation and the flat space of special relativity. We've already built into our derivation the condition that our initial metric (9.3.1) approximates the Minkowski metric as r increases (ie $e^{2X} \rightarrow 1$ and $e^{2Y} \rightarrow 1$ as $r \rightarrow \infty$). Now we need to ensure that the metric behaves as it ought to for weak gravitational fields.

In Section 8.3 we saw that at the Newtonian limit (where $g_{\mu\nu} = \eta_{\mu\nu} + h_{\mu\nu}$ and $|h_{\mu\nu}| \ll 1$) we have (8.3.8)

$$h_{00} = \frac{2\phi}{c^2},$$

and therefore

$$g_{00} = 1 + \frac{2\phi}{c^2},$$

where ϕ is the Newtonian gravitational potential at a point in a gravitational field and is given by $\phi = -\frac{GM}{r}$. We can therefore stipulate that at the Newtonian limit, b/r must equal $2\phi/c^2$. As $\phi = -\frac{GM}{r}$, we can therefore say

$$\frac{b}{r} = -\frac{2GM}{c^2 r}.$$

We can now define the exponential functions e^{2X} and e^{2Y}, using (9.3.10) and (9.3.9), as

$$e^{2X} = \left(1 - \frac{2GM}{c^2 r}\right)$$

and

$$e^{2Y} = \left(1 - \frac{2GM}{c^2 r}\right)^{-1},$$

and we can rewrite (9.3.1)

$$ds^2 = e^{2X} c^2 dt^2 - e^{2Y} dr^2 - r^2 d\theta^2 - r^2 \sin^2 \theta d\phi^2$$

as

$$ds^2 = \left(1 - \frac{2GM}{c^2 r}\right) c^2 dt^2 - \frac{dr^2}{\left(1 - \frac{2GM}{c^2 r}\right)} - r^2 d\theta^2 - r^2 \sin^2 \theta d\phi^2,$$

our goal – the Schwarzschild metric (9.2.1).

9.4 More on Schwarzschild spacetime

We'll now look more closely at some of the important properties and implications of Schwarzschild spacetime. We first consider a crucial quantity known as the Schwarzschild radius. We then need to investigate the meaning of time and distance as described by the metric, and say a quick hello to the geodesic equations of motion, which determine how objects and photons move in this spacetime. Also, in this section, in no particular chronological order, we discuss what are known as the four classical tests of general relativity, these being:

- gravitational redshift

- precession of the perihelion of Mercury

- gravitational deflection of light

- gravitational time delay of signals passing the Sun.

9.4.1 The Schwarzschild radius

Buried in the coefficients $\left(1 - \frac{2GM}{c^2 r}\right)$ and $\left(1 - \frac{2GM}{c^2 r}\right)^{-1}$ is an important quantity R_S known as the **Schwarzschild radius**, where

$$R_S = \frac{2GM}{c^2} = 2m, \qquad (9.4.1)$$

M being the mass of our spherically symmetric gravitational source, a star for example, and $m = GM/c^2$. We can rewrite the Schwarzschild metric in terms of R_S as

$$ds^2 = \left(1 - \frac{R_S}{r}\right)c^2 dt^2 - \frac{dr^2}{1 - \frac{R_S}{r}} - r^2 d\theta^2 - r^2 \sin^2\theta d\phi^2. \qquad (9.4.2)$$

If we magic away our star so that $M = 0$, then $R_S = 0$ and the Schwarzschild metric again reverts to the Minkowski metric of special relativity expressed in spherical coordinates. Only when $M = 0$ do the coordinates t and r represent real clock-time and radial distance. If we increase M we start to curve spacetime and we can no longer assume that the coordinates t and r represent directly measurable quantities of time and distance.

Problem 9.2. Assuming that they are both spherically symmetric bodies, what is the Schwarzschild radius of (a) the Earth, (b) the Sun?

(a) The mass of the Earth $= 5.97 \times 10^{24}$kg. The gravitational constant $G = 6.67 \times 10^{-11}\,\mathrm{N\,m^2\,kg^{-2}}$. Using (9.4.1)

$$R_S = \frac{2GM}{c^2}$$

$$R_S = \frac{2 \times 6.67 \times 5.97 \times 10^{13}}{9 \times 10^{16}}$$

$$R_S = 8.85 \times 10^{-3}\,\mathrm{m}$$

or about 9 millimetres. The actual radius of the Earth is about $6.37 \times 10^6\,\mathrm{m}$ (6370 km).

(b) The mass of the Sun $= 1.99 \times 10^{30}$ kg.

$$R_S = \frac{2GM}{c^2}$$

$$R_S = \frac{2 \times 6.67 \times 1.99 \times 10^{19}}{9 \times 10^{16}}$$

$$R_S = 2.95 \times 10^3\,\mathrm{m}$$

or about 3 km or just under 2 miles. The actual radius of the Sun is about $6.96 \times 10^8\,\mathrm{m}$ (696,000 km).

A strange property of the Schwarzschild metric is that if all the mass M could be squeezed inside a sphere of radius R_S, light would be unable to escape from the object and we would have created a **black hole**, one of the strangest objects in the universe. We'll be taking a closer look at black holes in the next chapter.

From our previous example we can see that if we wanted to transform the Earth into a black hole, we would need to squash it down into an $18\,\mathrm{mm}$ diameter ball – something about the size of a grape! The reason that planets and most stars do not shrink down into black holes of their own accord is that there are countervailing internal pressures and forces that act to prevent such a total gravitational collapse.

You may have noticed that as the value of r approaches R_S, the factor $1 - \frac{2GM}{c^2 r}$ approaches zero and the metric starts to go haywire. First, the g_{00} metric coefficient $\to 0$. Second, the g_{11} coefficient $\to \infty$. We say there is a **singularity** in the Schwarzschild metric at $r = R_S$. A singularity is a point where a mathematical object is undefined. For example, there's a singularity in the function $y = 1/x$ where $x = 0$. At that point the function 'blows up', and we cannot assign a value to y.

But does the singularity at $r = R_S$ have any physical significance? In other words, is it describing a real lower-limit boundary of r, or is it some kind of mathematical blip? It took some time following the publication of the Schwarzschild metric for physicists to realise that $r = R_S$ is not a physical singularity, but instead a consequence of the coordinates being used. Hence, $r = R_S$ is known as a **coordinate singularity** and can be removed by suitable replacement of coordinates that are valid for r. Because $r = R_S$ is a coordinate singularity, by careful choice of coordinates other than t, r, θ, ϕ, the Schwarzschild metric can describe the simplest form of non-rotating, electrically neutral black hole, where the radius of the central mass is less than the Schwarzschild radius, ie $r < R_S$. These are known as Schwarzschild black holes, which we look at in the next chapter.

Recall that the Schwarzschild metric is a vacuum solution to the Einstein field equations. This means it is only valid in the empty region of spacetime outside of a massive spherical object. For 'ordinary' celestial bodies such as planets, our Sun and most stars, the Schwarzschild radius is located deep within the object and $R_S < r$. This means that the singularity $r = R_S$ won't be a problem when using our usual t, r, θ, ϕ coordinates because this singularity will not occur in the empty space surrounding our object.

Because $R_S < r$, the $\left(1 - \frac{R_S}{r}\right)$ and $\left(1 - \frac{2GM}{c^2 r}\right)$ terms in the Schwarzschild metric will be less than 1 for 'ordinary' celestial bodies.

It's worth noting that there is a genuine **physical singularity** when $r = 0$, which cannot be transformed away by changing coordinates. At this point, which is thought to occur at the centre of a black hole, spacetime has infinite curvature, matter has infinite density and the laws of physics break down.

9.4.2 Gravitational time dilation

We want to see how time is measured in Schwarzschild spacetime. We'll consider the gravitational field outside of an imaginary planet Zog, which is standing alone in the middle of boundless empty space. An observer, let's call him Yuri, with a torch is standing on the surface of Zog. As well as a torch, he is also equipped with a super-accurate clock. Yuri points the torch into the sky and rapidly flashes it on and off twice. Each flash of light can be thought of as an event. He records the time interval between the two flashes/events using his clock.

In terms of the Schwarzschild metric, what do we know about these events?

We'll call the Schwarzschild coordinates of the first event $(t_{ZOG}, r_{ZOG}, \theta_{ZOG}, \phi_{ZOG})$ and those of the second event $(t_{ZOG} + dt_{ZOG}, r_{ZOG}, \theta_{ZOG}, \phi_{ZOG})$. We don't know the value of the $r_{ZOG}, \theta_{ZOG}, \phi_{ZOG}$ coordinates, but we know they are constant (because the torch is stationary relative to Zog), and therefore $dr = d\theta = d\phi = 0$. The two events are separated by a difference in coordinate time equal to dt_{ZOG}. We don't know how this coordinate time relates to the time recorded on Yuri's clock.

From our previous discussion, we know that the proper time (7.4.3) between the two events in terms of the spacetime line element is given by

$$c^2 d\tau^2 = ds^2,$$

which is the time measured by a stationary clock at the same position as the two events. Yuri's is such a clock. We can call the proper time measured by Yuri's clock $d\tau_{ZOG}$. Feeding all this information into the Schwarzschild metric (9.2.1)

$$ds^2 = \left(1 - \frac{2GM}{c^2 r}\right) c^2 dt^2 - \frac{dr^2}{1 - \frac{2GM}{c^2 r}} - r^2 d\theta^2 - r^2 \sin^2 \theta d\phi^2$$

we find that the small (we'll assume that it's infinitesimal) spacetime separation of the two events is

$$ds_{ZOG}^2 = c^2 d\tau_{ZOG}^2 = \left(1 - \frac{2GM}{c^2 r}\right) c^2 dt_{ZOG}^2 - \frac{(0)^2}{1 - \frac{2GM}{c^2 r}} - r^2 (0)^2 - r^2 \sin^2 \theta (0)^2$$

$$d\tau_{ZOG}^2 = \left(1 - \frac{2GM}{c^2 r}\right) dt_{ZOG}^2$$

giving

$$d\tau_{ZOG} = \left(1 - \frac{2GM}{c^2 r}\right)^{1/2} dt_{ZOG}. \tag{9.4.3}$$

Because $\left(1 - \frac{2GM}{c^2 r}\right)^{1/2}$ is less than 1, we can see that the proper time $d\tau_{ZOG}$ measured by Yuri between the two events is less than the coordinate time dt_{ZOG} separating the events, ie $d\tau_{ZOG} < dt_{ZOG}$.

We can rewrite (9.4.3) in general terms for any increment of proper time as

$$d\tau = \left(1 - \frac{2GM}{c^2 r}\right)^{1/2} dt \tag{9.4.4}$$

or, substituting $m = GM/c^2$, we can write

$$d\tau = \left(1 - \frac{2m}{r}\right)^{1/2} dt. \tag{9.4.5}$$

Looking at (9.4.3) we can see that if we travelled far, far away from Zog so $r \to \infty$, the term $\left(1 - \frac{2GM}{c^2 r}\right)^{1/2} \to 1$, and $d\tau$ will increasingly agree with dt. We can therefore say that the coordinate time t is the same as the proper time τ kept by a stationary clock at infinite distance from Zog or, in other words, $d\tau_\infty = dt$. Now we can rewrite (9.4.4) as

$$d\tau = \left(1 - \frac{2GM}{c^2 r}\right)^{1/2} d\tau_\infty$$

or

$$d\tau_\infty = \frac{d\tau}{\left(1 - \frac{2GM}{c^2 r}\right)^{1/2}}. \tag{9.4.6}$$

This equation tells us that the proper time $d\tau_\infty$ between two events as recorded by a stationary distant observer is more than the proper time $d\tau$ recorded by an observer located where the events occurred. We've actually slipped an assumption in here: that the coordinate time difference between two events in Schwarzschild spacetime is the same as that recorded by a distant observer. We'll show that this assumption is true in Section 10.4.1.

This phenomenon – that clocks run slower in a gravitational field as seen by a distant observer – is known as **gravitational time dilation**. In other words, clocks lower down in a gravitational field (nearer the surface of the star, planet, etc) run slower than clocks higher up in that field. To paraphrase – gravity makes time run slower.

Note that although we have stated that our distant observer is infinitely far away from the central body (Zog, in our example) all we are really asking is that the observer should be sufficiently distant so that $\frac{2GM}{c^2 r}$ is negligible, meaning $\left(1 - \frac{2GM}{c^2 r}\right) \approx 1$.

We can also see from (9.4.6) that if the two events are moved nearer the surface of the spherically symmetric body, then r decreases and $\left(1 - \frac{2GM}{c^2 r}\right)^{-1/2}$ increases. This means the distant observer would measure the clock recording proper time $d\tau$ next to the events to be running even slower compared to his own clock showing proper time $d\tau_\infty$.

Don't confuse the gravitational time dilation of general relativity with the time dilation of special relativity we met in Section 3.3.6. The differences are:

- In special relativity we are comparing two observers moving in relative uniform motion. In our example of gravitational time dilation both observers are stationary with respect to each other.

- In special relativity two observers in relative uniform motion measure each other's clocks to be running slow. This is not the case with gravitational time dilation where both observers agree that the clock in the stronger gravitational field runs slower than the one in the weaker gravitational field.

- In special relativity we use global reference frames (think of a spacetime diagram stretching on for ever) that can, in theory, extend to infinity. We assume that inertial observers in special relativity use synchronised clocks to time events wherever they happen in their reference frame. In general relativity we are dealing with curved spacetime and a global coordinate system isn't usually possible, so we tend to use local frames. In our example of gravitational time dilation the distant observer measured the time difference between Yuri's light flashes locally at his own location, not where they occurred where Yuri was standing on the surface of Zog.

9.4.3 Gravitational redshift – the first test of general relativity

First, some basic information about waves in general – light, water, sound, etc. Figure 9.2 shows two simple waves travelling from left to right across the page. The top wave has a shorter wavelength λ

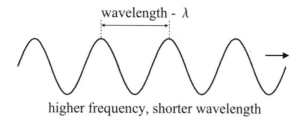

higher frequency, shorter wavelength

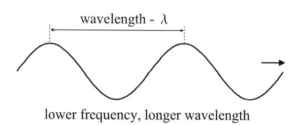

lower frequency, longer wavelength

Figure 9.2: Simple waves.

(distance between two peaks) than the bottom one. The frequency f of a wave is the number of cycles of a wave to pass some point in a second, peaks per second for example. There is an easy formula expressing the relationship between the wavelength, frequency and velocity v of a wave. This is

$$v = f\lambda. \tag{9.4.7}$$

So, for example, if a wave has a frequency of 10 cycles per second (or 10 hertz, to use the proper SI units) and a wavelength of 2 m, it will have a velocity of $10 \times 2 = 20\,\mathrm{m\,s^{-1}}$.

The period of a wave is the duration of one cycle, so is the reciprocal of the frequency. In our example, the period would therefore equal $1/10 = 0.1$ seconds – the time it would take each peak to pass a particular point.

Returning to Yuri. When he flicked his torch on and off he measured the proper time dt_{ZOG} between two events, ie two flashes of light. Another pair of events might be the successive peaks of the light wave leaving the torch. The proper time interval dt_{ZOG} then represents the period of the wave as it leaves the torch. We can use (9.4.6) to show the period of the light wave as measured by a distant observer $d\tau_\infty$ in terms of the period of the light wave as measured by Yuri (dt_{ZOG}):

$$d\tau_\infty = \frac{d\tau_{ZOG}}{\left(1 - \frac{2GM}{c^2 r}\right)^{1/2}}.$$

In general terms, we can write

$$d\tau_\infty = \frac{d\tau}{\left(1 - \frac{2GM}{c^2 r}\right)^{1/2}},$$

where $d\tau$ is the period of the wave measured where it is emitted. Because frequency is the reciprocal

of the period, we can say

$$f_\infty = f_{em} \left(1 - \frac{2GM}{c^2 r} \right)^{1/2} , \qquad (9.4.8)$$

where f_∞ is the frequency of the wave measured by a distant observer, and f_{em} is the frequency of the wave measured at the point of emission. This equation tells us that the frequency of a wave as recorded by a distant observer is less than the frequency recorded by an observer located where the events occurred. This phenomenon is known as **gravitational redshift**, because a reduction in frequency means a shift toward the longer wavelengths or 'red' end of the electromagnetic spectrum. We can think of the photons losing energy as they climb out of the gravitational field – loss of energy equating to a drop in frequency. Gravitational redshift doesn't only apply to visible light of course but to all electromagnetic waves. Figure 9.3 illustrates this phenomenon schematically.

Figure 9.3: Gravitational redshift of a light wave escaping from a massive object.

9.4.4 Using the equivalence principle to predict gravitational time dilation and gravitational redshift

Recall Einstein's 'happiest thought' (Section 7.2.1), when he realised the fundamental importance of the equivalence principle in his search for a gravitational theory of relativity. Using this principle, Einstein predicted a local version of gravitational time dilation and gravitational redshift in 1907, eight years before his full formulation of the general theory and the publication of the Schwarzschild metric. The following modern derivation is based on Schutz [29].

First, we need to mention a phenomenon known as the **Doppler effect**, which describes the observed change in frequency of waves relative to a moving observer. A police car siren, for example, appears to increase in pitch as the car speeds towards you and decrease in pitch as it moves away. Think of the 'yeeeoooow' sound a racing car makes as speeds past. The police officer driving the car will not of course notice any change in pitch of her siren. But if she knows how fast she is travelling, and how to do the physics, she could predict how the siren would sound to a passing observer.

The Doppler effect applies to light as well as sound. An observer moving away from a light source will measure an increase in wavelength, a decrease in frequency, and the light will appear redshifted. Conversely, an observer moving towards the light source will see a decrease in wavelength, an increase in frequency, and the light will appear blueshifted.

A useful approximation of the change in frequency of a light wave is given by

$$f_o = f_e \left(1 - v/c\right), \tag{9.4.9}$$

where f_e is the emitted frequency, and f_o is the measured frequency of an observer moving with velocity v away from the light source. This equation assumes that v is much smaller than c. Because the term $(1 - v/c)$ is less than 1, the observed frequency f_o will be less than the emitted frequency f_e and the light will appear redshifted. If the observer is moving towards the light source we change the sign from $-v/c$ to $+v/c$ and the light would appear blueshifted.

Now we recall the equivalence principle, which states that, locally, the physical behaviour of objects in a gravitational field cannot be distinguished by any experiment from the behaviour of uniformly accelerating objects. That means that if a beam of light is redshifted in a uniformly accelerating frame it must also be redshifted in a gravitational field. How can we show this?

Consider a thought experiment where an observer A is on top of a tower of height h. On the ground next to the tower is a laser shining a beam of light vertically upwards of frequency f_{bot} (measured on the ground, ie at the bottom of the tower). Observer A measures the frequency of the light beam when it reaches him to be f_{top}. What is the relationship between f_{bot} and f_{top}? In order to answer this question we introduce another (fanatically dedicated) observer B, who jumps off the top of the tower the moment the laser emits its beam of light. Observer B is momentarily stationary when he steps off the tower, which is the same time as when the light is emitted, and so he measures its frequency at that instant to be f_{bot}. In fact, because observer B is freely falling, and therefore constitutes an inertial frame of reference, he measures the frequency of the light beam to be constant at f_{bot} (think of observer B as being effectively in deep space measuring the speed of a passing light beam).

From falling observer B's point of view, observer A is moving away from the light source. He can therefore use the Doppler effect equation (9.4.9) to calculate the frequency f_{top} measured by observer A on top of a tower compared to the frequency f_{bot} measured by observer B. Just to be clear, observer B doesn't see any change in frequency in the light beam, but he can predict that observer A will measure a Doppler-caused redshift (just as the police officer doesn't hear the pitch of her own siren change, but can calculate that change for a roadside observer).

To calculate the Doppler redshift measured by observer A, we need the relative velocity of that observer compared to the laser. This is the same as the velocity v of observer B when the beam of light reaches the top of the tower. We find v using the relevant equation of uniform motion (1.10.15) we met earlier

$$v = u + at,$$

where u is the initial velocity of observer A, ie $u = 0$, a is the acceleration due to gravity, ie $a = g$, and t is the time it takes the light to reach the top of the tower, ie $t = h/c$. We can therefore say

$$v = 0 + gh/c.$$

We can now plug this into the Doppler effect equation (9.4.9) to get

$$f_{top} = f_{bot}\left(1 - v/c\right)$$

$$f_{top} = f_{bot}\left(1 - \frac{gh}{c^2}\right). \tag{9.4.10}$$

In our approximation, gh will be much smaller than c^2, therefore the term $\left(1 - \frac{gh}{c^2}\right)$ is less than 1, f_{top} will be smaller than f_{bot}, and we have shown the light is redshifted.

The principle of equivalence says that if some physical effect happens in a uniformly accelerating frame, it must also happen in a stationary frame in a gravitational field. From observer B's point of view the light is redshifted for observer A due to the Doppler effect. From observer A's point of view the light is red shifted because it climbed out of a gravitational field.

Because frequency is the reciprocal of period we can invert (9.4.10) to give an equation for gravitational time dilation:

$$t_{top} = \frac{t_{bot}}{\left(1 - \frac{gh}{c^2}\right)}, \tag{9.4.11}$$

where t_{bot} is the time measured at the bottom of our hypothetical tower, and t_{top} is the time measured at the top.

We have therefore used the principle of equivalence to derive approximations ((9.4.11) and (9.4.10)) of the equations of gravitational time dilation and redshift.

These equations are reasonable approximation for small distances h near the Earth. They are local versions of the gravitational time dilation and redshift equations we derived in the previous section using the Schwarzschild metric. As we saw when using our example of Yuri shining his torch into the sky on planet Zog, the Schwarzschild equations are not limited to a single local frame but can be used to take measurements in widely separated local frames (think of the distant observer far away from Zog, measuring the time intervals of Yuri's flashing torch).

In 1960 Pound and Rebka performed the first laboratory-based test of general relativity when they measured the gravitational redshift of photons travelling 22.5m vertically in a tower at Harvard University's Jefferson Laboratory. Their experiment, using gamma rays, showed a 10% deviation from the predictions of general relativity, later improved to better than 1% by Pound and Snider.

9.4.5 Proper distance

We'll now look at how to measure distance in Schwarzschild spacetime. When deriving the Schwarzschild metric we had to change the definition of coordinate r, meaning we could no longer assume that it represented a straightforward radial distance from the origin. We saw that only when $M = 0$, and the Schwarzschild metric reduces to the Minkowski metric, do the coordinates t and r represent real clock-time and radial distance. So, what does the coordinate r actually represent?

Consider two events that occur at the same coordinate time (ie $dt = 0$), but are separated by an infinitesimal spatial distance. Using the Schwarzschild metric (9.2.1) we can say

$$ds^2 = \left(1 - \frac{2GM}{c^2 r}\right)c^2 dt^2 - \frac{dr^2}{1 - \frac{2GM}{c^2 r}} - r^2 d\theta^2 - r^2 \sin^2\theta d\phi^2$$

$$ds^2 = \left(1 - \frac{2GM}{c^2 r}\right) c^2 \left(0\right)^2 - \frac{dr^2}{1 - \frac{2GM}{c^2 r}} - r^2 d\theta^2 - r^2 \sin^2 \theta d\phi^2$$

$$ds^2 = -\frac{dr^2}{1 - \frac{2GM}{c^2 r}} - r^2 d\theta^2 - r^2 \sin^2 \theta d\phi^2, \tag{9.4.12}$$

and we can say the **proper distance** (denoted by the Greek letter sigma σ) between the two events is given by

$$d\sigma^2 = -ds^2 = \frac{dr^2}{1 - \frac{2GM}{c^2 r}} + r^2 d\theta^2 + r^2 \sin^2 \theta d\phi^2 \tag{9.4.13}$$

or

$$d\sigma = \left(-ds^2\right)^{1/2} = \left(\frac{dr^2}{1 - \frac{2GM}{c^2 r}} + r^2 d\theta^2 + r^2 \sin^2 \theta d\phi^2\right)^{1/2}. \tag{9.4.14}$$

For a fixed value of r (ie $dr = 0$), (9.4.13) reduces to

$$d\sigma^2 = r^2 d\theta^2 + r^2 \sin^2 \theta d\phi^2, \tag{9.4.15}$$

which is the line element (4.3.1) for the surface of an ordinary three-dimensional sphere embedded in Euclidean space that we met in Section 4.3.

Problem 9.3. What is the proper circumference C of a circle of Schwarzschild coordinate radius r in the $\theta = \pi/2$ plane centred on $r = 0$?

The $\theta = \pi/2$ plane defines the equator of a sphere. The coordinates t, r, θ are constant, so $dt = dr = d\theta = 0$. We use (9.4.14)

$$d\sigma = \left(-ds^2\right)^{1/2} = \left(\frac{dr^2}{1 - \frac{2GM}{c^2 r}} + r^2 d\theta^2 + r^2 \sin^2 \theta d\phi^2\right)^{1/2},$$

which becomes

$$d\sigma = \left(r \sin\left(\pi/2\right) d\phi\right),$$

and we can integrate to give

$$C = \int_0^{2\pi} r \sin\left(\pi/2\right) d\phi.$$

$\sin\left(\pi/2\right) = 1$ and thus

$$C = r\left(2\pi - 0\right)$$

$$C = 2\pi r, \tag{9.4.16}$$

the same as the ordinary circumference of a circle in flat space.

Let's be clear what we are saying here. Constant Schwarzschild coordinate time t and coordinate distance r, define the surface of a Euclidean sphere. Equation (9.4.15) then tells us that the infinitesimal proper distance ($\sqrt{d\sigma^2}$) between any two points/events in Schwarzschild spacetime is the same as the separation of those two points on the surface of such a sphere. Effectively, for each constant value of coordinate time t, Schwarzschild spacetime can be thought of as a series of nested Euclidean spheres (think of a set of Russian dolls), each one representing a different value of r.

Problem 9.3 tells us that the proper circumference of a great circle drawn on one of these spheres is $2\pi r$, just as it is in Euclidean space. However, because we are dealing with curved spacetime (ie $M \neq 0$ for the spherically symmetric central star or planet etc) the coordinate distance r would not equal the proper radius of such a circle. In other words, r does not represent proper radial distance. This is shown in Figure 9.4 where a great circle is drawn around a central spherical mass. We could use a ruler to measure the proper circumference C_P and proper radius R_P of this circle. We could calculate the Schwarzschild coordinate radius to be

$$r = C_P/2\pi,$$

but we would also find that

$$r \neq R_P.$$

This discrepancy is due to the central spherical mass distorting spacetime (actually space not spacetime in this particular example, where t is constant) as dictated by the Schwarzschild metric.

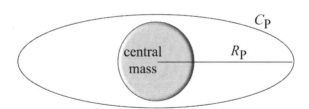

Figure 9.4: Measuring distance in Schwarzschild spacetime.

To quantify the relationship between proper distance σ and coordinate distance r, we can consider two events that not only occur at the same coordinate time t, but also at the same coordinate angles θ and ϕ (ie $d\theta = d\phi = 0$). Equation (9.4.14) now reduces to

$$d\sigma = \left(\frac{dr^2}{1 - \frac{2GM}{c^2 r}} \right)^{1/2} = \frac{dr}{\left(1 - \frac{2GM}{c^2 r}\right)^{1/2}}. \tag{9.4.17}$$

The $\left(1 - \frac{2GM}{c^2 r}\right)$ terms in the Schwarzschild metric are less than 1 for 'ordinary' celestial bodies. Therefore, the $\left(1 - \frac{2GM}{c^2 r}\right)^{-1/2}$ term in (9.4.17) will be more than 1, meaning $d\sigma$ is usually greater than dr (as long as r is greater than the Schwarzschild radius $\frac{2GM}{c^2}$).

Recall (9.4.2) that we can also express the Schwarzschild metric in terms of the Schwarzschild radius, in which case (9.4.17) becomes

$$d\sigma = \frac{dr}{\left(1 - \frac{R_S}{r}\right)^{1/2}} \tag{9.4.18}$$

or, substituting $m = GM/c^2$, we can write

$$d\sigma = \frac{dr}{\left(1 - \frac{2m}{r}\right)^{1/2}}. \tag{9.4.19}$$

Figure 9.5 helps us visualise the relationship between σ and r. The lower portion of the diagram shows flat space where, there is no central mass (ie $M = 0$). The upper portion of the diagram shows the same space, but now curved due to the introduction of a central mass ($M > 0$). The circles C_1 and C_4 represent a Euclidean sphere of coordinate radius r, the circles C_2 and C_3 a slightly larger sphere of coordinate radius $r + dr$. Circles C_1 and C_4 have a proper circumference of $2\pi r$. Circles C_2 and C_3 have a proper circumference of $2\pi (r + dr)$. The proper measured distance between the two spheres equals dr in flat space, but $d\sigma$ (measured along the curve) in curved space. We can see that:

- $d\sigma$ is larger than dr.

- As we move radially outwards $d\sigma$ increasingly approximates to dr.

- As we move radially inwards, approaching the Schwarzschild radius, $d\sigma$ becomes increasingly large compared to dr.

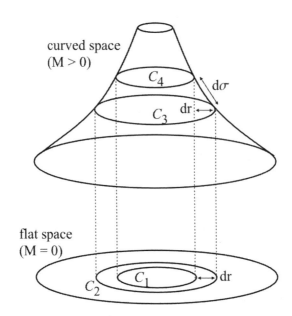

Figure 9.5: Radial distance in Schwarzschild spacetime.

Figure 9.5 is a type of **embedding diagram**, a frequently used model that helps us picture the spatial curvature of general relativity. Embedding diagrams take a slice through the equator ($\theta = \pi/2$,

$t = $ constant) of a star etc. The distance between nearby points on the curved surface then represent the proper distance $d\sigma$, as discussed above. It's important to realise that only the curved surface (the funnel-shaped bit) has any meaning in these diagrams. Points and distances away from the curved surface (ie in the three-dimensional Euclidean space in which the surface is embedded) have no physical meaning.

We can integrate 9.4.18 to obtain the proper radial distance $\Delta\sigma$ along a particular radial coordinate line

$$\Delta\sigma = \int_{r_1}^{r_2} d\sigma = \int_{r_1}^{r_2} dr \left(1 - \frac{R_S}{r}\right)^{-1/2}. \tag{9.4.20}$$

(This derivation and the following black hole examples are taken from Griest [13].) If we let $A_i = \sqrt{1 - \frac{2GM}{r_i c^2}}$, Griest gives

$$\Delta\sigma = r_2 A_2 - r_1 A_1 + \frac{R_S}{2} \ln\left(\frac{r_2 A_2 + r_2 - R_S/2}{r_1 A_1 + r_1 - R_S/2}\right), \tag{9.4.21}$$

where ln is the natural logarithm of what follows in the brackets. Recall that the natural logarithm of a number x is the power to which e would have to be raised to equal x. For example, $\ln(5) = 1.6094379$ because $e^{1.6094379} = 2.718281828^{1.6094379} = 5$.

Equation 9.4.21 looks unpleasantly complicated, both to derive and to solve for any particular values of radial coordinates r_1 and r_2. This is not a problem, however, because the process of finding both indefinite and definite integrals is made infinitely easier by using an online integral calculator. For example, using the WolframAlpha Calculus and Analysis Calculator [33], simply type 'integrate (1-R/r)^(-1/2)dr' (omit the quotes) into the input box and out pops an equivalent form of (9.4.21). This calculator also allows you to find definite integrals (ie we can solve for radial coordinates values r_1 and r_2), as we'll see shortly.

Griest uses an example of a black hole of three times the Sun's mass, giving a Schwarzschild radius of $R_S = 8.85$ km. He supposes that we fly a spaceship around the black hole and measure the circumference (using (9.4.16)) of our flight path to be $C = 2\pi \times 30$ km. Our radial coordinate distance is therefore $r = 30$ km We then fly to $r = 20$ km and might well assume that we have travelled a proper distance of 10 km. By integrating 9.4.18 between $r_1 = 20$ km and $r_2 = 30$ km, he shows that the proper distance we have covered is actually 12.51 km. Our measured circumference at $r = 20$ km would, however, be $2\pi \times 20$ km and not $2\pi \times (30 - 12.51)$ km.

This is easy enough using the WolframAlpha Calculus and Analysis Calculator [33]. Simply type 'integrate (1-8.850/r)^(-1/2)dr from r=20 to 30' (omit the quotes) into the input box, and out pops the answer: 12.5093.

Similarly, if we travelled from $r = 30$ km to $r = 10$ km we could cover a proper distance of 29.5 km, not 20 km Such weirdness is the result of spacetime being curved in the vicinity of a black hole.

Our own Sun also curves spacetime, but to a much smaller extent than a black hole. For example, assume that the Earth's radial coordinate distance from the centre of the Sun is $r = 15 \times 10^{10}$ m. If we fly from the Earth towards the Sun to $r = 14.9 \times 10^{10}$ m. we would have covered a coordinate distance of 1,000,000,000 m (a million kilometres), but a proper distance (ignoring all gravitational influences except the Sun's) of about 1,000,000,010 m – a 10 m difference.

The curvature of spacetime caused by the Earth is even smaller. If we assume that the radial coordinate height of Mount Everest is $r = 8848$ m, the proper height calculated by integrating 9.4.18 is around 8848.0000062 m. You'd hardly notice the difference!

Looking at (9.4.18)

$$d\sigma = \frac{dr}{\left(1 - \frac{R_S}{r}\right)^{1/2}},$$

such tiny discrepancies between proper distance σ and coordinate distance r are to be expected. Providing $R_S \ll r$ then $d\sigma \approx dr$, which is what astronomers can assume for most 'ordinary' celestial observations.

9.4.6 Geodesics in Schwarzschild spacetime

We saw in Section 7.4 that geodesic equations describe the paths of freely moving/falling particles in spacetime. These equations are in the form of parameterised curves. Time-like geodesics (where $ds^2 > 0$, and proper time $d\tau \neq 0$) describe the paths of massive objects ('massive' meaning anything with mass) and can use proper time as a parameter. Null geodesics (where $ds^2 = 0$ and proper time, $d\tau = 0$) describe the paths of (massless) photons and need to use another parameter (often denoted by λ) instead of proper time.

In order to understand how things (objects with mass as well as massless photons) move in Schwarzschild spacetime we therefore require the geodesic equations defined by the Schwarzschild metric. These are four complicated-looking differential equations that can be very difficult to solve. We'll give a partial derivation of one of them and simply state the other three. These are the important equations, after all, which allow physicists to describe how objects and photons move in Schwarzschild spacetime.

They are obtained using the basic geodesic equation (6.5.1)

$$\frac{d^2 x^\alpha}{d\lambda^2} + \Gamma^\alpha_{\gamma\beta} \frac{dx^\beta}{d\lambda} \frac{dx^\gamma}{d\lambda} = 0,$$

which means having to calculate the non-zero connection coefficients $\Gamma^\alpha_{\gamma\beta}$. We met the general form of these connection coefficients (in terms of X and Y and their derivatives X' and Y') when we were deriving the Schwarzschild metric in Section 9.3.3. For example, we found that $\Gamma^0_{01} = X'$. Now we know the actual Schwarzschild coefficients $g^{\mu\nu}$ and $g_{\mu\nu}$, we could plug these into (6.2.6)

$$\Gamma^\sigma_{\mu\nu} = \frac{1}{2} g^{\sigma\rho} \left(\frac{\partial g_{\rho\nu}}{\partial x^\mu} + \frac{\partial g_{\mu\rho}}{\partial x^\nu} - \frac{\partial g_{\mu\nu}}{\partial x^\rho} \right),$$

calculate the explicit non-zero connection coefficients $\Gamma^\alpha_{\gamma\beta}$ and hence find the geodesic equations. We'll now calculate one of the geodesic equations.

Problem 9.4. Calculate the appropriate Schwarzschild geodesic equation using the connection coefficient $\Gamma^\alpha_{\gamma\beta} = \Gamma^0_{01}$.

We saw in Section 9.3.3 that

$$\Gamma^0_{01} = X' = \Gamma^0_{10},$$

so we first need to find $X' = dX/dr$.

Our proposed metric (9.3.1) was

$$ds^2 = e^{2X} c^2 dt^2 - e^{2Y} dr^2 - r^2 d\theta^2 - r^2 \sin^2\theta d\phi^2.$$

We know from the Schwarzschild metric (9.2.1) that the metric coefficient

$$e^{2X} = \left(1 - \frac{2GM}{c^2 r}\right).$$

Using the definition of the natural logarithm we can say

$$X = \frac{1}{2}\ln\left(1 - \frac{2GM}{c^2 r}\right).$$

Using either the chain rule or an online derivative calculator ([33], for example) we find

$$X' = \frac{dX}{dr} = \frac{GM}{r^2 c^2 \left(1 - \frac{2GM}{c^2 r}\right)} = \Gamma^0_{01}.$$

We can now plug this value of the connection coefficient Γ^0_{01} into the geodesic equation (6.5.1)

$$\frac{d^2 x^\alpha}{d\lambda^2} + \Gamma^\alpha_{\gamma\beta}\frac{dx^\beta}{d\lambda}\frac{dx^\gamma}{d\lambda} = 0$$

and seeing that $x^\alpha = x^0 = t$, $x^\gamma = x^0 = t$ and $x^\beta = x^1 = r$ we can write

$$\frac{d^2 t}{d\lambda^2} + \frac{GM}{r^2 c^2 \left(1 - \frac{2GM}{c^2 r}\right)}\frac{dt}{d\lambda}\frac{dr}{d\lambda} = 0,$$

which is the required Schwarzschild geodesic equation.

The four Schwarzschild geodesic equations are:

$$\frac{d^2 t}{d\lambda^2} + \frac{GM}{r^2 c^2 \left(1 - \frac{2GM}{c^2 r}\right)}\frac{dt}{d\lambda}\frac{dr}{d\lambda} = 0, \tag{9.4.22}$$

$$\frac{d^2 r}{d\lambda^2} + \frac{GM}{r^2}\left(1 - \frac{2GM}{c^2 r}\right)\left(\frac{dt}{d\lambda}\right)^2 - \frac{GM}{r^2 c^2 \left(1 - \frac{2GM}{c^2 r}\right)}\left(\frac{dr}{d\lambda}\right)^2$$

$$- r\left(1 - \frac{2GM}{c^2 r}\right)\left(\left(\frac{d\theta}{d\lambda}\right)^2 + \sin^2\theta\left(\frac{d\phi}{d\lambda}\right)^2\right) = 0, \tag{9.4.23}$$

$$\frac{d^2\theta}{d\lambda^2} + \frac{2}{r}\frac{dr}{d\lambda}\frac{d\theta}{d\lambda} - \sin\theta\cos\theta\left(\frac{d\phi}{d\lambda}\right)^2 = 0, \tag{9.4.24}$$

$$\frac{d^2\phi}{d\lambda^2} + \frac{2}{r}\frac{dr}{d\lambda}\frac{d\phi}{d\lambda} + 2\frac{\cos\theta}{\sin\theta}\frac{d\theta}{d\lambda}\frac{d\phi}{d\lambda} = 0. \tag{9.4.25}$$

We can simplify these equations by introducing two quantities known as **constants of motion**. For an isolated system, these quantities are conserved throughout the motion of whatever is doing the moving. We are interested in two such constants of motion:

- Total energy per unit mass energy (E/mc^2) given by

$$\frac{E}{mc^2} = \left(1 - \frac{2GM}{c^2 r}\right)\frac{dt}{d\tau}. \tag{9.4.26}$$

- Objects (whether a spinning wheel or an orbiting planet) moving around a point possess a quantity known as **angular momentum**, analogous to the ordinary momentum we've already met. Angular momentum per unit mass (J/m) given by

$$\frac{J}{m} = r^2 \sin^2 \theta \frac{d\phi}{d\tau}. \tag{9.4.27}$$

Don't worry too much about what these equations actually mean. Their importance to us is, because they describe conserved quantities, we can use them to describe the motion of a freely falling particle of mass m in the constant plane $\theta = \pi/2$ in Schwarzschild spacetime. That motion is described by the **radial motion equation** (see Lambourne [17], for example, for a derivation):

$$\left(\frac{dr}{d\tau}\right)^2 + \frac{J^2}{m^2 r^2}\left(1 - \frac{2GM}{c^2 r}\right) - \frac{2GM}{r} = c^2 \left[\left(\frac{E}{mc^2}\right)^2 - 1\right]. \tag{9.4.28}$$

Although it looks pretty horrible, this is a very useful equation. Because it describes the motion of a freely falling object in Schwarzschild spacetime, it can be used to derive the equations of orbital motion (remember Newton's intuition that the Moon is 'freely falling' around the Earth), as well as describing the motion of objects in radial free fall – like a dropped cup, for example – as discussed in the next problem. We'll see that there are a couple of obvious ways (9.4.28) may be simplified. First, if the free fall motion of the object under consideration is radially inwards (ie radial free fall), then the object has no angular momentum and $J = 0$. Second, if the object is at rest and infinitely far away from the central mass M, the conserved energy given by (9.4.26) reduces to $E = mc^2$ and the right-hand side of (9.4.28) becomes zero.

Problem 9.5. Show that for an object in radial free fall, (9.4.28) reduces to

$$\frac{d^2 r}{d\tau^2} = -\frac{GM}{r^2}.$$

Radial free fall implies the object is moving 'straight down', ie ϕ is constant, therefore $d\phi/d\tau = 0$, and $J = 0$ in (9.4.27). Equation (9.4.28) thus simplifies to

$$\left(\frac{dr}{d\tau}\right)^2 = c^2 \left[\left(\frac{E}{mc^2}\right)^2 - 1\right] + \frac{2GM}{r}.$$

We can differentiate this with respect to τ. We do the left-hand side first using the product rule to give

$$\frac{d}{d\tau}\left(\frac{dr}{d\tau}\right)^2 = \frac{d^2 r}{d\tau^2}\frac{dr}{d\tau} + \frac{d^2 r}{d\tau^2}\frac{dr}{d\tau} = 2\frac{d^2 r}{d\tau^2}\frac{dr}{d\tau}. \tag{9.4.29}$$

Next, the right-hand side, which we'll call x, so

$$x = c^2 \left[\left(\frac{E}{mc^2}\right)^2 - 1\right] + \frac{2GM}{r}.$$

Using the chain rule, we can then say

$$\frac{dx}{d\tau} = \frac{dx}{dr} \times \frac{dr}{d\tau} = -\frac{2GM}{r^2}\frac{dr}{d\tau}. \tag{9.4.30}$$

We've differentiated both sides with respect to τ, so (9.4.29) must equal (9.4.30), giving

$$2\frac{d^2r}{d\tau^2}\frac{dr}{d\tau} = -\frac{2GM}{r^2}\frac{dr}{d\tau},$$

and dividing both sides by $2dr/d\tau$ gives

$$\frac{d^2r}{d\tau^2} = -\frac{GM}{r^2}, \tag{9.4.31}$$

which is what we were trying to show. This equation neatly approximates the Newtonian acceleration due to gravity (2.5.5)

$$a = -\frac{GM}{r^2}$$

we met in Section 2.5.3, but *only* when $dr/d\tau \ll c$, and r is sufficiently large (ie distant from the Schwarzschild radius) to give a weak (Newtonian) gravitational field and approximates to proper radial distance.

9.4.7 Perihelion advance – the second test of general relativity

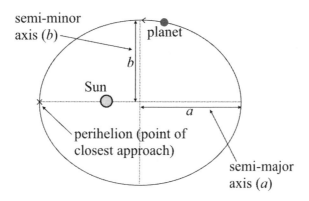

Figure 9.6: Newtonian orbit of isolated planet.

One of the classical tests of general relativity explains a small observed discrepancy in the orbit of Mercury. According to Newtonian mechanics, an isolated planet (one not affected by the gravitational fields of other planets) should follow an unchanging elliptical orbit around the Sun.

An ellipse (see Figure 9.6) is defined by the semi-major axis a, the eccentricity e (a measure of how 'squashed' the ellipse is), and the semi-minor axis b, where $b = a\sqrt{1 - e^2}$. If $e = 0$ then $b = a$ and the ellipse becomes a circle. Mercury, incidentally, has the most eccentric orbit ($e = 0.21$) of all the planets in the Solar System. The point where a planet's orbit is closest to the Sun is called the

perihelion. According to Newton, the position of the perihelion should be fixed in space, orbit after orbit.

But this isn't quite what astronomers actually observe with Mercury.

We now need to introduce an angular measurement called the **second of arc** (also known as an arcsecond or arcsec). There are 60 seconds of arc in a **minute of arc**, and 60 minutes of arc in a degree. In other words, 1 degree equals 3600 seconds of arc, and $180 \times 60 \times 60$ seconds of arc equals π radians. A second of arc is a tiny angle. If you look at a 1 cm diameter coin from a distance of 2.06 km, it will subtend (make an angle at your eye of) 1 second of arc.

Instead of being stationary in space, Mercury's perihelion is seen to advance, albeit very slowly, year after year. This phenomenon is known as the **precession of the perihelion** of Mercury (shown greatly exaggerated in Figure 9.7), and equals about 575 seconds of arc per century (that's right, *per century* – how on Earth do they measure that?). By the middle of the nineteenth century much of this movement had been meticulously explained in Newtonian terms, as being due to the gravitational interaction of the other planets on Mercury's orbit. However, a small residual and inexplicable discrepancy remained, amounting to 43 seconds of arc per century.

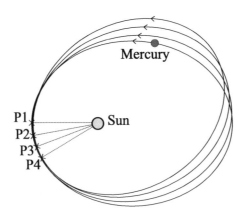

Figure 9.7: Precession of the perihelion of Mercury.

One suggestion was that this anomaly might be due to an unknown planet called Vulcan (Neptune had been discovered in 1846 based on similar discrepancies in the orbit of Uranus). There was no sign of planet Vulcan, however, and it wasn't until 1915 that Einstein was able to give an accurate explanation of the residual precession of Mercury in terms of his theory of general relativity.

Using proper time τ as the parameter λ, the Schwarzschild geodesic equations can be used to derive equations describing the orbital motion of an object of mass m, such as a planet, around a central mass M, such as the Sun. These general relativistic equations have an additional term compared to the tried and tested Newtonian orbital equations. This extra term $(-GMJ^2/m^2c^2r^3$, where J refers to the quantity known as angular momentum that we met in Equation 9.4.27) is negligible for large values of r, ie for larger orbits. However, for smaller values of r (Mercury's orbit, for example) the term becomes significant and has the effect of rotating the orbit through the $\theta = \pi/2$ plane, creating the pretty flower-petal effect shown in Figure 9.8. Dividing the number of seconds of arc in a circle $(360 \times 60 \times 60 = 1{,}296{,}000)$ by the perihelion advance in seconds of arc per century (575) tells us it takes 225,000 years for the perihelion point to trace out one complete orbit of the Sun.

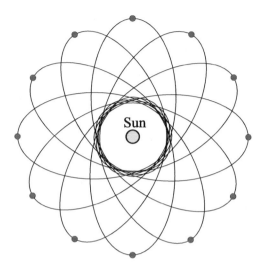

Figure 9.8: Rotation of orbit in its own plane.

It can be shown that the angle of general relativistic perihelion advance $\Delta\phi$ per orbit of a planet orbiting in the constant $\theta = \pi/2$ plane is given by

$$\Delta\phi = \frac{6\pi GM}{a\left(1 - e^2\right)c^2}, \qquad (9.4.32)$$

where M is the total mass of the system (which we can approximate in the case of Mercury to the mass of the Sun), a is the semi-major axis, and e is the orbit's eccentricity.

Problem 9.6. Using (9.4.32) calculate the general relativistic portion of Mercury's perihelion advance in seconds of arc per century. Assume the following:

Mercury has a period (ie the time for one complete orbit) of 87.97 days,

the semi-major axis of Mercury's orbit $a = 5.79 \times 10^{10}$ m,

eccentricity of Mercury's orbit $e = 0.207$,

the mass of the Sun $M = 1.99 \times 10^{30}$ kg,

the gravitational constant $G = 6.67 \times 10^{-11}$ N m^2 kg^{-2},

one year equals 365.25 days.

Using (9.4.32) we can write

$$\Delta\phi = \frac{6\pi \times 6.67 \times 10^{-11} \times 1.99 \times 10^{30}}{5.79 \times 10^{10} \times \left(1 - (0.207)^2\right) \times (3 \times 10^8)^2}. \qquad (9.4.33)$$

The angle $\Delta\phi$ is in radians. We want the angle expressed in seconds of arc. There are $180 \times 60 \times 60$ seconds of arc in π radians so we need to multiply (9.4.33) by $(180 \times 3600)/\pi$ to give $\Delta\phi$ in seconds of arc, ie

$$\Delta\phi = \frac{6\pi \times 6.67 \times 10^{-11} \times 1.99 \times 10^{30} \times 180 \times 3600}{5.79 \times 10^{10} \times \left(1 - (0.207)^2\right) \times (3 \times 10^8)^2 \times \pi}$$

$$\Delta\phi = \frac{516.065 \times 10^{24}}{49.877 \times 10^{26}} = 0.1035$$

seconds of arc per orbit. We want seconds of arc per century, so we need to multiply our answer by the number of orbits in one century, which equals the number of days in a century (365.25×100) divided by the period 87.97. We can therefore say

$$\Delta\phi = 0.1035 \times \frac{365.25 \times 100}{87.97}$$

$$\Delta\phi = 42.97 \text{ seconds of arc per century.}$$

Relativistic precession of the perihelion also affects other planets and objects in our Solar System, as shown in Table 9.1 (from Lambourne [17]).

Planet	Predicted relativistic precession – seconds of arc per century	Observed relativistic precession – seconds of arc per century
Mercury	43.0	43.1 ± 0.5
Venus	8.6	8.4 ± 4.8
Earth	3.8	5.0 ± 1.2
Icarus	10.3	9.8 ± 0.8

Table 9.1: Predicted and observed rates of relativistic precession of perihelion of planets and minor body Icarus.

9.4.8 Gravitational deflection of light – the third test of general relativity

In 1911 consideration of the equivalence principle led Einstein to believe that light would be deflected in a gravitational field. It wasn't until 1915, however, after he had successfully incorporated curved spacetime into a gravitational theory of relativity, that he was able to make an accurate prediction as to the magnitude of such a deflection.

Figure 9.9: Arthur Stanley Eddington (1882–1944).

Einstein suggested that the bending of light from stars appearing close to the Sun (normally, of course, these stars are hidden by the Sun's glare) could be measured during a total eclipse. In 1919 British astrophysicist Arthur Eddington (Figure 9.9) led the famous expedition to observe a solar eclipse and test general relativity. Eddington set up camp on the island of Principe, in the Gulf of Guinea off the coast of west Africa. Another team, led by Andrew Crommelin, observed the eclipse in northern Brazil. Eventually, after returning to England and analysing the data, Eddington confirmed that starlight was deflected as predicted by general relativity. In September 1919 Einstein received a cable from Hendrik Lorentz telling him the good news. Einstein wrote to his mother:

> 'Dear Mother – Good news today. H.A. Lorentz has wired me that the British expeditions have actually proved the light deflection near the sun.'

Later that year, the news was announced publicly at a joint meeting of the Royal Society and the Royal Astronomical Society in London, and Einstein became a worldwide celebrity.

That's the meeting, incidentally, where as he was leaving, Eddington was (allegedly) asked whether it was true that only three people in the world understood the theory of general relativity. When Eddington refrained from answering, his questioner said, 'Don't be modest Eddington.' Eddington replied, 'Not at all. I was wondering who the third one might be.'

Although optical measurements of light deflection are in broad agreement with general relativity there remain significant experimental difficulties with this method, and the results are not totally conclusive. With the advent of large radio telescopes and the discovery of quasars (very distant emitters of electromagnetic radiation), a new technique known as **radio interferometry** has been developed. This involves two widely spaced radio telescopes comparing radiation from a quasar as it passes behind the Sun. These results have shown only a 0.04% deviation from the predictions of general relativity.

Returning to the mathematics – using an alternative parameter to proper time, the Schwarzschild geodesic equations can be used to derive null geodesic equations that describe the path of a light ray in spacetime. The merest outline of this quite difficult (well, I thought it was) derivation is that we start with the geodesic equation (9.4.25)

$$\frac{d^2\phi}{d\lambda^2} + \frac{2}{r}\frac{dr}{d\lambda}\frac{d\phi}{d\lambda} + 2\frac{\cos\theta}{\sin\theta}\frac{d\theta}{d\lambda}\frac{d\phi}{d\lambda} = 0,$$

and limit motion to the $\theta = \pi/2$ plane, enabling us to simplify to

$$\frac{d^2\phi}{d\lambda^2} + \frac{2}{r}\frac{dr}{d\lambda}\frac{d\phi}{d\lambda} = 0,$$

from which, setting $u = 1/r$, and some mathematical hand waving, can be obtained

$$\frac{d^2u}{d\phi^2} + u = 3mu^2,$$

where $m = GM/c^2$. A few more passes of the magic wand and a final abracadabra yields the useful equation

$$\delta = \frac{4GM}{c^2b}, \tag{9.4.34}$$

where δ is the angle of deflection (in radians) of a light ray passing close to a spherically symmetric body of mass M, and b is the distance of closest approach to the origin (which we can approximate to the distance from the centre of the central body) – see Figure 9.10.

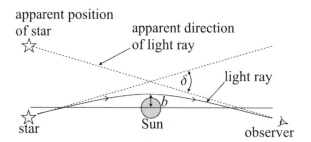

Figure 9.10: The deflection of light due to the curvature of spacetime close to the Sun.

We can see from (9.4.34) that the angle of deflection δ will be greatest when b is least, which will be when the ray of light just grazes the surface of the massive body. Equation (9.4.34) gives twice the deflection angle to that predicted by Newtonian gravitational theory.

Problem 9.7. Calculate the angle of deflection of a ray of light (in seconds of arc) of a ray of light just grazing the Sun. Assume the following:

the radius of the Sun $= 6.96 \times 10^8$ m,

the mass of the Sun $M = 1.99 \times 10^{30}$ kg,

the gravitational constant $G = 6.67 \times 10^{-11}$ N m^2 kg^{-2}.

We use (9.4.34), and let $b =$ the radius of the Sun for a grazing ray:

$$\delta = \frac{4GM}{c^2 b}$$

$$\delta = \frac{4 \times 6.67 \times 10^{-11} \times 1.99 \times 10^{30}}{6.96 \times 10^8 \times (3 \times 10^8)^2}$$

$$\delta = \frac{53.093 \times 10^{-5}}{62.64} = 8.478 \times 10^{-6} \text{radians.} \qquad (9.4.35)$$

We want the answer in seconds of arc. There are $180 \times 60 \times 60$ seconds of arc in π radians so we need to multiply (9.4.35) by $(180 \times 3600)/\pi$ to give δ in seconds of arc,

$$\delta = \frac{8.478 \times 10^{-6} \times 180 \times 60 \times 60}{3.142}$$

$$\delta = 1.75 \text{ seconds of arc.}$$

9.4.9 Gravitational time delay of signals passing the Sun – the fourth test of general relativity

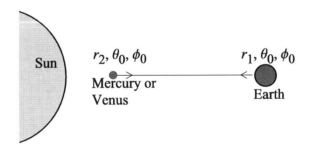

Figure 9.11: Gravitational time delay.

In 1964 Irwin I. Shapiro suggested that the curvature of spacetime could be measured by bouncing a high-powered radar beam off a planet or satellite as it passed behind the Sun. The transit time (ie the beam's journey time from Earth to the target and then back to Earth) should be slightly longer in spacetime curved by the influence of the Sun's gravitational field than in flat space. This is known as the **Shapiro time delay experiment**.

The experimental details are complex, but we can get a basic feel for the principle behind the experiment by considering the simple configuration as shown in Figure 9.11 – taken from Foster and Nightingale [9].

We want to compare two transit time intervals as measured by our observer on Earth:

1. the proper time interval $\Delta\tau$, calculated from the coordinate time equation (9.4.5);

2. the proper time interval $\Delta\tau'$, calculated from the proper distance equation (9.4.19).

If spacetime were flat, we would expect these two time intervals to be equal. The fact that they aren't is a consequence of the slight curvature of spacetime caused by the Sun.

The spatial coordinates of the observer on Earth are r_1, θ_0, ϕ_0, and of the object r_2, θ_0, ϕ_0. Therefore, $d\theta = d\phi = 0$, and r_2 is obviously less than r_1.

First, we measure the proper time taken by the there and back journey of the radar pulse based on the Schwarzschild coordinate time taken.

We saw in Section 7.4 that null geodesics (where $ds^2 = c^2d\tau^2 = 0$) describe the paths of light rays. Because a radar pulse travels at the speed of light we can use (9.2.2)

$$ds^2 = \left(1 - \frac{2m}{r}\right)c^2dt^2 - \frac{dr^2}{1 - \frac{2m}{r}} - r^2d\theta^2 - r^2\sin^2\theta d\phi^2$$

and write

$$0 = \left(1 - \frac{2m}{r}\right)c^2dt^2 - \frac{dr^2}{1 - \frac{2m}{r}}$$

$$\left(1 - \frac{2m}{r}\right)^2 c^2dt^2 = dr^2$$

$$\frac{dr}{dt} = \pm c\left(1 - \frac{2m}{r}\right). \tag{9.4.36}$$

Which is the coordinate speed of light in the radial direction, ie the speed of light measured using the r and t Schwarzschild coordinates. The coordinate time for both legs of the journey can be found by integrating (9.4.36) to give

$$\Delta t = -\frac{1}{c}\int_{r_1}^{r_2} \frac{dr}{1 - 2m/r} + \frac{1}{c}\int_{r_2}^{r_1} \frac{dr}{1 - 2m/r}$$

or

$$\Delta t = \frac{2}{c}\int_{r_2}^{r_1} \frac{dr}{1 - 2m/r}. \tag{9.4.37}$$

This is the coordinate time taken. We need the proper time $\Delta\tau$ for the trip as measured by an observer on Earth at r_1. We find this by substituting (9.4.37) into (9.4.5)

$$d\tau = \left(1 - \frac{2m}{r}\right)^{1/2} dt$$

to give

$$\Delta\tau = \left(1 - \frac{2m}{r_1}\right)^{1/2} \Delta t$$

$$\Delta\tau = \frac{2}{c}\left(1 - \frac{2m}{r_1}\right)^{1/2} \int_{r_2}^{r_1} \frac{dr}{1 - 2m/r}. \tag{9.4.38}$$

Now, we measure the proper time taken by the radar pulse based on the proper distance it travels, given by (9.4.19)

$$d\sigma = \frac{dr}{\left(1 - \frac{2m}{r}\right)^{1/2}}.$$

We can integrate this equation to give the proper distance σ travelled by the radar pulse. If we then multiply that distance by 2 to account for the there and back journey, then divide the result by c we obtain the total transit time $\Delta\tau'$, which is

$$\Delta\tau' = \frac{2}{c} \int_{r2}^{r1} \left(1 - \frac{2m}{r}\right)^{-1/2} dr. \qquad (9.4.39)$$

For curved spacetime $\Delta\tau \neq \Delta\tau'$, that difference being the time delay caused by spacetime curvature.

In practice, the time delay is too short to measure when the target planet is between the Earth and the Sun (known as inferior conjunction). Shapiro measured the delay when the planets – he used Mercury and Venus – were behind the Sun (superior conjunction), which requires a more sophisticated analysis than the one we've used in our example. For Venus the measured time delay was about $200\,\mu s$ (microseconds), which agreed with the theoretical prediction to within 5%.

More accurate results (not depending on the terrain of the planet) were later achieved by bouncing signals off the Viking and Voyager space probes, launched in the 1970's, with time delays in agreement with theoretical predictions to an accuracy of one part in one thousand. The most precise measurement of gravitational time delay to date used signals from the Cassini spacecraft (see Figure 9.12), launched in 1997, as it journeyed to Saturn. In 2003, results from Cassini showed only a 0.002% deviation from the predictions of general relativity.

Figure 9.12: High-precision test of general relativity by the Cassini space probe (NASA artist's impression).

10 Schwarzschild black holes

The black hole epitomizes the revolution wrought by general relativity. It pushes to the extreme – and therefore tests to the limit – the features of general relativity (the dynamics of curved spacetime) that set it apart from special relativity (the physics of static, "flat" spacetime) and the earlier mechanics of Newton.

JOHN WHEELER

10.1 Introduction

A black hole is a region of spacetime that has undergone gravitational collapse to such an extent that nothing, not even light, can escape. General relativity predicts the formation of black holes through the distortion of spacetime when a central mass, such as a collapsing star, becomes sufficiently dense. A black hole consists of a mathematically defined surface, known as an **event horizon**, surrounding a central singularity. A singularity is a point where the curvature of spacetime becomes infinite. Roger Penrose proved in 1965 that once an event horizon forms, a singularity must form inside it. The event horizon is a one-way boundary. Objects and light that cross the event horizon must then fall into the singularity. This means that events occurring within the event horizon – ie inside the black hole – cannot be seen by an external observer.

The term 'black hole' was first used by John Wheeler in 1967. However, the idea of light being unable to escape from a super-dense star was suggested in the eighteenth century, independently, by British physicist John Michell (1724–1793) and French mathematician and physicist Pierre-Simon Laplace (1749–1827). Their analysis was of course based on Newtonian gravitation and used the notion of escape speed (2.5.3) that we referred to earlier. We saw that for an object trying to 'break free' of a gravitational field the escape speed v_e is given by

$$v_e = \sqrt{\frac{2GM}{R}}.$$

If we let $R = 2GM/c^2$, then

$$v_e = \sqrt{\frac{2GMc^2}{2GM}} = c.$$

Therefore, if $R < 2GM/c^2$, light cannot escape from the object. There are fundamental differences between the Michell-Laplace and modern concepts of a black hole. For example, the speed of light has no particular relevance in non-relativistic mechanics, so objects moving faster than c could escape from a Michell-Laplace black hole. However, the eighteenth century physicists did calculate the correct radius of such a modern simple black hole – the quantity we have already met, now known as the Schwarzschild radius $R_S = \frac{2GM}{c^2}$.

We've previously mentioned that it took some time following the publication of the Schwarzschild metric for physicists to realise that not all that metric's singularities were physically significant, opening up the possibility that certain objects might be able to undergo a total gravitational collapse. In the 1920's it was suggested that the small dense stars known as **white dwarfs** (typically, think of a star with the mass of the Sun, but the diameter of the Earth) were supported against self-collapse by internal quantum electron effects known as **degeneracy pressure**. In 1931 the Indian astrophysicist Subrahmanyan Chandrasekhar (1910–1995) proposed an upper limit (about 1.4 times the mass of the Sun) for white dwarfs beyond which degeneracy pressure would be unable to resist gravity and the result would be gravitational collapse.

After the discovery of the neutron in 1932, it was suggested there might exist super-dense **neutron stars**. (take our solar mass white dwarf and shrink it to about 10 km across, for example). It was thought that the degeneracy pressure of the neutrons that are the main constituents of these stars would allow them to avoid gravitational collapse, and therefore exceed the upper limit that Chandrasekhar had proposed for white dwarfs.

In 1939 J. Robert Oppenheimer (1904–1967) and H. Snyder (1913–1962) suggested that neutron stars above approximately three solar masses would also collapse into black holes. Oppenheimer and Snyder showed, using general relativity, that for a distant observer the collapse of any star into a black hole takes an infinitely long time. For such an observer, the star's surface would appear to slow down and stop as it shrinks towards the event horizon. Because of gravitational redshift, the star would also become redder and dimmer as it contracted.

An observer unlucky enough to fall into a black hole, however, would measure a *finite* time to cross the event horizon and fall inwards to the central singularity. Just as the equator is an invisible line circling the Earth, the event horizon is an invisible surface surrounding a black hole. A falling observer (assuming that he hadn't been ripped apart by tidal forces, a process aptly known as **spaghettification**) wouldn't notice anything significant when passing through the event horizon, though it would of course mark his point of no return as he plunged towards the singularity.

Lambourne [17] states that many scientists regard Oppenheimer and Snyder's work, 'With its acceptance of complete gravitational collapse and recognition of the coordinate nature of the singularity at $r = R_S$, as the true birth of the black hole concept.'

The popular image of a black hole is of some huge celestial vacuum cleaner sucking up everything in the universe. This is incorrect. As long as you don't get too close, a black hole is perfectly well behaved, with a gravitational field identical to that of any other body of the same mass. If the Sun, for example, were to be replaced by a black hole of equivalent mass (ie shrunk down to its Schwarzschild radius of about 3 km) the sky would look very different, but the Earth and other planets would remain in their current orbits.

John Wheeler came out with the memorable quote that, 'Black holes have no hair.' This strange observation refers to the idea that black holes have only three externally measurable properties: mass, angular momentum and electric charge. All black holes have mass, so there are four different metrics that uniquely describe black holes with and without angular momentum and electric charge. The complete list is:

- Schwarzschild metric – describes a black hole with mass only.

- Kerr metric – describes a black hole with mass and angular momentum only.

- Reissner-Nordström metric – describes a black hole with mass and electric charge only.

- Kerr-Newman metric – describes a black hole with mass, angular momentum and electric charge.

Black holes may also be classified by size, for example ($M_\odot$ equals one solar mass):

Mini black holes	$0 - 0.1\,M_\odot$
Stellar mass black holes	$0.1 - 100\,M_\odot$
Intermediate mass black holes	$100 - 10^5\,M_\odot$
Supermassive black holes	$10^5 - 10^{10}\,M_\odot$

There is a theoretical, but unlikely, possibility that very low mass mini black holes could be produced by the Large Hadron Collider at CERN. Fortunately for all of us, CERN claim these hypothetical black holes would be short lived and harmless.

Figure 10.1: NASA artist's impression of Cygnus X-1 stellar mass black hole.

Although stellar mass black holes cannot be observed directly, material that falls towards them is thought to produce huge amounts of X-rays, which are detectable (there is a nice analogy of a black hole being a 'messy eater' – not everything it tries to shovel into its mouth actually gets eaten). X-ray emitting matter spiralling towards the black hole forms what is known as an **accretion disc**. Indirect evidence for this type of black hole comes predominantly from binary star systems, where a companion star sheds matter that falls into the supposed neighbouring black hole. Cygnus X-1 (see Figure 10.1) is one of the strongest X-ray sources seen from Earth, and is widely accepted as a likely stellar mass black hole. So far, around twenty binary systems thought to contain black holes have been found.

The ultimate fate of a star depends on its mass. Most average sized stars (including our Sun) will become white dwarfs. Larger stars will end up as neutrons stars. It is now thought that stellar mass black holes are the evolutionary end point of massive (several times larger than the Sun) stars.

Though theoretically possible, the existence of intermediate mass black holes remains an open question. Some candidates have been proposed, but none are widely accepted.

Most, maybe all, galaxies are thought to contain a supermassive black hole at their centre. There is strong evidence, for example, that our own Milky Way galaxy contains a huge central black hole with a mass of about $2.5 \times 10^6 M_\odot$ and a radius of no more than 6.25 light-hours (about the diameter of Uranus' orbit). NGC 4261, a giant elliptical galaxy in the Virgo galaxy cluster, is thought to contain a 400 million solar mass black hole, complete with two 'jets' of material ejected from the inner region of the accretion disk (see Figure 10.2).

Figure 10.2: NASA images of the Active Galaxy NGC 4261.

Because the Schwarzschild radius of an object is proportional to its mass, the event horizon of a supermassive black hole is much larger than that of stellar mass black hole. Tidal forces (the difference in gravitational force between your head and feet if you are falling vertically) in the vicinity of the event horizon surrounding a supermassive black hole are correspondingly much smaller than those around a stellar mass black hole. Lewis and Kwan [18] estimate that your survival time if you fell into a stellar mass black hole would be a fraction of a second, compared to several hours for a supermassive black hole. No matter what type of black hole you've encountered, once you've crossed the event horizon, there's nothing you can do to stop yourself fatally falling into the singularity. However, assuming that (a) you survive crossing the event horizon, and (b) you are in control of a powerful enough spaceship, you may be able to marginally prolong your survival time by judicious firing of your rocket engines. Lewis and Kwan show that if you fall from rest at the event horizon your best strategy is not to use your engines at all. This is because firing your rocket in *any* direction can only shorten your survival time compared to allowing yourself to free fall. The situation is different if you are moving through the event horizon. In that case the authors provide helpful equations (involving the mass of the black hole, how powerful your rocket is and how fast you crossed the event horizon) for you to calculate the optimum time to fire your engines.

10.2 Falling into a black hole

We now take a closer look at the simplest kind of black hole: one with mass, but no electric charge and no spin. Because these black holes are described using the Schwarzschild metric they are known as **Schwarzschild black holes.**

Our path to understanding spacetime in the vicinity of a black hole is via our old friend, the freely falling observer. We want to consider the time taken for such an observer to fall into a black hole. As is usual in relativity, our definition of 'observer' is a loose one encompassing a falling clock or unfortunate human space traveller dramatically plummeting towards an existing singularity or, less fancifully, a hypothetical clock 'sitting' on the surface of a collapsing star. By 'fall into a black hole' we mean how long does it take the observer to first reach the event horizon and then travel onwards to the central singularity. As usual, we are interested in proper time – two measurements of proper time, in fact. First, that recorded by the freely falling observer himself, using his own clock. Second, that measured by a stationary distant observer watching the fall and using *his* own clock. We already know that the answer to these questions was provided by Oppenheimer and Snyder: the distant observer sees the fall taking an infinite time, the falling observer records the fall taking a finite time. Now we'll put a little mathematical flesh onto the bare framework of those conclusions.

10.3 Falling – seen from up close

We are considering the motion of a radially free falling observer. Because radial free fall implies the object is moving 'straight down' toward the centre of the black hole, ϕ is constant. Therefore, from (9.4.27)

$$\frac{J}{m} = r^2 \sin^2 \theta \frac{d\phi}{d\tau} \tag{10.3.1}$$

$d\phi/d\tau = 0$ and $J = 0$. The radial motion equation (9.4.28)

$$\left(\frac{dr}{d\tau}\right)^2 + \frac{J^2}{m^2 r^2}\left(1 - \frac{2GM}{c^2 r}\right) - \frac{2GM}{r} = c^2\left[\left(\frac{E}{mc^2}\right)^2 - 1\right]$$

thus simplifies to

$$\left(\frac{dr}{d\tau}\right)^2 - \frac{2GM}{r} = c^2\left[\left(\frac{E}{mc^2}\right)^2 - 1\right]. \tag{10.3.2}$$

If we assume that our observer starts freely falling at rest from infinitely far away ($r = \infty$), then the conserved energy E is only equal to mc^2. The right-hand side of (10.3.2) therefore equals zero and we have

$$\left(\frac{dr}{d\tau}\right)^2 = \frac{2GM}{r} \tag{10.3.3}$$

$$\frac{dr}{d\tau} = \pm\sqrt{\frac{2GM}{r}}. \tag{10.3.4}$$

We take the negative square root of (10.3.4) (because freely falling implies the observer's r is decreasing) and integrate between an arbitrary point r_1 on the observer's descent and another lower point r_2 (where, obviously, $r_2 < r_1$), giving us

$$\int_{r_1}^{r_2} r^{1/2} dr = -\int_0^\tau \sqrt{2GM} d\tau,$$

where τ is the observer's proper time measured between r_2 and r_1. Evaluating this integral gives

$$\left[\frac{2}{3}r^{3/2}\right]_{r_1}^{r_2} = -\tau\sqrt{2GM},$$

so

$$\tau = \frac{2}{3\sqrt{2GM}}\left(r_1^{3/2} - r_2^{3/2}\right). \tag{10.3.5}$$

To make our calculations a little easier, we can rewrite (10.3.5) in terms of the Schwarzschild radius $R_S = 2GM/c^2$. When we do this, we are in effect multiplying the right-hand side by $\sqrt{c^2}$, so we also need to divide by c to keep the equation balanced:

$$\tau = \frac{2}{3c\sqrt{R_S}}\left(r_1^{3/2} - r_2^{3/2}\right). \tag{10.3.6}$$

This equation tells us the proper time, as measured by the observer, for a radially free falling observer to fall from r_1 to r_2. The essential thing to note about (10.3.6) is that, because it is well behaved and doesn't blow up for any value of r, the freely falling observer measures a finite time to both pass through the event horizon at $r = R_S$ and then travel onwards to the singularity at $r = 0$. This journey, in terms of proper time τ and radial coordinate r is shown in Figure 10.3.

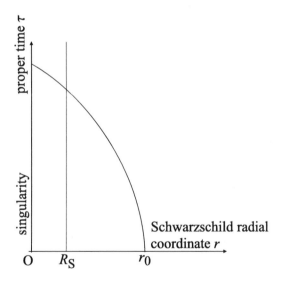

Figure 10.3: Falling into a black hole – proper time τ and radial coordinate r.

Problem 10.1. An observer starts to free fall from infinity towards a three solar mass black hole of Schwarzschild radius $R_S = 8.85\,\text{km}$. How long does it take the observer to fall (a) from $r_1 = 30\,\text{km}$ to the event horizon, (b) to fall from $r_1 = 30\,\text{km}$ to the singularity, and (c) to fall from the event horizon to the singularity? Assume that (10.3.6) is valid inside as well as outside the black hole.

(a) Using 10.3.6, the event horizon $r_2 = 8.85\,\text{km}$, and converting all distances to metres we obtain

$$\tau = \frac{2}{3c\sqrt{R_S}}\left(r_1^{3/2} - r_2^{3/2}\right)$$

$$\tau = \frac{2}{3 \times 3 \times 10^8 \sqrt{8.85 \times 10^3}} \left(\left(30 \times 10^3\right)^{3/2} - \left(8.85 \times 10^3\right)^{3/2} \right)$$

$$\tau = \left(\frac{2 \times \left(\left(5.20 \times 10^6\right) - \left(8.33 \times 10^5\right)\right)}{8.47 \times 10^{10}} \right) = 1.03 \times 10^{-4}\,\mathrm{s} = 0.103\,\mathrm{ms}.$$

(b) We use $r_2 = 0\,\mathrm{km}$ in the above calculation to find the time to fall to the singularity:

$$\tau = \frac{2}{3 \times 3 \times 10^8 \sqrt{8.85 \times 10^3}} \left(\left(30 \times 10^3\right)^{3/2} - \left(0\right)^{3/2} \right)$$

$$\tau = \frac{2 \times 5.20 \times 10^6}{8.47 \times 10^{10}} = 0.123\,\mathrm{ms}.$$

(c) Therefore, the observer takes $0.123 - 0.103 = 0.020$ milliseconds to fall from the event horizon to the singularity.

The above calculations are based on the observer freely falling at rest from infinity. Now let's ask what happens if the observer starts his free fall not from infinity but at rest from a point closer to the black hole, from $r_1 = 30\,\mathrm{km}$, for example. In this case we can no longer assume that the conserved energy E equals mc^2. To calculate the conserved energy we use (10.3.2)

$$\left(\frac{dr}{d\tau}\right)^2 - \frac{2GM}{r} = c^2 \left[\left(\frac{E}{mc^2}\right)^2 - 1 \right].$$

At $\tau = 0$, when the observer starts along his geodesic from rest, $dr/d\tau = 0$ and $r = r_1$. Plugging those values into the above equation to find the conserved energy E gives

$$\left(0\right)^2 - \frac{2GM}{r_1} = c^2 \left[\left(\frac{E}{mc^2}\right)^2 - 1 \right]$$

$$E = \pm mc^2 \sqrt{1 - \frac{2GM}{c^2 r_1}},$$

this being the energy conserved along the observer's geodesic. This value of E is then plugged into (10.3.2), and after some fairly heavy mathematics and a few working approximations (see Lambourne [17], for example) it is possible to derive an equation for the proper time taken to reach the singularity starting free fall at rest from r_1. That equation is

$$\tau_{sing} = \frac{\pi \left(r_1\right)^{3/2}}{2c\sqrt{R_S}}. \tag{10.3.7}$$

Problem 10.2. An observer starts to free fall from rest at $r_1 = 30\,\mathrm{km}$ towards a three solar mass black hole of Schwarzschild radius $R_S = 8.85\,\mathrm{km}$. How long does it take the observer to reach the singularity?

We use (10.3.7) and obtain

$$\tau_{sing} = \frac{3.142 \left(30 \times 10^3\right)^{3/2}}{2 \times 3 \times 10^8 \sqrt{8.85 \times 10^3}} = \frac{1.63 \times 10^7}{5.64 \times 10^{10}}$$

$$\tau_{sing} = 0.29\,\text{ms}.$$

As would be expected, it takes slightly longer (0.29 milliseconds) to fall into the black hole starting at rest from $r = 30\,\text{km}$, compared to the 0.103 milliseconds from $r = 30\,\text{km}$ when the free fall starts from infinity. The crucial point to note is that no matter where the observer starts from, he records his journey into the black hole as taking a finite time (and, in our example, an extremely short time!).

Next, we'll look at what a stationary distant observer sees when watching something fall into a black hole.

10.4 Falling – seen from far, far away

Our stationary distant observer is carefully watching some object falling into a black hole or, perhaps, watching a star undergoing total gravitational collapse. What does he see? First, we'll consider how long a light signal from the freely falling object takes to reach the distant observer. Then we'll consider how the distant observer measures the relationship between coordinate time t and the position r of the freely falling object. We saw when looking at gravitational time dilation (Section 9.4.2) that the coordinate time t of an event in Schwarzschild spacetime is the same as the proper time τ measured by a stationary distant observer, ie $d\tau_\infty = dt$. This means that if we can find the coordinate time taken for a light signal, or for a change in position of the freely falling object, we have automatically found the distant observer's proper time measurement for those two events. That's what we're now going to do.

10.4.1 Journey time of a light signal

First, we'll find the journey time for a light signal/photon emitted by an object freely falling into a black hole and received by a distant observer. To keep things as simple as possible, we assume that both observers are on the same radial line, ie $d\theta = d\phi = 0$.

We saw in Section 9.4.9 that the coordinate speed of light in Schwarzschild spacetime in a radial direction ($d\theta = d\phi = 0$) is given by (9.4.36)

$$\frac{dr}{dt} = \pm c \left(1 - \frac{2m}{r}\right),$$

which we can rearrange to give

$$\frac{dt}{dr} = \pm \frac{1}{c} \left(\frac{1}{1 - 2m/r}\right),$$

which, in terms of the Schwarzschild radius (9.4.1) $R_S = \frac{2GM}{c^2} = 2m$, can be written as

$$\frac{dt}{dr} = \pm\frac{1}{c}\left(\frac{1}{1 - R_S/r}\right), \tag{10.4.1}$$

which we can integrate to find the total time taken for the photon

$$t_2 - t_1 = \int_{t_1}^{t_2} dt = \frac{1}{c}\int_{r_1}^{r_2} \frac{dr}{1 - R_S/r},$$

where t_1 and r_1 are the coordinate time and radial coordinate of the photon emitted by the object freely falling into the black hole, and t_2 and r_2 are the coordinate time and radial coordinate when the same photon is received by the distant observer. The integral of $dr/\left(1 - R_S/r\right)$, found either by hand or by using an online integral calculator ([33], for example), is

$$\int \frac{dr}{1 - R_S/r} = R_S \ln\left(r - R_S\right) + r + \text{constant}.$$

So, we can say

$$t_2 - t_1 = \frac{r_2 - r_1}{c} + \frac{R_S}{c}\ln\left(\frac{r_2 - R_S}{r_1 - R_S}\right). \tag{10.4.2}$$

What does this equation tell us?

First, if there were no central mass (ie $M = 0$, meaning $R_S = 0$), the right-hand term would disappear, we would be describing flat space, and the photon journey time would be simply distance divided by speed $(r_2 - r_1)/c$.

Second, as the falling object approaches the event horizon $(r_1 \to R_S)$, the right-hand term $\to \infty$, and therefore the photon journey time also approaches infinity. Because coordinate time is the same as a distant observer's proper time, such an observer will never quite see the object reach the event horizon.

Third, we've justified the assumption we made in Section 9.4.2 (see just after (9.4.6)) that the coordinate time difference between two signals sent in Schwarzschild spacetime is the same as that recorded by a stationary distant observer. Equation (10.4.2) tells us that the journey time of a photon only depends on the coordinate positions of the signal's emitter and receiver. To see this, let's assume for the moment that (10.4.2) is describing a stationary object in Schwarzschild spacetime, and that shortly after the t_1 signal is emitted, another signal t_3 is broadcast and received by the distant observer at t_4. Equation (10.4.2) tells us that

$$t_4 - t_3 = t_2 - t_1,$$

which we can juggle about to give

$$t_3 - t_1 = t_4 - t_2,$$

which tells us the coordinate time interval Δt between two emitted signals in Schwarzschild spacetime is the same as the coordinate time interval Δt between those signals being received by a distant observer.

10.4.2 Relationship between coordinate time t and coordinate distance r

We know that in Schwarzschild spacetime, coordinate time t is the same as proper time τ as measured by a distant stationary observer. Therefore, it will be helpful for us to find the relationship between coordinate time and coordinate distance r of a massive object (not, as in the previous section, a photon) freely falling into a black hole as measured by a such an observer. In other words, we want to find the function linking t to r. So, first we need to find dt/dr.

We start with (10.3.3), which describes a freely falling something or other starting its journey to the black hole at rest from infinitely far away ($r = \infty$), with consequent conserved energy $E = mc^2$:

$$\left(\frac{dr}{d\tau}\right)^2 = \frac{2GM}{r}.$$

We take the negative square root of (10.3.3) (freely falling implies the observer's r is decreasing) to give

$$\frac{dr}{d\tau} = -\sqrt{2GM/r},$$

which we can invert and rewrite in terms of the Schwarzschild radius $R_S = 2GM/c^2$ as

$$\frac{d\tau}{dr} = -\frac{1}{c\sqrt{R_S/r}} = -\frac{1}{c\left(R_S/r\right)^{1/2}}. \tag{10.4.3}$$

Next we consider the conserved energy E equation (9.4.26)

$$\frac{E}{mc^2} = \left(1 - \frac{2GM}{c^2 r}\right)\frac{dt}{d\tau},$$

which, as we've already assumed $E = mc^2$, becomes

$$\frac{dt}{d\tau} = \left(1 - \frac{2GM}{c^2 r}\right)^{-1}$$

or, in terms of the Schwarzschild radius $R_S = 2GM/c^2$,

$$\frac{dt}{d\tau} = \left(1 - \frac{R_S}{r}\right)^{-1}. \tag{10.4.4}$$

Using the chain rule we know

$$\frac{dt}{dr} = \frac{d\tau}{dr} \times \frac{dt}{d\tau}$$

so we can multiply (10.4.3) by (10.4.4) to get

$$\frac{dt}{dr} = -\frac{1}{c\left(R_S/r\right)^{1/2}\left(1 - \frac{R_S}{r}\right)}$$

$$\frac{dt}{dr} = -\frac{(r)^{1/2}}{c\left(R_S\right)^{1/2}\left(1 - \frac{R_S}{r}\right)}. \tag{10.4.5}$$

This looks complicated, but we can see that as $r \to R_S$, $(1 - R_S/r) \to 0$, meaning $dt/d\tau \to -\infty$. In other words, a tiny decrease in r as the falling object approaches the event horizon results in an ever larger (approaching infinite) increase in t. The distant observer never actually sees the falling object or collapsing star reach the event horizon as it takes an infinite coordinate time (proper time as measured by the distant observer) to do so. The situation is illustrated in Figure 10.4, which also shows the finite proper time for the fall (as recorded by the freely falling observer) that we saw in Figure 10.3.

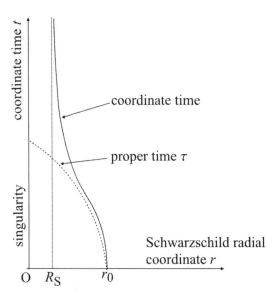

Figure 10.4: Falling into a black hole – coordinate time t and radial coordinate r.

In order to plot the coordinate time curve shown in Figure 10.4, it would be necessary to integrate (10.4.5), which gives (from d'Inverno [5])

$$t_1 - t_0 = \frac{1}{c} \left(\frac{-2}{3 \left(R_S\right)^{1/2}} \left(r^{3/2} - r_0^{3/2} + 3R_S r^{1/2} - 3R_S r_0^{1/2} \right) \right) +$$

$$\frac{R_S}{c} \ln \frac{\left(r^{1/2} + R_S^{1/2} \right) \left(r_0^{1/2} - R_S^{1/2} \right)}{\left(r_0^{1/2} + R_S^{1/2} \right) \left(r^{1/2} - R_S^{1/2} \right)}. \tag{10.4.6}$$

Although this equation looks very different to (10.3.6)

$$\tau = \frac{2}{3c\sqrt{R_S}} \left(r_1^{3/2} - r_2^{3/2} \right),$$

(the one that tells us the proper time according to a freely falling observer), they produce remarkably similar results ($t \approx \tau$) *until* r gets very close to R_S. Only then does the coordinate time tend to shoot off to infinity, as shown in Figure 10.4.

10.4.3 Gravitational redshift and luminosity

In Section 9.4.3, as part of our earlier general discussion of Schwarzschild spacetime, we met the equation for gravitational redshift (9.4.8)

$$f_\infty = f_{em} \left(1 - \frac{2GM}{c^2 r}\right)^{1/2}.$$

The relationship between velocity v, wavelength λ and frequency f is given by (9.4.7)

$$v = f\lambda$$

or, in the case of light,

$$c = f\lambda, \tag{10.4.7}$$

which we can use to rewrite (9.4.8) as

$$\frac{c}{f_\infty} = \frac{c}{f_{em} \left(1 - \frac{2GM}{c^2 r}\right)^{1/2}},$$

and therefore

$$\lambda_\infty = \frac{\lambda_{em}}{\left(1 - \frac{2GM}{c^2 r}\right)^{1/2}}$$

or, in terms of the Schwarzschild radius R_S,

$$\lambda_\infty = \frac{\lambda_{em}}{\left(1 - \frac{R_S}{r}\right)^{1/2}}. \tag{10.4.8}$$

We can see from this equation that as the falling object or collapsing star's surface approaches the event horizon, the wavelength of the emitted radiation λ_∞, as seen by a distant observer, increases, ie as $r \to R_S$, $\lambda_\infty \to \infty$. In other words, the radiation becomes infinitely redshifted.

It can also be shown that as $r \to R_S$, the brightness or luminosity of the emitted radiation as seen by a distant observer $\to 0$.

10.4.4 Distant observer – summary

Now we can (literally) bring our discussion back down to Earth and summarise what an astronomer, on our planet for example, would see when observing a fall into a Schwarzschild black hole, such as a distant star in the process of total gravitational collapse. First, as $r \to R_S$ (ie the star's surface shrinks toward the event horizon), light signals emitted from the star take an increasing time to reach the astronomer. Second, the contracting surface takes an infinite amount of coordinate time to reach the event horizon. Third, as the surface approaches the event horizon, emitted radiation becomes infinitely redshifted and infinitely dim. In short, our astronomer would see the star redden and dim and fade from view into, well, into a black hole. Anything falling into an existing black hole would appear to stop, redden and fade from sight at the event horizon, frozen for eternity in spacetime, even though the object itself had long since plunged into the singularity.

10.5 Coordinates and lightcones

We've seen that a freely falling observer records a finite proper time τ for the journey to the singularity. We now ask, what does he measure in terms of coordinate time t? We've already answered the first part of that question when looking at how a distant observer records the fall. Because coordinate time is the same for both observers, the observer freely falling towards the event horizon will calculate the same coordinate time as he moves towards the event horizon as shown in Figure 10.3, ie coordinate time will shoot off to infinity. However, for him, unlike the distant observer, coordinate time does not equal proper time. Furthermore, as he seamlessly crosses the event horizon, coordinate time will go into reverse and begin to *decrease* as he falls towards the singularity, as shown in Figure 10.5.

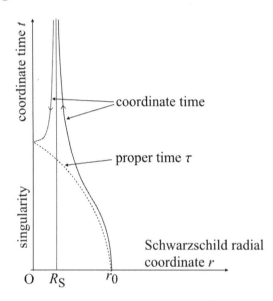

Figure 10.5: Coordinate time t for an observer freely falling into a black hole.

Does this mean that time is now running backwards for the observer? No, it doesn't. The observer's proper time, the time recorded by his super-accurate wristwatch for example, is still ticking away normally. What Figure 10.5 demonstrates are the limitations of Schwarzschild coordinates in visualising the fall into a black hole from the event horizon to the singularity. As we already know, only when $M = 0$ and the Schwarzschild metric reduces to the Minkowski metric, do the coordinates t and r represent real clock-time and real radial distance.

We can further explore the use (and inappropriateness) of Schwarzschild coordinates, by returning to the concept of lightcones. We first came across lightcones when looking at special relativity, where they helped us understand the possible causal relationships between two events. Specifically, we saw that only events that lie within each other's lightcones can be causally related. Recall that the sides of the lightcone are formed from light rays passing through the origin, and that the cone above the origin contains events that may have *been caused* by the event at the origin.

In the flat spacetime of special relativity, light rays travel in straight lines and lightcones can therefore be extended indefinitely. This is not the case in curved spacetime, where light rays do not travel in straight lines. Nevertheless, we can still use small, local lightcones to explore the causal structure of curved spacetime. We know that freely falling massive particles follow time-like geodesics (where $ds^2 > 0$ – see Section 7.4). Although we are particularly interested in freely falling particles, *all* massive particles can be thought of as moving on time-like world-lines through the origin of a succession of local lightcones, and will be constrained (not being able to travel faster than light) by the sides of that lightcone (see Figure 10.6).

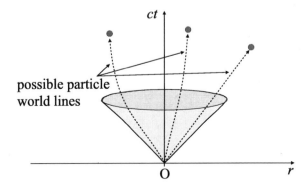

Figure 10.6: Lightcone constraining the world-lines of moving particles.

In Minkowski space, lightcones are defined at the intersection of incoming and outgoing light rays, ie light rays travelling in the negative and positive x direction. Similarly, in Schwarzschild spacetime, we can also draw a spacetime diagram showing the paths of incoming and outgoing light rays, each defined by a null geodesic (where $ds^2 = 0$ – see Section 7.4). We could then draw lightcones at the intersection of these ingoing and outgoing null geodesics.

And what are the equations for null geodesics or photon paths in Schwarzschild spacetime? Recall that we found them, at least for radially moving photons, in Section 10.4.1, when finding the journey time of a photon from a falling to a distant observer. On a spacetime diagram, the gradient of these null geodesic paths would be equal to $d\left(ct\right)/dr$ (using our usual ct time units), which we obtain by multiplying (10.4.1)

$$\frac{dt}{dr} = \pm \frac{1}{c} \left(\frac{1}{1 - R_S/r} \right)$$

by c to give

$$\frac{dct}{dr} = \pm \left(\frac{1}{1 - R_S/r} \right), \tag{10.5.1}$$

and which we could integrate (as we did in Section 10.4.1) to give a slight variation of (10.4.2), ie

$$ct = \pm \left((r - r_1) + R_S \ln \left(\frac{r - R_S}{r_1 - R_S} \right) \right), \tag{10.5.2}$$

the plus sign in front of the right-hand side of (10.5.2) describing the paths of outgoing photons, the minus sign ingoing photons. If we plot (10.5.2) using different values of r_1 (corresponding to where the outgoing and ingoing photons cross the r axis) we obtain the spacetime diagram shown in Figure 10.7.

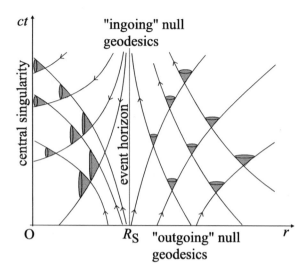

Figure 10.7: Ingoing and outgoing null geodesics in Schwarzschild coordinates.

What does this diagram tell us?

First, at sufficiently large values of r, the gradient of the ingoing and outgoing null geodesics $\to \pm 1$. Equation (10.5.1) confirms this: as $r \to \infty$ then $(1 - R_S/r)^{-1} \to 1$ and $dct/dr \to 1$, meaning the lightcones revert to those of flat space, making angles of $45°$ with the coordinate axes.

Second, as $r \to R_S$ the lightcones become narrower, when $r < R_S$ they broaden and flip over abruptly before narrowing again, indicating that inside the event horizon both photons and massive particles must relentlessly move toward the central singularity at $r = 0$ (see Figure 10.8).

The reversal of coordinate time and strange behaviour of the lightcones illustrates the inappropriateness of Schwarzschild coordinates for the region $r \leq R_S$. If we want to smoothly show the path of an in-falling particle or photon we need to use other coordinates, ones that can cope with the fact that $r = R_S$ is a coordinate and not a physical singularity.

One solution to this problem is a coordinate system known as **advanced Eddington-Finkelstein coordinates**. The new coordinate ct' is defined by

$$ct' = ct + R_S \ln\left(\frac{r}{R_S} - 1\right), \tag{10.5.3}$$

which we can differentiate (see Section 1.10.1.5) to give

$$cdt' = cdt + \left(\frac{r}{R_S} - 1\right)^{-1} dr,$$

and then substitute into the Schwarzschild metric line element to give

$$ds^2 = \left(1 - \frac{R_S}{r}\right)c^2 dt'^2 - 2\frac{R_S}{r}cdt'dr - \left(1 + \frac{R_S}{r}\right)dr^2 - r^2\left(d\theta^2 + \sin^2\theta d\phi^2\right), \tag{10.5.4}$$

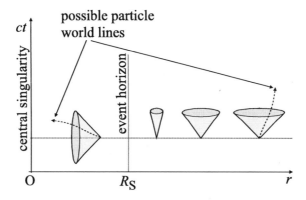

Figure 10.8: Flipping lightcones in Schwarzschild coordinates.

which looks pretty horrible, but crucially is non-singular at $r = R_S$. We can therefore use advanced Eddington-Finkelstein coordinates to illustrate smoothly what happens when objects or photons cross the event horizon. The spacetime diagram for advanced Eddington-Finkelstein coordinates is shown in Figure 10.9.

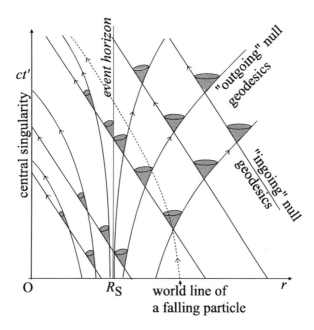

Figure 10.9: Ingoing and outgoing null geodesics in advanced Eddington-Finkelstein coordinates.

What does this diagram tell us?

As with Schwarzschild coordinates, at sufficiently large values of r, the lightcones revert to those of flat space, making angles of $45°$ with the coordinate axes.

The important thing to note here is that ingoing particles and photons are well behaved – coordinate time doesn't reverse or shoot off to infinity as it does with Schwarzschild coordinates. Consequently, the lightcones don't abruptly flip about as they cross $r = R_S$, but gently tilt toward the ct' axis, showing that anything inside the event horizon must move to the singularity (see Figure 10.10). Photons emitted at $r = R_S$ stay on the event horizon.

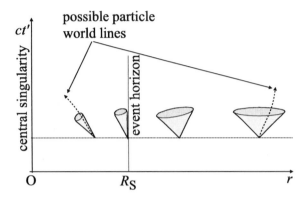

Figure 10.10: Changing lightcones in advanced Eddington-Finkelstein coordinates.

Ingoing null geodesics are represented straight lines. However, this can be reversed by changing the sign in (10.5.3) to give

$$ct' = ct - R_S \ln \left(\frac{r}{R_S} - 1 \right),$$

which would straighten out the *outgoing* radial null geodesics, a coordinate system known as **retarded Eddington-Finkelstein coordinates**. We can do this because, unlike Schwarzschild coordinates, Eddington-Finkelstein coordinates are not time-symmetric.

Although an improvement on Schwarzschild coordinates, Eddington-Finkelstein coordinates are not perfect. The advanced form are good for describing in-falling particles, but not out-falling; the retarded type are good for describing out-falling particles, but not in-falling. These difficulties provide the motivation for other, more complicated, coordinate systems used to describe Schwarzschild black holes such as **Kruskal coordinates** that are non-singular for all $r \neq 0$.

We started this chapter with a quotation from John Wheeler, who has been called 'the father of the black hole'. It's fitting to end with another:

> '[The black hole] teaches us that space can be crumpled like a piece of paper into an infinitesimal dot, that time can be extinguished like a blown-out flame, and that the laws of physics that we regard as "sacred", as immutable, are anything but.'

11 Cosmology

Space is big. You just won't believe how vastly, hugely, mind-bogglingly big it is. I mean, you may think it's a long way down the road to the chemist's, but that's just peanuts to space.

DOUGLAS ADAMS

11.1 Introduction

Cosmology is the study of the nature, origin and evolution of the universe. Gravity dominates our understanding of the universe, and general relativity, the best current theory of gravitation, thus underpins modern cosmology. Newtonian theory is fine for describing gravitational phenomena – such as the Solar System, galaxies, even clusters of galaxies – where the system's mass is small in relation to its size. But compared to the size of the known universe these structures are insignificant verging on tiny. On a cosmological scale, as we consider larger and larger volumes of space the ratio of mass to size increases, and general relativity becomes the essential theoretical framework.

We here give only a brief introduction to this (literally) huge subject. We start with four key observed properties of the universe that underpin modern cosmology:

- the dark night sky
- the cosmological principle
- Hubble's law
- Cosmic microwave background radiation.

11.2 Key observed properties of the universe

11.2.1 The dark night sky

The sky at night is dark. So what, you might say. The importance of the observed fact that the night sky is dark is that it conflicts with the once common assumption of an infinite and static universe. This contradiction, known as **Olbers' paradox**, – after Heinrich Olbers (1758–1840), a German astronomer – or the **dark night sky paradox**, occurs because if there were an infinite number of infinitely old stars, every line of sight should end at a star, the light from that star would have had sufficient time to reach us, and the night sky should therefore be as bright as the surface of the average star. This would still be true even if the stars were obscured by hypothetical clouds of cosmic matter – eventually, the starlight would heat the matter up so that it emits as much light as it absorbs.

The principle reason that the night sky is dark is because the universe has a finite age (about 13.7 billion years), so the light from more distant stars hasn't had time to yet reach us. A secondary reason is that because the universe is expanding, the light from more distant stars is redshifted into obscurity.

11.2.2 The cosmological principle

This principle, based on a wealth of diverse evidence, states that on a sufficiently large scale and at any given time, the universe looks the same to all observers wherever they are. Specifically this means the universe is:

- **Homogeneous** – uniform throughout space, ie the number of stars per unit volume is roughly the same everywhere.

- **Isotropic** – uniform in all directions, eg the number of stars per unit solid angle is roughly the same in all directions.

At first thought, these assumptions are questionable to say the least. The Solar System, for example, contains just one star and is definitely not homogeneous or isotropic. Nor is our Galaxy, the Milky Way, a disc-like structure approximately 100,000 light-years across, with the greatest concentration of its 200 billion stars towards the centre of the disc. Galaxies tend to be concentrated in groups and clusters. The Milky Way is part of the Local Group of 40 or so galaxies. Clusters of galaxies form superclusters separated by vast emptier regions known as voids – think of a sponge or holey cheese. Even at these huge scales the universe is 'lumpy', with superclusters denser than the voids between them. However, on an even larger scale than superclusters and voids, deep sky galaxy surveys such as the Sloan Digital Sky Survey or the 2dF Galaxy Redshift Survey indicate that the universe finally becomes homogeneous.

This is illustrated in Figure 11.1, which shows a slice of the universe surveyed by the 2dF (Two-degree-Field) galaxy redshift survey conducted by the Anglo-Australian Observatory. The image contains over 100,000 galaxies stretching out more than 4 billion light-years.

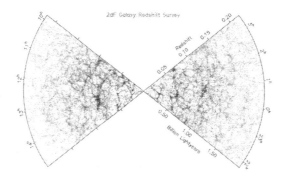

Figure 11.1: 2dF map of the distribution of galaxies.

The astronomical unit of distance known as the **parsec** (symbol – pc) is approximately equal to 3.26 light-years ($\approx 3.09 \times 10^{16}$ m). Cosmological evidence indicates that in any part of the universe, the average density of stuff in a sphere radius 100 Mpc (recall that M equals mega, so 1 Mpc equals one million parsecs) is the same as in any other sphere anywhere else of the same size. On a similar scale, the universe also becomes isotropic, looking more or less the same in all directions.

Figure 11.2 shows the temperature (about 2.725 K) of all-sky cosmic background radiation taken by the Wilkinson Microwave Anisotropy Probe (WMAP) in 2003. Anisotropy is the property of being directionally dependent, ie the opposite of isotropy. The different colours (hard to see in black and white, but in the original coloured image red is warmer, blue is cooler) show tiny temperature variations or anisotropies of less than one part in ten thousand – further evidence for the isotropy of the universe. The radiation recorded in this image was emitted almost 14 billion years ago, when the universe was in its infancy, long before the formation of stars and galaxies. The small fluctuations of matter density indicated by the different colour patterns eventually evolved to form galaxies.

Figure 11.2: WMAP image of the cosmic microwave background radiation (CMBR).

Incidentally, homogeneity is not equivalent to isotropy. A universe that is isotropic for all observers is necessarily homogeneous, but the reverse isn't true. For example, a universe permeated by a uniform magnetic field would be homogeneous, but not isotropic as the field would have a definite direction at each point. Luckily for cosmologists, it just so happens that our universe is both isotropic *and* homogeneous

11.2.3 Cosmic microwave background radiation

Mentioned above as evidence for the cosmological principle, cosmic microwave background radiation, a faint, diffuse glow coming from all directions in the sky, was discovered in 1965 by Arno Penzias and Robert Wilson, who won the Nobel prize for their efforts.

In physics, a **black body** is an idealised body that perfectly absorbs and emits radiation. The importance of black body radiation is that its spectrum only depends on temperature, not the composition of the radiating body. Spectral analysis of CMBR shows an almost perfect fit to a black body spectrum with a temperature of 2.725 K.

The existence of CMBR is excellent evidence for a **Big Bang** as opposed to a Steady State model of the universe. CMBR is thought to have been emitted about 13.7 billion years ago, 400,000 or so years after the Big Bang. The Big Bang theory neatly predicts an initially very hot and dense universe of highly energised subatomic particles that expands and cools for several hundred thousand years until reaching a temperature of about 3000 K, when it emits huge amounts of radiation. The expansion of the universe since that time has redshifted that radiation – now detected as CMBR – to its current temperature of 2.725 K.

Ryden [27] makes the point that if Olbers' eyes could somehow have seen CMBR, he would never have formulated his eponymous paradox. She says, 'Unknown to Olbers, the night sky actually *is* uniformly bright – it's just uniformly bright at a temperature of ... 2.725 K rather than at a [visible to the human eye] temperature of a few thousand degrees Kelvin.'

Incidentally, CMBR, an echo of the Big Bang, is what causes some of the static on a non-digital television.

11.2.4 Hubble's law

Figure 11.3: Part of the emission spectrum of hydrogen.

We've already met redshift – a shift in the lines of the spectra toward the 'red' end of the electromagnetic spectrum, ie the lengthening of wavelength – in the form of gravitational time dilation and the Doppler effect. A third type of redshift is **cosmological redshift**, caused by the expansion of space itself, a concept we'll be discussing shortly. Both Doppler redshift and cosmological redshift can be used to calculate the relative motion of observer and light source.

Different elements have different spectral signatures, a bit like product bar codes, which can be studied and catalogued in the laboratory (a small part of the emission spectrum of hydrogen is shown in Figure 11.3). Stars containing these elements (mainly hydrogen and helium) emit light that can then be analysed to (a) identify those elements, and (b) measure the extent to which the spectral lines are shifted: usually to the red end of the spectrum (redshift), occasionally to the blue (blueshift). A star or galaxy has a redshift z given by the formula

$$z = \frac{\lambda_{ob} - \lambda_{em}}{\lambda_{em}},$$

where λ_{ob} and λ_{em} are, respectively, the wavelengths of the observed and emitted radiation. A blueshift is described if $z < 0$, and a redshift if $z > 0$. For small redshifts of less than about 0.1, z is related to the velocity v of the receding object by the simple formula $v = cz$.

In 1929 Edwin Hubble used this relationship to calculate the recessional velocity of a sample of nearby galaxies. After estimating their distance (a much more difficult measurement) he then proposed a simple law that describes the expansion of the universe. **Hubble's law** states that the apparent recessional velocity of a galaxy is proportional to its distance from the observer, ie

$$v = cz = H_0 d, \qquad\qquad (11.2.1)$$

where H_0 is known as **Hubble's constant**. The fundamental importance of Hubble's law is that it describes an **expanding universe**.

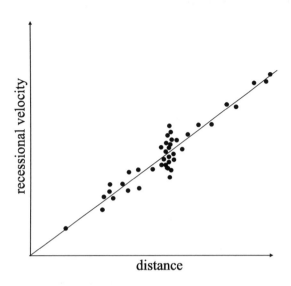

Figure 11.4: Hubble's law: recessional velocity plotted against distance for a sample of galaxies.

Figure 11.4 shows a typical plot of recessional velocity versus distance for a sample of galaxies. The gradient of the straight line, ie recessional velocity divided by distance, gives the value of Hubble's constant. Hubble's original estimate of this constant was $H_0 = 500\,\mathrm{km\,s^{-1}\,Mpc^{-1}}$. However, he seriously underestimated the distance of the galaxies from Earth. Recent estimates give a value of $H_0 = 70.4\,\mathrm{km\,s^{-1}\,Mpc^{-1}}$. Recalling that one parsec equals about 3.26 light-years, this means that the universe is expanding at a rate of around $70.4\,\mathrm{km\,s^{-1}}$ per 3.26 million light-years.

This overall, large-scale motion of galaxies due to the expansion of the universe is known as the **Hubble flow**. At any one time, the fact that the Hubble flow can be described using a single rate of expansion (the Hubble constant) is due to the validity of the cosmological principle.

A good way to visualise the Hubble flow is to imagine dots painted on the surface of a spherical balloon (see Figure 11.5). As the balloon is inflated any two dots will separate at a rate proportional to their distance apart. Think about it – if they all moved with the same velocity, after time t they would have all moved the same distance, which obviously doesn't happen. An observer A on a particular dot would see all the other dots moving away from him and might well think he's at the centre of the 'balloon universe'. But so would another observer B on a different dot. As with our universe, there is no 'centre' from which everything expands away from. This idea is known as the **Copernican principle**, namely that there is no central, privileged location in the universe.

(If the balloon was sufficiently large, with lots of more or less evenly spaced dots, we would have an even better model because any observer on the balloon's surface would see the dot distribution obeying the cosmological principle, ie being both homogeneous and isotropic.)

Hubble's law describes the large-scale uniform motion of galaxies due to the expansion of the universe. Galaxies may also be influenced by the gravitational effects of other galaxies. This component of a

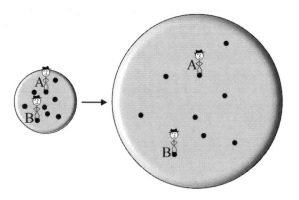

Figure 11.5: Hubble flow on a balloon.

galaxy's or star's velocity, which is not explained by Hubble's law, is described by the term **peculiar motion** and may need to taken into account when carrying out cosmological calculations.

11.3 Robertson-Walker spacetime

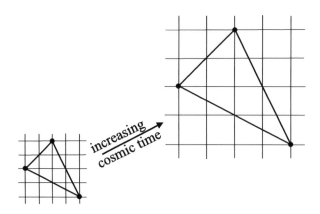

Figure 11.6: Co-moving coordinates.

In terms of general relativity, the obvious question to ask is: what is the metric that describes a spacetime that is both homogeneous and isotropic? In the 1930s the American mathematician and cosmologist Howard Robertson (1903–1961) and the British mathematician Arthur Walker (1909–2001) independently derived such a metric. The Robertson-Walker metric is the most general possible metric describing a spacetime that conforms to the cosmological principle. The metric's most common form is

$$ds^2 = c^2dt^2 - R^2(t)\left[\frac{dr^2}{1-kr^2} + r^2d\theta^2 + r^2\sin^2\theta d\phi^2\right]. \tag{11.3.1}$$

The time coordinate t is known as **cosmic time** and is the time measured by an observer whose peculiar motion is negligible, ie whose only motion is due to the expansion or contraction of homogeneous, isotropic spacetime. These observers, who all share the same cosmic time, are sometimes called **fundamental observers**. In an expanding universe such as ours, fundamental observers would all be moving with the Hubble flow.

The spatial coordinates (in this case r, θ, ϕ) assigned by a fundamental observer are known as **co-moving coordinates** and remain constant with time for any point. Figure 11.6 shows a triangle in expanding spacetime. Although the coordinate grid has increased in size, the co-moving coordinates describing the position of the triangle remain the same.

At any particular value of cosmic time, all fundamental observers will be measuring the same spatial slice of three-dimensional space, known as a **space-like hypersurface**. This is shown schematically in Figure 11.7 where each space-like hypersurface represents all of space at a particular moment of cosmic time t_1, t_2, t_3, etc. Also shown are the diverging world-lines of fundamental observers moving through cosmic time in an expanding universe.

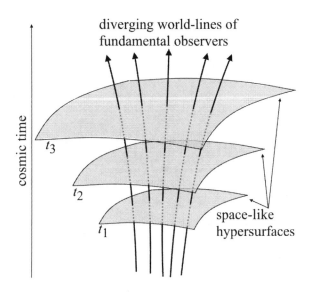

Figure 11.7: Hypersurfaces and diverging world-lines of fundamental observers.

Apart from the spacetime coordinates t, r, θ, ϕ there are two other quantities in the Robertson-Walker metric – the **curvature parameter** k and the **scale factor** $R(t)$. Notice that the scale factor is a function of time. This is the first metric we've met where the spatial coordinates may be time dependent. A scale factor $R(t)$ that increases with time describes an expanding universe; one that decreases with time describes a contracting universe. Trying to find $k, R(t)$ and R_0 (the current value of $R(t)$) is a major preoccupation of cosmologists trying to understand and model the universe.

11.3.1 The curvature parameter

The cosmological principle demands that if space is curved then it must be constantly curved at every point. This simplifies matters considerably as there are only three types of space exhibiting constant curvature: flat space, **positively curved** or **closed** space, and **negatively curved** or **open** space. This aspect of the spatial r, θ, ϕ geometry described by the Robertson-Walker metric is determined by the curvature parameter k. By rescaling the coordinate r, k can take one of the three discrete values $0, +1$ or -1, corresponding to flat, positively curved or negatively curved three-dimensional space-like hypersurfaces. The internal angles of a triangle in flat space add up to $180°$, in positively curved space to more than $180°$, and in negatively curved space to less than $180°$. Although we can't draw three-dimensional curved spaces, we can get a feel for their characteristics by considering two-dimensional surfaces. Figure 11.8 shows flat, open and closed two-dimensional curved surfaces corresponding to values of $k = 0, +1$ and -1. Just like a great circle on the surface of a sphere, a rocket following a geodesic in positively curved space would eventually return to its starting point.

Incidentally, it's not possible to construct a two-dimensional surface of constant negative curvature in three-dimensional space. The $k = -1$ saddle-shaped open surface illustrated in Figure 11.8 is a good approximation, but will only have constant curvature in the centre of the saddle region.

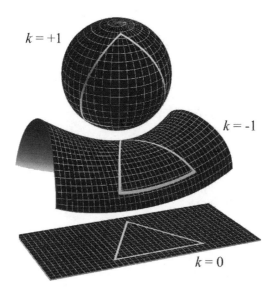

Figure 11.8: Positively curved, negatively curved and flat two-dimensional surfaces.

The overall curvature of a constant t space-like hypersurface depends on both k and $R(t)$ and can be shown to be equal to $k/R^2(t)$.

11.3.2 Proper distance

How might we determine a spatial distance between two points in the universe, the distance to a far away galaxy, for example? In practice, cosmologists often estimate distance by measuring the luminosity and brightness of a star or galaxy to calculate a quantity known as the **luminosity distance**. Theoretically, in terms of the Robertson-Walker metric, we need to determine the relationship between the constant co-moving coordinates and the actual distance measured with a hypothetical line of rulers. This 'measured' distance between two points is known as the **proper distance** and is equal to the length of a geodesic between those points on a space-like hypersurface at a fixed time t. Finding this proper distance (theoretically, don't forget – we can't practically construct a line of rulers to a galaxy) is easier than it sounds. We start with the Robertson-Walker metric (11.3.1)

$$ds^2 = c^2 dt^2 - R^2(t) \left[\frac{dr^2}{1 - kr^2} + r^2 d\theta^2 + r^2 \sin^2 \theta d\phi^2 \right].$$

For an observer at time t trying to determine the distance σ to a galaxy along a radial coordinate r, the coordinates θ and ϕ are constant. Therefore, $dt = d\theta = d\phi = 0$ and we can write

$$d\sigma = R(t) \left[\frac{dr^2}{1 - kr^2} + r^2 d\theta^2 + r^2 \sin^2 \theta d\phi^2 \right]^{1/2}$$

$$d\sigma = R(t) \left[\frac{dr^2}{1 - kr^2} \right]^{1/2}. \tag{11.3.2}$$

Recall that spacetime is modelled using a pseudo-Riemannian manifold, where ds^2 may be positive, zero or negative. If we let the radial coordinate r of the galaxy be $r = \chi$ (χ is the Greek letter chi) we can find σ by integrating (11.3.2), ie

$$\sigma(t) = \int_0^\chi R(t) \frac{dr}{(1 - kr^2)^{1/2}}, \tag{11.3.3}$$

where, for convenience, we have designated the observer's coordinates as the origin $r = 0$, and we write $\sigma(t)$ to acknowledge that proper distance is a function of time. Integrating either by hand or using an online integral calculator ([33], for example), we can solve (11.3.3) for curvature parameter values of $k = 0, +1$ and -1 to give

$$\sigma(t) = R(t) \sin^{-1} \chi \quad \text{if } k = +1, \tag{11.3.4}$$

$$\sigma(t) = R(t) \chi \quad \text{if } k = 0, \tag{11.3.5}$$

$$\sigma(t) = R(t) \sinh^{-1} \chi \quad \text{if } k = -1, \tag{11.3.6}$$

(where $\sinh^{-1}$ is a function called the hyperbolic sine, a type of hyperbolic function). We don't need to worry too much about what a hyperbolic sine actually is (we get rid of it in a few lines time). Instead, we introduce a quantity known as the **proper radial velocity** $d\sigma/dt$, which simply equals the rate of change of proper distance with respect to cosmic time. We can then rewrite the above three equations as

$$\frac{d\sigma}{dt} = \frac{dR}{dt} \sin^{-1} \chi \quad \text{if } k = +1, \tag{11.3.7}$$

$$\frac{d\sigma}{dt} = \frac{dR}{dt} \chi \quad \text{if } k = 0, \tag{11.3.8}$$

$$\frac{d\sigma}{dt} = \frac{dR}{dt} \sinh^{-1} \chi \quad \text{if } k = -1. \tag{11.3.9}$$

We can rewrite equations (11.3.4), (11.3.5) and (11.3.6) to give

$$\frac{\sigma}{R} = \sin^{-1}\chi \ \text{ if } k = +1,$$

$$\frac{\sigma}{R} = \chi \ \text{ if } k = 0,$$

$$\frac{\sigma}{R} = \sinh^{-1}\chi \ \text{ if } k = -1,$$

which we can substitute into (11.3.7), (11.3.8) and (11.3.9) to give

$$\frac{d\sigma}{dt} = \frac{1}{R}\frac{dR}{dt}\sigma \qquad (11.3.10)$$

for all three equations. We can rewrite (11.3.10) in the more conventional form of

$$v_p = H(t)\, d_p, \qquad (11.3.11)$$

known as the **velocity-distance law**, where v_p is the proper radial velocity, d_p is the proper distance, and

$$H(t) = \frac{1}{R}\frac{dR}{dt} \qquad (11.3.12)$$

defines a quantity known as the **Hubble parameter**. The velocity-distance law tells us that in Robertson-Walker spacetime (ie spacetime that conforms to the cosmological principle) any fundamental observer is moving away from every other fundamental observer with a proper radial velocity that is proportional to the proper distance that separates them. This is a *theoretical* relationship derived from the Robertson-Walker metric. Note both the similarity and difference to Hubble's law (11.2.1)

$$v = cz = H_0 d$$

that we met earlier, where H_0 is the Hubble constant. Hubble's law is an *observational* relationship that calculates recessional velocity v from redshift z measurements using the formula $v = cz$. Because this formula is only valid for small redshifts, Hubble's law is only accurate for relatively nearby galaxies. The term 'Hubble constant' is actually a bit of a misnomer as it isn't constant but changes with time. The Hubble constant H_0 is actually the $t = 0$ (ie the present value) of the time-dependent function $H(t)$, the Hubble parameter.

It doesn't take an Einstein to see that there's nothing in the velocity-distance law (11.3.11) prohibiting the recessional velocity v_p exceeding the speed of light (known as **superluminal speeds**). If the proper distance d_p is set to

$$d_p = d_H = c/H_0,$$

the quantity d_H is known as the **Hubble length** or **distance**. Galaxies further away than d_H are therefore currently moving away from us at superluminal speeds. The current value of the Hubble distance is about 4200 Mpc, or 13.7 billion light-years. But doesn't special relativity prohibit superluminal speeds? It does, but only in an observer's inertial frame. Galaxies receding from us at superluminal speeds are doing so because space itself is expanding. They are not travelling faster than light in our or any other observer's inertial frame. In this context, superluminal speeds do not contradict special relativity.

Using the velocity-distance law, we can do a back-of-the-envelope calculation of the age of the universe. First, we assume that the recessional velocity of two galaxies currently moving away from each other has been constant since the expansion of the universe started. Then, knowing that for an object moving with constant velocity, time equals distance divided by that velocity, we can rearrange the velocity-distance law (11.3.11) to find elapsed time t_0 since the galaxies were in contact, ie the age of the universe. This is given by

$$t_0 = \frac{d_p}{v_p} = \frac{1}{H_0}. \tag{11.3.13}$$

We known $H_0 = 70.4 \, \text{km s}^{-1} \, \text{Mpc}^{-1}$ and $1 \, \text{pc} \approx 3.09 \times 10^{16} \, \text{m}$. Therefore,

$$\frac{1}{H_0} = \frac{3.09 \times 10^{22}}{7.04 \times 10^4} = 4.39 \times 10^{17} \, \text{s},$$

which, converted to years, is

$$\frac{4.39 \times 10^{17}}{365 \times 24 \times 3600} \approx 13,900,000,000 \, \text{years}.$$

This calculation is based on the assumption of a constantly expanding universe. The current estimate of the age of the universe is about 13.7×10^9 years. The reason the two figures are so close (what's 200,000,000 or so years between friends?) is that it is now thought that for most of the universe's history the expansion has been more or less linear.

A final point regarding Robertson-Walker spacetime. We need to note the distinction between the spatial r, θ, ϕ curvature and the spacetime t, r, θ, ϕ curvature described by the Robertson-Walker metric. It's possible to have a flat space-like universe (where $k = 0$), but the full Robertson-Walker spacetime will not be flat as long as $R(t)$ isn't constant. This is because the proper distance between any two points in that universe will change as $R(t)$ changes.

11.4 Introducing the Friedmann equations

Our aim is to model a universe that obeys the cosmological principle. Introducing the Robertson-Walker metric was the first step. Next we need to make an educated guess regarding the nature of the energy-momentum tensor $T^{\mu\nu}$ that describes the large-scale distribution and flow of energy and momentum in the universe. Then it's possible (we don't go into details) to feed the Robertson-Walker metric and $T^{\mu\nu}$ into the field equations to produce a set of differential equations called the Friedmann equations. By changing the parameters of these equations we can construct various Friedmann-Robertson-Walker models that show how the scale factor $R(t)$, and thus the universe, may change with time. Finally, we ask which of these models, based on the available evidence, best describes our own universe.

11.4.1 The cosmic energy-momentum tensor

Cosmologists make the wonderfully simplifying assumption that the whole universe can be treated as a perfect fluid, characterised by proper density $\rho(t)$ and pressure $p(t)$ – (to preserve homogeneity both proper density and pressure may only be functions of cosmic time). A fundamental observer is by definition travelling with the flow of this cosmic perfect fluid. Such a fluid is therefore described by the energy-momentum tensor of a perfect fluid in its momentarily comoving rest frame (MCRF) that we met earlier (7.5.6)

$$[T^{\mu\nu}] = \begin{pmatrix} \rho c^2 & 0 & 0 & 0 \\ 0 & p & 0 & 0 \\ 0 & 0 & p & 0 \\ 0 & 0 & 0 & p \end{pmatrix}.$$

In Section 7.7 we introduced what Einstein referred to as his 'greatest blunder', the cosmological constant Λ and its associated dark energy, a sort of 'anti-gravity' repulsive force or negative pressure. Confronted with the evidence of an accelerating universe, cosmologists have now dusted off that apparent 'blunder' and reintroduced dark energy into their equations. Specifically, they treat the cosmic perfect fluid as being a mixture of three constituent perfect fluids representing matter, radiation and the source (whatever that is) of dark energy. The cosmic density can then be defined as

$$\rho(t) = \rho_m(t) + \rho_r(t) + \rho_\Lambda \tag{11.4.1}$$

and the cosmic pressure as

$$p(t) = p_m(t) + p_r(t) + p_\Lambda. \tag{11.4.2}$$

The matter-component of the cosmic perfect fluid consists of the radiation-emitting matter that we can detect (stars, galaxies etc), and other invisible matter whose existence is inferred because of its gravitational effects on the stuff we can see. This mysterious substance, which is thought to account for the majority of matter in the universe, is called **dark matter** (and is no relation to dark energy).

Let's look at the cosmological constant and dark energy in a little more detail. Equation (7.7.1) gives the modified field equations with the cosmological constant:

$$R_{\mu\nu} - \frac{1}{2}Rg_{\mu\nu} + \Lambda g_{\mu\nu} = -\kappa T_{\mu\nu},$$

which we can rearrange to give

$$R_{\mu\nu} - \frac{1}{2}Rg_{\mu\nu} = -\kappa \left(T_{\mu\nu} + \frac{\Lambda}{\kappa}g_{\mu\nu} \right).$$

As the $\frac{\Lambda}{\kappa}g_{\mu\nu}$ term contributes energy and momentum to the right-hand side of the equation we can regard it as representing another energy-momentum tensor that we'll call $\overline{T}_{\mu\nu}$ and write

$$R_{\mu\nu} - \frac{1}{2}Rg_{\mu\nu} = -\kappa \left(T_{\mu\nu} + \overline{T}_{\mu\nu} \right).$$

We've already assumed that density ρ_Λ and pressure p_Λ constitute a perfect fluid. Equation (7.5.8) tells us the components of the energy-momentum tensor of a perfect fluid are given by

$$T^{\mu\nu} = \left(\rho + p/c^2 \right) U^\mu U^\nu - pg^{\mu\nu},$$

which we can thus rewrite using ρ_Λ and p_Λ as

$$\overline{T}_{\mu\nu} = \left(\rho_\Lambda + p_\Lambda/c^2 \right) U_\mu U_\nu - p_\Lambda g_{\mu\nu}. \tag{11.4.3}$$

We've defined $\overline{T}_{\mu\nu} = \frac{\Lambda}{\kappa}g_{\mu\nu}$, so we can write

$$\frac{\Lambda}{\kappa}g_{\mu\nu} = \left(\rho_\Lambda + p_\Lambda/c^2 \right) U_\mu U_\nu - p_\Lambda g_{\mu\nu}$$

or, in the MCRF,

$$\frac{\Lambda}{\kappa}\eta_{\mu\nu} = \left(\rho_\Lambda + p_\Lambda/c^2\right)U_\mu U_\nu - p_\Lambda \eta_{\mu\nu}. \qquad (11.4.4)$$

In the MCRF, when $\eta_{00} = 1$, $U_{00} = c^2$, and therefore

$$\frac{\Lambda}{\kappa} = \left(\rho_\Lambda + p_\Lambda/c^2\right)c^2 - p_\Lambda$$

giving

$$\rho_\Lambda = \frac{\Lambda}{\kappa c^2}. \qquad (11.4.5)$$

We know from Section 7.6 that $\kappa = 8\pi G/c^4$, so

$$\rho_\Lambda = \frac{\Lambda c^2}{8\pi G}. \qquad (11.4.6)$$

In the MCRF, when $\eta_{ii} = -1$, $U_{ii} = 0$, and therefore, from (11.4.4)

$$-\frac{\Lambda}{\kappa} = p_\Lambda$$

$$p_\Lambda = -\frac{\Lambda}{\kappa}. \qquad (11.4.7)$$

Substituting $\Lambda = \rho_\Lambda \kappa c^2$ from (11.4.5) into (11.4.7) we obtain

$$-p_\Lambda = \rho_\Lambda c^2,$$

which tells us that a positive dark energy density ρ_Λ results in a negative dark energy pressure $-p_\Lambda$. The modified field equations then become

$$R_{\mu\nu} - \frac{1}{2}Rg_{\mu\nu} = -\kappa\left(T_{\mu\nu} + \rho_\Lambda c^2 g_{\mu\nu}\right).$$

General relativity states that ordinary, positive pressure components of $T^{\mu\nu}$ contribute to a system's 'gravitational attraction'. Negative pressure (whatever that is) has the opposite effect, tending to drive things apart. We've now shown how introducing a cosmological constant into the field equations results in negative pressure.

11.4.2 More on density and pressure

As well as assuming that the universe can be described as a perfect fluid, cosmologists also generally assume that the matter component of such a universe (stars, galaxies, dark matter etc) has zero pressure $(p_m(t) = 0)$, ie that it can be treated as dust. That leaves three density components (matter, radiation and dark energy) plus two pressure components (radiation and dark energy) that may need to be taken into consideration when modelling the universe. The question we now ask is: how do these different density and pressure components change with time?

Imagine a cube of space containing matter and radiation. What happens to the density of that matter and radiation in an expanding or contracting universe? First, we'll consider matter. If the lengths of the sides of the cube double (from 1 to 2, for example) the volume of the cube will increase by a factor of 8, and the density ρ_m will decrease by a factor of 8 (the same amount of matter is now in a larger cube). In terms of the scale factor $R(t)$ we can say

$$\rho_m \propto \frac{1}{R^3},\tag{11.4.8}$$

where the $\propto$ symbol means 'proportional to'. In the case of radiation, Planck's law states the energy of each particle is given by

$$E = hf,$$

where h is Planck's constant and f is frequency, ie a radiation particle's energy is proportional to its frequency, or inversely proportional to its wavelength λ. If, in an expanding universe, the length of the sides of the cube of space double, the wavelength will also double. Combined with the above $\propto 1/R^3$ relationship for particle energy density, this means that the density of radiation ρ_r in terms of the scale factor $R(t)$ is

$$\rho_r \propto \frac{1}{R^4}.\tag{11.4.9}$$

In other words, in an expanding universe, radiation density decreases more quickly than matter density and both decline relative to dark energy density, which is assumed to be time invariant. These evolving density contributions are illustrated schematically in Figure 11.9. Over time, matter density must therefore overtake radiation density. Dark energy density must similarly overtake both matter and radiation density. That dynamic is now thought to describe the history of our (expanding) universe. First, there was a relatively short-lived radiation-dominated era lasting about 50,000 years. That was followed by a matter-dominated era lasting about 9.8 billion years. 13.7 billion years after the Big Bang, we are now living in a dark energy dominated universe.

At an arbitrary time $t = 0$ (often taken to be the present time) let $\rho_m = \rho_{m,0}$ and $R(t) = R_0$. At time $t = t$, let $\rho_m = \rho_m(t)$ and $R(t) = R(t)$. Using (11.4.8), we can then divide $\rho_m(t) \propto 1/(R(t))^3$ by $\rho_{m,0} \propto 1/(R_0)^3$ to give

$$\rho_m(t) = \rho_{m,0}\left(\frac{R_0}{R(t)}\right)^3.$$

Similarly, we can say

$$\rho_r(t) = \rho_{r,0}\left(\frac{R_0}{R(t)}\right)^4.$$

We can then use the cosmic density equation (11.4.1) to write

$$\rho(t) = \rho_{m,0}\left(\frac{R_0}{R(t)}\right)^3 + \rho_{r,0}\left(\frac{R_0}{R(t)}\right)^4 + \rho_\Lambda.\tag{11.4.10}$$

In order to calculate how cosmic pressure changes with time, it is necessary to use something called an **equation of state**, which in relation to cosmology is given by

$$p = \omega\rho c^2,$$

where $\omega = 0$ for dust, $1/3$ for radiation, and -1 for dark energy. Using these equations of state, a corresponding cosmic pressure equation can also be derived (see, for example, Lambourne [17] for details):

$$p(t) = \frac{\rho_{r,0}c^2}{3}\left(\frac{R_0}{R(t)}\right)^4 - \rho_\Lambda c^2.\tag{11.4.11}$$

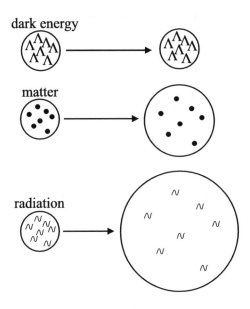

dark energy

matter

radiation

Figure 11.9: Evolving density contributions of dark energy, matter and radiation in an expanding universe.

Equations (11.4.10) and (11.4.11) demonstrate the fundamental importance of the cosmic density components $\rho_{m,0}, \rho_{r,0}$ and ρ_Λ. If cosmologists know (a) these three values at a certain cosmic time, and (b) the scale factor function $R(t)$, they can calculate the cosmic density and cosmic pressure at any other cosmic time. We'll use the above relationships to expand and simplify the Friedmann equations, which we look at next.

11.4.3 The Friedmann equations

The non-zero components of the Robertson-Walker metric (11.3.1) can be used to calculate the connection coefficients $\Gamma^\gamma_{\alpha\beta}$, which can then be used to find the Riemann curvature tensor $R^\rho{}_{\sigma\mu\nu}$, which can then be used to find the Ricci tensor $R_{\mu\nu}$ components and the Ricci scalar R (not to be confused with the scale factor $R(t)$), which can then be fed (phew!) into the field equations to give a pair of independent equations known as the **Friedmann equations**, after the Russian mathematical physicist Alexander Friedmann (Figure 11.10).

The Friedmann equations are some of the most important equations in cosmology. Textbooks give various forms of these equations. The two we start off with are

$$\left[\frac{1}{R}\frac{dR}{dt}\right]^2 = \frac{8\pi G}{3}\rho - \frac{kc^2}{R^2}, \tag{11.4.12}$$

which is sometimes called *the* Friedmann equation, and

$$\frac{1}{R}\frac{d^2R}{dt^2} = -\frac{4\pi G}{3}\left(\rho + \frac{3p}{c^2}\right), \tag{11.4.13}$$

Figure 11.10: Alexander Friedmann (1888–1925).

which is sometimes called the **Friedmann acceleration equation**.

We'll also make use of two expanded and simplified forms of these equations. We know that cosmic density ρ has three components: ρ_m, ρ_r and ρ_Λ. Using (11.4.10), we can expand the first Friedmann equation in terms of these components to give:

$$\left[\frac{1}{R}\frac{dR}{dt}\right]^2 = \frac{8\pi G}{3}\left[\rho_{m,0}\left(\frac{R_0}{R(t)}\right)^3 + \rho_{r,0}\left(\frac{R_0}{R(t)}\right)^4 + \rho_\Lambda\right] - \frac{kc^2}{R^2}. \tag{11.4.14}$$

Using (11.4.10) and (11.4.11), we can also eliminate cosmic pressure p in the second Friedmann equation to give

$$\frac{1}{R}\frac{d^2R}{dt^2} = -\frac{4\pi G}{3}\left[\rho_{m,0}\left(\frac{R_0}{R(t)}\right)^3 + 2\rho_{r,0}\left(\frac{R_0}{R(t)}\right)^4 - 2\rho_\Lambda\right]. \tag{11.4.15}$$

You will also commonly see the Friedmann equations written in terms of a **normalised scale factor** $a(t)$. Normalised means $a(t_0) = 1$, ie the scale factor at the present time t_0 is defined to be 1. This is done by letting

$$a(t) = R/R_0 \tag{11.4.16}$$

so

$$a(t_0) = \frac{R_0}{R_0} = 1.$$

Friedmann equation (11.4.12) then changes from

$$\left[\frac{1}{R}\frac{dR}{dt}\right]^2 = \frac{8\pi G}{3}\rho - \frac{kc^2}{R^2}$$

to

$$\left[\frac{1}{aR_0}\frac{d\left(aR_0\right)}{dt}\right]^2 = \frac{8\pi G}{3}\rho - \frac{kc^2}{R_0^2 a^2}$$

$$\left[\frac{1}{a}\frac{da}{dt}\right]^2 = \frac{8\pi G}{3}\rho - \frac{kc^2}{R_0^2 a^2}. \tag{11.4.17}$$

From the above, we've shown that

$$\frac{1}{R}\frac{dR}{dt} = \frac{1}{a}\frac{da}{dt},$$

and from the definition of the Hubble parameter (11.3.12), we can say

$$H\left(t\right) = \frac{1}{R}\frac{dR}{dt} = \frac{1}{a}\frac{da}{dt}.$$

At t_0, $H\left(t\right) = H_0$ and $a\left(t_0\right) = 1$, so

$$H_0 = \frac{da}{dt}. \tag{11.4.18}$$

In other words, the rate of change of the normalised scale factor equals the Hubble constant.

Equation (11.3.13) allowed us to calculate $1/H_0$, which was found to equal 4.39×10^{17} s. H_0 is the reciprocal of that value, ie

$$\frac{da}{dt} = H_0 = \frac{1}{4.39 \times 10^{17}\,\text{s}} = 2.27 \times 10^{-18}\,\text{s}^{-1}.$$

By definition, the normalised scale factor is currently 1, so every second from now it increases by a factor of 2.3×10^{-18}.

You may also see the Friedmann equations written not in terms of proper density ρ (with units of $\text{kg}\,\text{m}^{-3}$), but in terms of proper energy density $\epsilon = \rho c^2$ (with units of $\text{J}\,\text{m}^{-3}$ – where ϵ is the Greek letter epsilon).

Now we've met the Friedmann equations we can consider some simple cosmological models. First, we'll look at a particularly basic universe, an empty one containing no matter, radiation or dark energy. Next, we'll look at the metaphorical source of dark energy, Einstein's original 1917 static universe, his famous 'greatest blunder', where he attempted to 'fix' the field equations with the addition of a cosmological constant. This is a $k = 1$ model. Then we'll look at three $k = 0$ models, each having just one non-zero value of the three constituent perfect fluids $\rho_{m,0}$, $\rho_{r,0}$ and ρ_Λ, ie we are considering spatially flat universes containing only mass or radiation or dark energy. These five examples are not now thought to accurately represent the universe as it currently exists. However, as well as being a useful introduction to the subject of cosmological modelling they may also represent stages in the evolutionary history of our universe. The radiation only model, for example, is believed to reasonably well describe the radiation dominated early universe. Cosmological models based on the Friedmann equations and the Robertson-Walker metric are known as **Friedmann-Robertson-Walker (FRW) models**.

11.4.4 The empty universe model

If we turn all the dials to zero, we have an empty universe with no matter, radiation or dark energy, ie $\rho_m = \rho_r = \rho_\Lambda = 0$. Plugging $\rho = 0$ into the first Friedmann equation (11.4.12)

$$\left[\frac{1}{R} \frac{dR}{dt} \right]^2 = \frac{8\pi G}{3} \rho - \frac{kc^2}{R^2}$$

gives

$$\frac{dR}{dt} = \sqrt{-kc^2}.$$

In order to avoid taking square roots of negative numbers, k must equal 0 or -1. If $k = 0$, R integrates to a constant, and this solution describes empty, static, flat Minkowski space. If $k = -1$

$$\frac{dR}{dt} = \pm c,$$

which integrates to give (ignoring the constant of integration)

$$R = \pm ct, \tag{11.4.19}$$

or $R = \pm t$ if we let $c = 1$. In an empty, negatively curved, expanding universe $R = t$, meaning the scale factor increases linearly with time (see Figure 11.11).

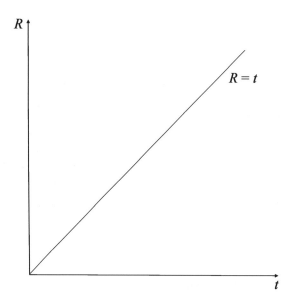

Figure 11.11: An empty, expanding universe.

In terms of the normalised scale factor 11.4.16 $a(t) = R/R_0$, 11.4.19 can be written as

$$a(t) = \frac{ct}{R_0}. \tag{11.4.20}$$

At time t_0, this becomes (recall that $a(t)$ is normalised, ie is defined to equal 1 at time t_0)

$$a_0 = 1 = \frac{ct_0}{R_0}. \tag{11.4.21}$$

And dividing (11.4.20) by (11.4.21) gives

$$a(t) = \frac{t}{t_0}. \tag{11.4.22}$$

11.4.5 The static Einstein model

In search of a static universe, Einstein added a cosmological constant term to his original field equations. By a static universe we of course mean one where the scale factor is constant, ie

$$\frac{dR}{dt} = \frac{da}{dt} = 0.$$

Let's look at the implications of using a cosmological constant to achieve a constant scale factor. Equation (7.7.1) gives the modified field equations with the cosmological constant:

$$R_{\mu\nu} - \frac{1}{2}Rg_{\mu\nu} + \Lambda g_{\mu\nu} = -\kappa T_{\mu\nu}.$$

If the Friedmann equations are derived from these field equations they become

$$\left[\frac{1}{R}\frac{dR}{dt}\right]^2 = \frac{8\pi G}{3}\rho - \frac{kc^2}{R^2} + \frac{\Lambda c^2}{3} \tag{11.4.23}$$

and

$$\frac{1}{R}\frac{d^2R}{dt^2} = -\frac{4\pi G}{3}\left(\rho + \frac{3p}{c^2}\right) + \frac{\Lambda c^2}{3}, \tag{11.4.24}$$

where now, in both equations, ρ equals $\rho_m + \rho_r$ and not $\rho_m + \rho_r + \rho_\Lambda$ as in the original (11.4.12) and (11.4.13) Friedmann equations. In other words, the dark energy chunk of the universe is provided by the cosmological constant term $\frac{\Lambda c^2}{3}$ in (11.4.23) and (11.4.24), and by the equivalent term dark energy term ρ_Λ buried in (11.4.12) and (11.4.13). Effectively, we are saying the same thing in two different ways.

We can easily check this equivalence with respect to the first Friedmann equation (11.4.12), by using (11.4.6) that we derived for dark energy

$$\rho_\Lambda = \frac{\Lambda c^2}{8\pi G}.$$

We can divide this by 3 and rearrange to give

$$\frac{8\pi G}{3}\rho_\Lambda = \frac{\Lambda c^2}{3},$$

and we can see that adding a $\frac{\Lambda c^2}{3}$ term is equivalent to adding a ρ_Λ component to (11.4.12) to give

$$\left[\frac{1}{R}\frac{dR}{dt}\right]^2 = \frac{8\pi G}{3}(\rho + \rho_\Lambda) - \frac{kc^2}{R^2}. \tag{11.4.25}$$

In order for the universe to be static, both $\frac{dR}{dt}$ and $\frac{d^2R}{dt^2}$ must equal zero, as must pressure p. Friedmann equation (11.4.24) then neatly reduces to

$$0 = -\frac{4\pi G}{3}(\rho + 0) + \frac{\Lambda c^2}{3}$$

or

$$\Lambda = \frac{4\pi G \rho}{c^2},$$

which is the value Einstein had to give his cosmological constant. Friedmann equation (11.4.23) reduces to

$$0 = \frac{8\pi G}{3}\rho - \frac{kc^2}{R^2} + \frac{\Lambda c^2}{3}$$

$$0 = \frac{8\pi G}{3}\rho - \frac{kc^2}{R^2} + \frac{4\pi G \rho c^2}{3c^2}$$

$$\frac{kc^2}{R^2} = 4\pi G \rho,$$

meaning k must equal $+1$, ie this static universe has to be positively curved. We can now solve for R_0 (the scale factor – constant in the case of a static universe), to give

$$R_0 = \frac{c}{\sqrt{4\pi G \rho}} = \frac{1}{\Lambda^{1/2}}.$$

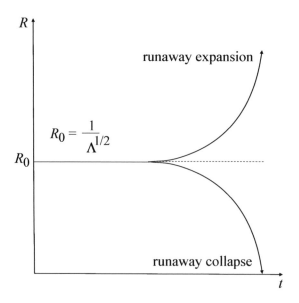

Figure 11.12: Einstein's static universe.

The problem with Einstein's static universe is that it is inherently unstable, a bit like trying to balance a plank on a knife-edge. At equilibrium, the attractive force of ρ does indeed exactly balance the repulsive force of Λ. But the slightest increase in size of the universe leads to runaway expansion, and the slightest contraction leads to runaway collapse (see Figure 11.12).

11.4.6 The de Sitter model

Proposed in 1917 by the Dutch astronomer Willem de Sitter (1872–1934), this model has $k = 0$, and contains no matter or radiation, only dark energy, ie $\rho_{m,0} = \rho_{r,0} = 0$. This model was the first to describe an expanding universe. Equation (11.4.14) becomes

$$\frac{dR}{dt} = \sqrt{\frac{8\pi G \rho_\Lambda}{3}} R. \tag{11.4.26}$$

We can solve this first-order differential equation as follows. First, divide both sides by R

$$\frac{\frac{dR}{dt}}{R} = \sqrt{\frac{8\pi G \rho_\Lambda}{3}},$$

then integrate both sides with respect to t

$$\int \frac{\frac{dR}{dt}}{R} dt = \int \sqrt{\frac{8\pi G \rho_\Lambda}{3}} dt$$

$$\int \frac{dR}{R} = \int \sqrt{\frac{8\pi G \rho_\Lambda}{3}} dt,$$

which gives

$$\ln R = \left(\sqrt{\frac{8\pi G \rho_\Lambda}{3}} \right) t + K,$$

where K is a constant of integration. From the definition of the natural logarithm $e^{\ln(x)} = x$, we obtain

$$R = e^{\left(\sqrt{\frac{8\pi G \rho_\Lambda}{3}} \right) t + K}$$

$$R = K e^{\left(\sqrt{\frac{8\pi G \rho_\Lambda}{3}} \right) t}. \tag{11.4.27}$$

(We've changed the constant). If we let $t = t_0$ when $R = R_0$, this becomes

$$R_0 = K e^{\left(\sqrt{\frac{8\pi G \rho_\Lambda}{3}} \right) t_0}. \tag{11.4.28}$$

Divide (11.4.27) by (11.4.28) (thus cancelling the constant K) gives

$$\frac{R}{R_0} = e^{\left(\sqrt{\frac{8\pi G \rho_\Lambda}{3}} (t - t_0) \right)},$$

and we can say

$$R(t) = R_0 e^{\left(\sqrt{\frac{8\pi G \rho_\Lambda}{3}} (t - t_0) \right)}. \tag{11.4.29}$$

From (11.4.26), we can see the Hubble parameter (11.3.12)

$$H(t) = \frac{1}{R}\frac{dR}{dt} = \sqrt{\frac{8\pi G \rho_\Lambda}{3}}$$

is a constant. Letting $H_0 = H(t_0)$, (11.4.29) can then be rewritten as

$$R(t) = R_0 e^{H_0(t-t_0)}$$

or, in terms of the normalised scale factor (11.4.16),

$$a(t) = e^{H_0(t-t_0)}. \tag{11.4.30}$$

To simplify the above, we can say

$$R(t) \sim e^{H_0(t-t_0)}$$

as shown in Figure 11.13. However, note that the de Sitter curve shown is a plot of $R(t) = e^{(t-t_0)}$, with a consequently much exaggerated slope. A more accurate plot of $R(t) = e^{H_0(t-t_0)}$ would still grow exponentially, but at a much smaller rate (because $H_0 = 2.27 \times 10^{-18}\,\mathrm{s}^{-1} \ll 1$) and would appear almost parallel to the horizontal time axis.

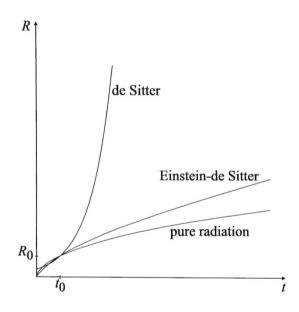

Figure 11.13: de Sitter, Einstein-de Sitter and pure radiation models – not to scale.

11.4.7 The radiation only model

This model is believed to reasonably well describe the radiation dominated early universe. The parameters are $k = 0$, $\rho_{m,0} = \rho_\Lambda = 0$. Equation (11.4.14) then becomes

$$\frac{dR}{dt} = \sqrt{\frac{8\pi G}{3}\rho_{r,0}\frac{R_0^2}{R}},$$

which we can solve as follows. First, multiply both sides by R

$$\frac{dR}{dt}R = \sqrt{\frac{8\pi G}{3}\rho_{r,0}R_0^2},$$

then multiply both sides by dt and integrate the left side with respect to R and the right side with respect to t

$$\int R dR = \int \sqrt{\frac{8\pi G}{3}\rho_{r,0}R_0^2}\,dt. \tag{11.4.31}$$

We can simplify this by seeing that the Hubble parameter (11.3.12) is given by

$$H(t) = \frac{1}{R}\frac{dR}{dt} = \sqrt{\frac{8\pi G\rho_{r,0}}{3}\frac{R_0^2}{R^2}}$$

so for $H_0 = H(t_0)$ and $R = R_0$, we can write

$$H_0 = \sqrt{\frac{8\pi G\rho_{r,0}}{3}},$$

which is substituted into (11.4.31) to give

$$\int R dR = \int H_0 R_0^2\,dt.$$

Evaluating this integral gives

$$\frac{R^2}{2} = H_0 R_0^2 t + K,$$

where K is a constant of integration, which we can set to zero by letting $R = 0$ when $t = 0$ to give

$$R(t) = R_0\sqrt{2H_0 t}. \tag{11.4.32}$$

If we again let $t = t_0$ when $R = R_0$, this becomes

$$R_0 = R_0\sqrt{2H_0 t_0}. \tag{11.4.33}$$

Dividing (11.4.32) by (11.4.33) gives

$$R(t) = R_0\left(\frac{t}{t_0}\right)^{1/2}$$

or, in terms of the normalised scale factor (11.4.16),

$$a(t) = \left(\frac{t}{t_0}\right)^{1/2}. \tag{11.4.34}$$

To simplify the above, we can say

$$R(t) \sim t^{1/2}$$

as shown in Figure 11.13.

11.4.8 The Einstein-de Sitter model

In 1932 Einstein and de Sitter proposed this mass-only (for mass, think zero pressure dust) model. This is an example of a **Friedmann universe**, where the dark energy/cosmological constant component is assumed to be zero. For many years, this model was thought to be a credible description of our universe. Why? First, because the universe is expanding. Second, recall that in an expanding universe, radiation density decreases more quickly than matter density. That means that a radiation dominated universe is inherently unstable. Sooner or later, matter density must overtake radiation density and become the dominant component. Hence the widespread support for the Einstein-de Sitter model, which was only finally rejected in the late 1990s when mounting evidence for an accelerating universe persuaded cosmologists to turn to models dominated by dark energy.

The parameters of this model are $k = 0$, $\rho_{r,0} = \rho_\Lambda = 0$. Equation (11.4.14) then becomes

$$\frac{dR}{dt} = \sqrt{\frac{8\pi G\rho_{m,0}}{3}}\frac{R_0^{3/2}}{R^{1/2}},$$

which we can solve as follows. First, multiply both sides by $R^{1/2}$

$$\frac{dR}{dt}R^{1/2} = \sqrt{\frac{8\pi G\rho_{m,0}}{3}}R_0^{3/2},$$

then multiply both sides by dt and integrate the left side with respect to R and the right side with respect to t

$$\int R^{1/2}dR = \int \sqrt{\frac{8\pi G\rho_{m,0}}{3}}R_0^{3/2}dt. \tag{11.4.35}$$

We can simplify this by seeing that the Hubble parameter (11.3.12) is given by

$$H(t) = \frac{1}{R}\frac{dR}{dt} = \sqrt{\frac{8\pi G\rho_{m,0}}{3}}\frac{R_0^{3/2}}{R^{3/2}},$$

so for $H_0 = H(t_0)$ and $R = R_0$, we can write

$$H_0 = \sqrt{\frac{8\pi G\rho_{m,0}}{3}}, \tag{11.4.36}$$

which is substituted into (11.4.35) to give

$$\int R^{1/2}dR = \int H_0 R_0^{3/2}dt.$$

Evaluating this integral gives

$$\frac{2}{3}R^{3/2} = H_0 R_0^{3/2}t + K,$$

where K is a constant of integration, which we can again eliminate by letting $R = 0$ when $t = 0$ to give

$$R(t) = R_0\left(\frac{3}{2}H_0 t\right)^{2/3}. \tag{11.4.37}$$

If we again let $t = t_0$ when $R = R_0$, this becomes

$$R_0 = R_0 \left(\frac{3}{2} H_0 t_0 \right)^{2/3}.$$
(11.4.38)

Dividing (11.4.37) by (11.4.38) gives

$$R(t) = R_0 \left(\frac{t}{t_0} \right)^{2/3}$$

or, in terms of the normalised scale factor (11.4.16),

$$a(t) = \left(\frac{t}{t_0} \right)^{2/3}.$$
(11.4.39)

To simplify the above, we can say

$$R(t) \sim t^{2/3},$$

as shown in Figure 11.13.

As well as being a one-time contender for representing our universe, this model is significant because it describes a universe with an expansion rate that is set 'just right' to prevent collapse. The Einstein-de Sitter model is a Friedmann universe with just the right amount of mass, so that the universe will expand forever (but with a continually decreasing expansion rate). Turn the mass dial up a fraction in an Einstein-de Sitter universe, and the expansion will eventually halt and the universe contract into a Big Crunch.

We can see why this is by plugging $k = 0$ (corresponding to a spatially flat universe) into (11.4.12) to obtain

$$\left[\frac{1}{R} \frac{dR}{dt} \right]^2 = \frac{8\pi G}{3} \rho,$$

which we can write in terms of the Hubble parameter (11.3.12)

$$H^2(t) = \left(\frac{1}{R} \frac{dR}{dt} \right)^2 = \frac{8\pi G}{3} \rho(t)$$

or

$$\rho(t) = \frac{3H^2(t)}{8\pi G}.$$

This quantity, known as the **critical density**, is denoted by $\rho_c(t)$

$$\rho_c(t) = \frac{3H^2(t)}{8\pi G},$$
(11.4.40)

currently thought to be about 9×10^{-27} kg m^{-3} (which you can confirm by plugging in the value of H_0 we found using (11.4.18)). Equation 11.4.1 is the same density as the density $\rho_{m,0}$ of the Einstein-de Sitter universe given by (11.4.36)

$$H_0 = \sqrt{\frac{8\pi G \rho_{m,0}}{3}}.$$

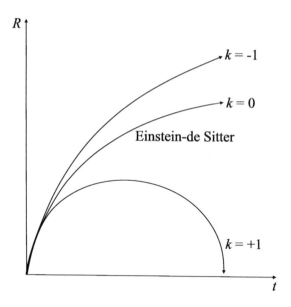

Figure 11.14: Three possible fates of a Friedmann universe.

We can give the Friedmann equation (11.4.12) in terms of the Hubble parameter (11.3.12)

$$\left[\frac{1}{R}\frac{dR}{dt}\right]^2 = H^2(t) = \frac{8\pi G}{3}\rho - \frac{kc^2}{R^2}. \tag{11.4.41}$$

If the actual density of a Friedmann universe $\rho < \rho_c(t)$, then k must equal -1 and the spatial universe is negatively curved. In that case, the right-hand side of (11.4.41) will never equal zero, the Hubble constant will therefore also never equal zero, and the universe will expand for ever. It's a different story if $\rho > \rho_c(t)$. In that case, k must equal $+1$, the spatial universe is positively curved, and it is possible for the two terms on the right-hand side of (11.4.41) to cancel each other out and the Hubble constant to equal zero. In fact, this must happen because matter density (which is proportional to $1/R^3$) will eventually be less than the k term (which is proportional to $1/R^2$). A $k = +1$ Friedmann universe must therefore eventually collapse. Finally, if $\rho = \rho_c(t)$ then k must equal 0, meaning the spatial universe is flat.

In truth, there's been a hefty chunk of mathematical handwaving in this explanation. To derive $R(t)$ for even a relatively simple matter-only Friedmann universe involves the use of $\sin\theta$ and $\cos\theta$ parametric equations, where the parameter θ varies from 0 to 2π. See Ryden [27], for example.

The three possible fates of a Friedmann universe, corresponding to $k = 0$ and ± 1 are shown in Figure 11.14.

11.4.9 Density parameters

Instead of proper density $\rho(t)$, cosmologists often use quantities known as **density parameters**, which are the ratios of the density components $\rho_m(t)$, $\rho_r(t)$ and ρ_Λ to the critical density $\rho_c(t)$. These density parameters are given the symbol Ω (the Greek letter Omega) and are defined as follows:

$$\Omega_m(t) = \frac{\rho_m(t)}{\rho_c(t)},$$

$$\Omega_r(t) = \frac{\rho_r(t)}{\rho_c(t)},$$

$$\Omega_\Lambda(t) = \frac{\rho_\Lambda}{\rho_c(t)}.$$

The total density parameter $\Omega(t)$ is just the sum of the individual parameters, ie

$$\Omega(t) = \Omega_m(t) + \Omega_r(t) + \Omega_\Lambda(t).$$

Using these density parameters, the first Friedmann equation can be rewritten as

$$\frac{c^2 k}{H^2(t) R^2(t)} = \Omega_m(t) + \Omega_r(t) + \Omega_\Lambda(t) - 1$$

or, more concisely,

$$\frac{c^2 k}{H^2(t) R^2(t)} = \Omega(t) - 1. \tag{11.4.42}$$

Because k cannot change its sign as a universe evolves, the right-hand side of (11.4.42) cannot change its sign either. The total density parameter $\Omega(t)$ therefore determines the sign of k, ie if

$$\Omega(t) < 1, \text{ then } k = -1,$$

$$\Omega(t) = 1, \text{ then } k = 0,$$

$$\Omega(t) > 1, \text{ then } k = +1.$$

With much substituting and jiggling of the density parameters and the Friedmann equations it can be shown that (in terms of the normalised scale factor $a(t) = R/R_0$)

$$\frac{da}{dt} = H_0 \left[\frac{\Omega_{r,0}}{a^2} + \frac{\Omega_{m,0}}{a} + \Omega_{\Lambda,0} a^2 + (1 - \Omega_0) \right]^{1/2}. \tag{11.4.43}$$

Solving this differential equation for given values of H_0, $\Omega_{r,0}$, $\Omega_{m,0}$ and $\Omega_{\Lambda,0}$ will show how the normalised scale factor $a(t)$ changes with time, and thus the evolution of that particular universe. Very simple models, such as the single component ones we've looked at, can be solved exactly using (11.4.43). For example, recall the empty universe model (Section 11.4.4) with no radiation, matter or dark energy, ie $\rho = \Omega(t) = 0$. Equation 11.4.43 then simplifies to

$$\frac{da}{dt} = H_0 \left[\frac{0}{a^2} + \frac{0}{a} + \left(0 \times a^2\right) + (1 - 0) \right]^{1/2}$$

$$\frac{da}{dt} = H_0.$$

The Hubble parameter (11.3.12) for an empty ($k = -1$) universe is given by

$$H(t) = \frac{1}{R}\frac{dR}{dt} = \frac{c}{R},$$

and therefore

$$\frac{da}{dt} = \frac{c}{R_0},$$

which integrates to give

$$a(t) = \frac{ct}{R_0},$$

which is the same equation as (11.4.20) that we found earlier. This is an example of an analytic or exact solution to equation (11.4.43). However, more complicated models, with multiple components, can usually only be solved numerically (ie approximately) using computer power.

11.5 More FRW models, and our universe

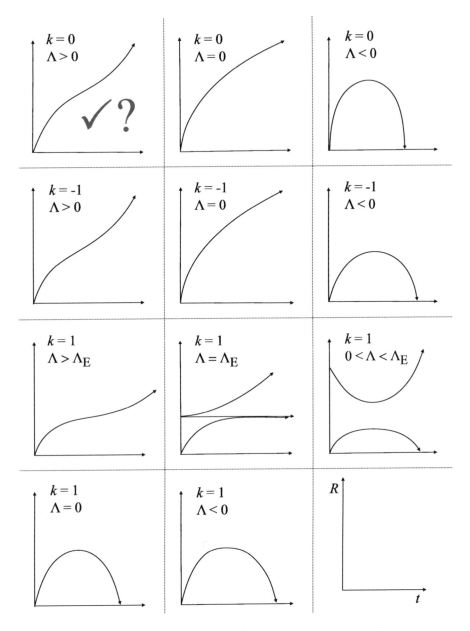

Figure 11.15: A range of FRW cosmological models.

Based on solutions to (11.4.43), cosmologists have proposed many different Friedmann-Robertson-Walker models using different permutations of H_0, $\Omega_{r,0}$, $\Omega_{m,0}$ and $\Omega_{\Lambda,0}$. An illustrative range of these models (taken from Harrison [10]) is shown in Figure 11.15 (Λ_E refers to the value of the cosmological constant in the static Einstein model we met earlier).

Most, but not all of these possible universes start with a Big Bang of some sort. Some expand forever, a scenario known as the **Big Chill**. Some expand initially before collapsing into a **Big Crunch**. One suggestion, known as the **Big Bounce** model, describes an oscillating universe stuck in an infinitely repeating cycle of Big Bang, expansion, contraction, Big Crunch.

The tick in the top left box indicates the model that is currently thought to best describe our own universe. This $k = 0$ universe is sometimes known as an **accelerating model**. The acceleration is due to the changing proportions of radiation, mass and dark energy density. We now appear to be living in a dark energy dominated, and therefore accelerating, universe. This model of the universe is known as the **Lambda-CDM** model, where CDM stands for cold dark matter. The title therefore neatly includes the two big mysteries at the heart of current cosmology: dark energy and dark matter. The Lambda-CDM model is also known as the **standard model** of Big Bang cosmology.

Cosmologists are able to make use of various techniques to estimate the key cosmological parameters of our universe. At the time of writing these are thought to be:

$$\Omega_{m,0} \approx 0.27, \Omega_{r,0} \approx 0, \Omega_{\Lambda,0} \approx 0.73, H_0 \approx 70.4 \, \text{km}\,\text{s}^{-1}\,\text{Mpc}^{-1}.$$

Note that $\Omega_{m,0} + \Omega_{r,0} + \Omega_{\Lambda,0} \approx 1$, which, from (11.4.42), tells us the spatial geometry of the universe is approximately flat. As previously mentioned, the history of the universe can be roughly divided into three periods:

- A relatively short-lived radiation-dominated era lasting about 50,000 years after the Big Bang. During this time the radiation only model discussed in Section 11.4.7 would be a reasonable approximation and, from (11.4.34), $a(t) = \left(\frac{t}{t_0}\right)^{1/2}$.

- A matter-dominated era lasting about 9.8 billion years. During this time the matter only Einstein-de Sitter model discussed in Section 11.4.8 would be a reasonable approximation and, from (11.4.39), $a(t) = \left(\frac{t}{t_0}\right)^{2/3}$.

- The current dark energy dominated universe. During this time the dark energy only de Sitter model discussed in Section 11.4.6 would be a reasonable approximation and, from (11.4.30), $a(t) = e^{H_0(t-t_0)}$.

Our final, nifty little calculation estimates the age of the universe t_0 based on the above cosmological parameters. Ignoring the contribution of radiation, and assuming that $\Omega_{m,0} + \Omega_{\Lambda,0} = \Omega_0 = 1$ (ie $k = 0$), (11.4.43)

$$\frac{da}{dt} = H_0 \left[\frac{\Omega_{r,0}}{a^2} + \frac{\Omega_{m,0}}{a} + \Omega_{\Lambda,0}a^2 + (1 - \Omega_0) \right]^{1/2}$$

can be rearranged to give

$$\int_0^1 \frac{a^{1/2}da}{\sqrt{((1 - \Omega_{\Lambda,0}) + \Omega_{\Lambda,0}a^3)}} = H_0 t_0.$$

By using an online integral calculator ([33], for example) this evaluates to

$$H_0 t_0 = \frac{2}{3\sqrt{\Omega_{\Lambda,0}}} \ln \left[2 \left(\Omega_{\Lambda,0}\sqrt{a^3} + \sqrt{\Omega_{\Lambda,0}}\sqrt{\Omega_{\Lambda,0}(a^3 - 1) + 1} \right) \right]_0^1,$$

and after some work,

$$t_0 = \frac{2}{H_0 3 \sqrt{\Omega_{\Lambda,0}}} \ln \left(\frac{\sqrt{\Omega_{\Lambda,0}} + 1}{\sqrt{1 - \Omega_{\Lambda,0}}} \right). \tag{11.5.1}$$

By plugging in $1/H_0 = 4.39 \times 10^{17}$ s and $\Omega_{\Lambda,0} = 0.73$, we calculate the age of the universe t_0 to be about 13.78 billion years. NASA's current estimate of the age of the universe [22] is 13.7 ± 0.13 billion years.

We finish this chapter with a beautiful NASA graphic (Figure 11.16) showing a representation of the timeline of the universe. I can't better NASA's own description of this image. Here it is, taken from their website [22]:

'A representation of the evolution of the universe over 13.7 billion years. The far left depicts the earliest moment we can now probe, when a period of "inflation" produced a burst of exponential growth in the universe ... For the next several billion years, the expansion of the universe gradually slowed down as the matter in the universe pulled on itself via gravity. More recently, the expansion has begun to speed up again as the repulsive effects of dark energy have come to dominate the expansion of the universe. The afterglow light seen by WMAP was emitted about 380,000 years after inflation and has traversed the universe largely unimpeded since then. The conditions of earlier times are imprinted on this light; it also forms a backlight for later developments of the universe.'

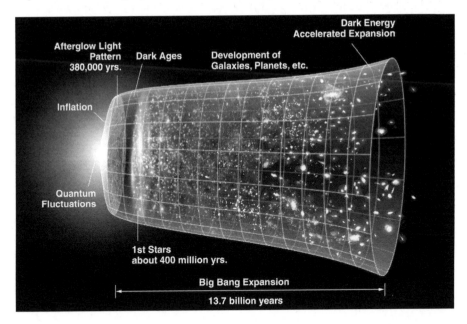

Figure 11.16: NASA representation of the timeline of the universe.

12 Gravitational waves

Ladies and gentlemen, we have detected gravitational waves. We did it!

DAVID REITZE – 11 FEBRUARY 2016

12.1 Introduction

When I originally wrote this book, space was tight and I only briefly mentioned (at the tail end of Section 7.3) the phenomenon of **gravitational waves**, a form of radiation whose existence was predicted by Einstein in 1916. Gravitational waves can be thought of as ripples in the curvature of spacetime, travel at the speed of light, but are not a form of electromagnetic radiation. I ended that short paragraph by saying, 'To date, despite a world-wide effort to detect gravitational waves, there is strong, but only indirect, evidence for their existence.'

Tempus fugit. Gravitational waves have now been detected. On 11 February 2016, the Advanced Laser Interferometer Gravitational-Wave Observatory (LIGO) team announced they had detected gravitational waves from the merger of two black holes. The press release stated:

> 'For the first time, scientists have observed ripples in the fabric of spacetime called gravitational waves, arriving at the earth from a cataclysmic event in the distant universe. This confirms a major prediction of Albert Einstein's 1915 general theory of relativity and opens an unprecedented new window onto the cosmos.'

Space is still tight, but this momentous discovery warrants more coverage than my original scant lines. Hence this short additional chapter on the subject.

12.2 Linearized gravity

In Chapter 8, when looking at the Newtonian limit of general relativity, we assumed a weak gravitational field and made use of the linearized gravity equation (8.3.3)

$$g_{\mu\nu} = \eta_{\mu\nu} + h_{\mu\nu}.$$

Here, in nearly flat spacetime, $g_{\mu\nu}$ equals the Minkowski metric $\eta_{\mu\nu}$ plus a small perturbation (an extra little bit of metric) $h_{\mu\nu}$, which is due to the weak gravitational field. The components of $h_{\mu\nu}$ are small compared to $\eta_{\mu\nu}$, meaning $|h_{\mu\nu}| \ll 1$. Although gravitational waves are created by some of the most violent events in the cosmos, linear theory is sufficiently accurate to describe them both in empty space far away from their source and also, to a good approximation, by the time they reach us on Earth.

Hidden in the Einstein field equations (7.6.4)

$$R_{\mu\nu} - \frac{1}{2}Rg_{\mu\nu} = -\kappa T_{\mu\nu}$$

are the wave equations that describe gravitational waves. To winkle them out we need to write the field equations in terms of $h_{\mu\nu}$.

We first find the appropriate connection coefficients, using – with renamed indices – (6.2.6)

$$\Gamma^{\mu}_{\nu\sigma} = \frac{1}{2}g^{\mu\beta}\left(\frac{\partial g_{\beta\sigma}}{\partial x^{\nu}} + \frac{\partial g_{\nu\beta}}{\partial x^{\sigma}} - \frac{\partial g_{\nu\sigma}}{\partial x^{\beta}}\right),$$

and substituting $g_{\mu\nu} = \eta_{\mu\nu} + h_{\mu\nu}$ and (as we found in Section 8.3) $g^{\mu\nu} = \eta^{\mu\nu} - h^{\mu\nu}$. When we discussed the Newtonian limit we assumed a static gravitational field and could thus ignore time derivatives in the connection coefficients. We don't have that luxury when considering gravitational waves. We are also now raising and lowering indices by using the Minkowski metric $\eta^{\mu\nu}$ and $\eta_{\mu\nu}$, not $g^{\mu\nu}$ and $g_{\mu\nu}$. Finally, because $h_{\mu\nu}$ and its derivatives are very small, we can ignore all products of these. By following this recipe we obtain

$$\Gamma^{\mu}_{\nu\sigma} = \frac{1}{2}\eta^{\mu\beta}\left(\frac{\partial h_{\beta\sigma}}{\partial x^{\nu}} + \frac{\partial h_{\nu\beta}}{\partial x^{\sigma}} - \frac{\partial h_{\nu\sigma}}{\partial x^{\beta}}\right)$$

$$\Gamma^{\mu}_{\nu\sigma} = \frac{1}{2}\left(\partial_{\nu}h^{\mu}_{\sigma} + \partial_{\sigma}h^{\mu}_{\nu} - \partial^{\mu}h_{\nu\sigma}\right).$$

The notation $\partial_{\nu}h^{\mu}_{\sigma}$ is equivalent to $\frac{\partial h^{\mu}_{\sigma}}{\partial x^{\nu}}$. The upper μ on the ∂ is a consequence of multiplying by $\eta^{\mu\beta}$ and indicates that that term's spatial components have a reversed sign. Using these connection coefficients, we can go on to find the Ricci tensor

$$R_{\mu\nu} = \frac{1}{2}\left(\partial_{\mu\nu}h^{\alpha}_{\alpha} - \partial_{\nu\alpha}h^{\alpha}_{\mu} - \partial_{\mu\alpha}h^{\alpha}_{\nu} + \partial^{\alpha}_{\alpha}h_{\mu\nu}\right)$$

$$= \frac{1}{2}\left(\partial_{\mu\nu}h - \partial_{\nu\alpha}h^{\alpha}_{\mu} - \partial_{\mu\alpha}h^{\alpha}_{\nu} + \partial^{\alpha}_{\alpha}h_{\mu\nu}\right),$$

(where $h = h^{\mu}_{\mu} = \eta^{\mu\nu}h_{\mu\nu}$) and Ricci scalar

$$R = \partial^{\alpha}_{\alpha}h - \partial_{\alpha\beta}h^{\alpha\beta}.$$

We can then plug these into the field equations to get

$$\frac{1}{2}\left(\partial_{\mu\nu}h - \partial_{\nu\alpha}h^{\alpha}_{\mu} - \partial_{\mu\alpha}h^{\alpha}_{\nu} + \partial^{\alpha}_{\alpha}h_{\mu\nu}\right) - \frac{1}{2}\left(\partial^{\alpha}_{\alpha}h - \partial_{\alpha\beta}h^{\alpha\beta}\right)\left(\eta_{\mu\nu} + h_{\mu\nu}\right) = -\kappa T_{\mu\nu},$$

which, ignoring higher order terms, gives the linearized field equations

$$\partial_{\mu\nu}h - \partial_{\nu\alpha}h^{\alpha}_{\mu} - \partial_{\mu\alpha}h^{\alpha}_{\nu} + \partial^{\alpha}_{\alpha}h_{\mu\nu} - \eta_{\mu\nu}\left(\partial^{\alpha}_{\alpha}h - \partial_{\alpha\beta}h^{\alpha\beta}\right) = -2\kappa T_{\mu\nu}. \qquad (12.2.1)$$

This can be 'simplified' (the term is relative, but things do look less messy be the time we've finished – honestly) in several steps. First, we introduce something called the **D'Alembertian operator** (symbol $\Box$), also known as the **wave operator**, where

$$\Box = \partial_\alpha^\alpha = \frac{1}{c^2}\frac{\partial^2}{\partial t^2} - \nabla^2 = \frac{1}{c^2}\frac{\partial^2}{\partial t^2} - \frac{\partial^2}{\partial x^2} - \frac{\partial^2}{\partial y^2} - \frac{\partial^2}{\partial z^2}.$$

This operator defines waves travelling at the speed of light. Plugging this into (12.2.1) gives

$$\partial_{\mu\nu}h - \partial_{\nu\alpha}h_\mu^\alpha - \partial_{\mu\alpha}h_\nu^\alpha + \Box h_{\mu\nu} - \eta_{\mu\nu}\left(\Box h - \partial_{\alpha\beta}h^{\alpha\beta}\right) = -2\kappa T_{\mu\nu}.$$

Next we define a quantity

$$\overline{h}_{\mu\nu} = h_{\mu\nu} - \frac{1}{2}\eta_{\mu\nu}h,$$

and tweak the coordinates (known as imposing a **gauge condition**) by assuming

$$\partial_\mu \overline{h}_{\mu\nu} = 0.$$

The upshot is that we end up with the much simplified field equations

$$\Box \overline{h}_{\mu\nu} = -2\kappa T_{\mu\nu}.$$

In empty space, where $T_{\mu\nu} = 0$, the ripples in spacetime caused by gravitational waves travelling at the speed of light are described by the basic wave equation

$$\Box \overline{h}_{\mu\nu} = 0.$$

This equation can be further analysed to yield more useful wave equations, such as those for plane waves.

In theory, any accelerating mass, providing it isn't perfectly spherically symmetric, should produce gravitational waves. In practice, they are so ridiculously feeble that only the most heavyweight cosmic events (such as binary systems, spinning neutron stars, supernovae and maybe the Big Bang) generate detectable gravitational waves. Even if the Sun fell into a solar mass black hole, the resulting gravitational wave would pass through your body without you noticing it.

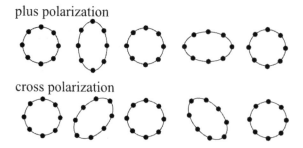

plus polarization

cross polarization

Figure 12.1: Effect of plus and cross polarized gravitational waves on a ring of test particles.

Gravitational waves are polarized in such a way that they will cause an object to expand in one direction and contract in the perpendicular direction. One way or another, this is the property utilized when trying to detect gravitational waves. Gravitational waves come in two independent polarizations, called 'plus' and 'cross'. The distorting effect of plus and cross polarized gravitational waves (travelling perpendicularly into the page) on a ring of test particles is shown in Figure 12.1.

12.3 The 2015 LIGO detection

Since the 1960s, various increasingly sensitive experiments have been conducted to detect gravitational waves. The phenomenon has been observed indirectly – the orbital decay of the Hulse-Taylor binary pulsar system is attributed to energy loss through the emission of gravitational waves. In recent years there have also been significant advances in **numerical relativity** – solving the field equations on a computer for extreme spacetimes where the metric isn't known, such as for closely orbiting black holes. Increasingly sophisticated computer simulations mean the characteristic gravitational waveform (shape of the wave) can be accurately predicted for a range of high-energy astrophysical events. The total mass

$$M = (m_1 + m_2),$$

(where m_1 and m_2 are the masses of the two bodies) and **chirp mass**, for example, are two of the key parameters used for modelling the merger of a pair of black holes, a pair of neutron stars or a black hole and neutron star (the chirp is the final burst of gravitational radiation as the two bodies coalesce). The chirp mass is given by

$$\mathcal{M} = \frac{(m_1\, m_2)^{3/5}}{(m_1 + m_2)^{1/5}} = \frac{c^3}{G}\left[\frac{5}{96}\,\pi^{-8/3}\,f^{-11/3}\,\dot{f}\right]^{3/5},$$

where the observables f and $\dot{f}$ are the gravitational wave frequency and its time derivative (a little dot above a symbol indicates a time derivative).

At last, in 2015, came the first direct observation. In September of that year the two widely separated detectors at the Laser Interferometer Gravitational-Wave Observatory in the US simultaneously observed a gravitational wave signal emitted by a pair of merging black holes. (The LIGO detectors, each having two 4 km long vacuum tubes arranged in an L-shape, are the most sensitive scientific instruments ever built, able to measure a displacement 10,000 times smaller than the diameter of a proton.)

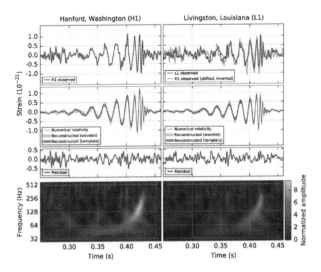

Figure 12.2: GW150914 observed by the LIGO Hanford (left) and Livingston (right) detectors.

Figure 12.2 shows the gravitational wave event (named GW150914) observed by the Hanford, Washington and Livingston, Louisiana observatories. The top three rows show strain (a measure of signal strength) plotted against time. The top row shows the received signal. The second row compares that signal with the 'numerical relativity' predicted waveform. The bottom row shows frequency plotted against time. Using the f and $\dot{f}$ data from the observed signal gives a chirp mass of about 30 solar masses, implying a pre-merger total mass of about 70 solar masses in the *detector* frame. After compensating for redshift, this corresponds to a physical total mass of 65 solar masses in the *source* frame. Further analysis very strongly suggests the source was the merger of two black holes 1.3 billion light-years away. 'Cataclysmic' scarcely does justice to the scale of the event that took place over a billion years ago. A 36 solar-mass black hole merged with a 29 solar mass black hole to give a 62 solar mass black hole. The missing 3 solar masses of energy was radiated as gravitational waves. That is a lot of energy. In the fraction of a second that the collision occurred, more power was generated than from 50 times the combined power output of all the stars in the observable universe.

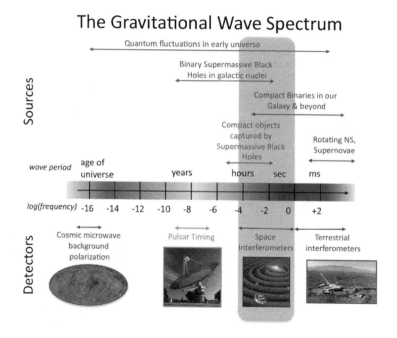

Figure 12.3: The gravitational wave spectrum.

Gravitational waves hold out the promise of providing a completely new way to study the universe. Many interesting cosmic events, such as merging black holes, the very early universe and who knows what else, are not observable using electromagnetic radiation. Figure 12.3 shows a NASA representation of the gravitational wave spectrum, which at present is more or less totally unexplored. A network of detectors searching for gravitational waves of widely different frequencies from a multitude of sources could therefore revolutionise our understanding of the cosmos.

Appendix: The Riemann curvature tensor

Derivation

Partial derivatives commute, in other words it doesn't matter in which order they are taken, the result is still the same. So, for a function $f(x, y)$

$$\frac{\partial}{\partial x}\left(\frac{\partial f}{\partial y}\right) = \frac{\partial}{\partial y}\left(\frac{\partial f}{\partial x}\right).$$

We say the commutator

$$\frac{\partial}{\partial x}\left(\frac{\partial f}{\partial y}\right) - \frac{\partial}{\partial y}\left(\frac{\partial f}{\partial x}\right) = 0.$$

In general, covariant derivatives do not commute, so

$$\nabla_b \nabla_c V_a \neq \nabla_c \nabla_b V_a.$$

Or, equivalently, the commutator

$$\nabla_b \nabla_c V_a - \nabla_c \nabla_b V_a \neq 0.$$

This property can be used to derive the Riemann curvature tensor In Section 6.6 we sketched out a derivation of the Riemann tensor in terms of the parallel transport of a vector around an infinitesimal loop on a manifold. An equivalent operation, and the derivation we look at in detail here, involves the above commutator of two covariant derivatives. In the words of Carroll [4]:

> 'The covariant derivative of a tensor in a certain direction measures how much the tensor changes relative to what it would have been if it had been parallel transported (since the covariant derivative of a tensor in a direction along which it is parallel transported is zero). The commutator of two covariant derivatives, then, measures the difference between parallel transporting the tensor first one way and then the other, versus the opposite ordering.'

So, our aim is to derive the Riemann tensor by finding the commutator

$$\nabla_c \nabla_b V_a - \nabla_b \nabla_c V_a,$$

or, in semi-colon notation,

$$V_{a;bc} - V_{a;cb}.$$

The covariant derivative of V_a is given by

$$\nabla_b V_a = \partial_b V_a - \Gamma_{ab}^d V_d.$$

(Where the notation $\partial_b V_a$ is equivalent to $\frac{\partial V_a}{\partial x^b}$). Now take another covariant derivative (remember that $\nabla_b V_a$ is itself a rank-2 tensor) to get

$$\nabla_c \nabla_b V_a = \partial_c \left(\nabla_b V_a\right) - \Gamma^e_{ac} \nabla_b V_e - \Gamma^e_{bc} \nabla_e V_a.$$

As ever, notice how the indices must balance on both sides of the equation and also that the derivative index c is the last one on Γ. Taking a term at a time, we find for the first right-hand side term

$$\partial_c \left(\nabla_b V_a\right) = \partial_c \partial_b V_a - \partial_c \left(\Gamma^d_{ab} V_d\right).$$

Then using the product rule we get

$$\partial_c \left(\nabla_b V_a\right) = \partial_c \partial_b V_a - \partial_c \left(\Gamma^d_{ab}\right) V_d - \partial_c \left(V_d\right) \Gamma^d_{ab}.$$

For the second right-hand side term

$$-\Gamma^e_{ac} \nabla_b V_e = -\Gamma^e_{ac} \left(\partial_b V_e - \Gamma^d_{eb} V_d\right).$$

And for the third right-hand side term

$$-\Gamma^e_{bc} \nabla_e V_a = -\Gamma^e_{bc} \left(\partial_e V_a - \Gamma^d_{ae} V_d\right).$$

Then put all these together to get

$$\nabla_c \nabla_b V_a = \partial_c \partial_b V_a - \partial_c \left(\Gamma^d_{ab}\right) V_d - \partial_c \left(V_d\right) \Gamma^d_{ab} - \Gamma^e_{ac} \left(\partial_b V_e - \Gamma^d_{eb} V_d\right) - \Gamma^e_{bc} \left(\partial_e V_a - \Gamma^d_{ae} V_d\right). \quad \text{(A.1)}$$

Interchanging b and c gives

$$\nabla_b \nabla_c V_a = \partial_b \partial_c V_a - \partial_b \left(\Gamma^d_{ac}\right) V_d - \partial_b \left(V_d\right) \Gamma^d_{ac} - \Gamma^e_{ab} \left(\partial_c V_e - \Gamma^d_{ec} V_d\right) - \Gamma^e_{cb} \left(\partial_e V_a - \Gamma^d_{ae} V_d\right). \quad \text{(A.2)}$$

Subtract equation (A.2) from equation (A.1):

$$\nabla_c \nabla_b V_a - \nabla_b \nabla_c V_a = \partial_c \partial_b V_a - \partial_c \left(\Gamma^d_{ab}\right) V_d - \partial_c \left(V_d\right) \Gamma^d_{ab} - \Gamma^e_{ac} \left(\partial_b V_e - \Gamma^d_{eb} V_d\right) - \Gamma^e_{bc} \left(\partial_e V_a - \Gamma^d_{ae} V_d\right) -$$
$$\left(\partial_b \partial_c V_a - \partial_b \left(\Gamma^d_{ac}\right) V_d - \partial_b \left(V_d\right) \Gamma^d_{ac} - \Gamma^e_{ab} \left(\partial_c V_e - \Gamma^d_{ec} V_d\right) - \Gamma^e_{cb} \left(\partial_e V_a - \Gamma^d_{ae} V_d\right)\right).$$

Multiplying out the brackets in the second line gives

$$\partial_c \partial_b V_a - \partial_c \left(\Gamma^d_{ab}\right) V_d - \partial_c \left(V_d\right) \Gamma^d_{ab} - \Gamma^e_{ac} \partial_b V_e + \Gamma^e_{ac} \Gamma^d_{eb} V_d - \Gamma^e_{bc} \partial_e V_a + \Gamma^e_{bc} \Gamma^d_{ae} V_d -$$
$$\left(\partial_b \partial_c V_a - \partial_b \left(\Gamma^d_{ac}\right) V_d - \partial_b \left(V_d\right) \Gamma^d_{ac} - \Gamma^e_{ab} \partial_c V_e + \Gamma^e_{ab} \Gamma^d_{ec} V_d - \Gamma^e_{cb} \partial_e V_a + \Gamma^e_{cb} \Gamma^d_{ae} V_d\right).$$

Note that $\Gamma^e_{ac} \partial_b V_e = \Gamma^e_{ab} \partial_c V_e$ and $\partial_c \left(V_d\right) \Gamma^d_{ab} = \partial_b \left(V_d\right) \Gamma^d_{ac}$ (because of the definition of $\frac{\partial e_x}{\partial x^z} = \Gamma^y_{xz} e_y$). Also, in the standard formulations of general relativity the connection coefficients are assumed to be symmetric in their lower indices, ie $\Gamma^e_{bc} = \Gamma^e_{cb}$, and therefore $\Gamma^e_{bc} \partial_e V_a = \Gamma^e_{cb} \partial_e V_a$. We then end up with

$$\nabla_c \nabla_b V_a - \nabla_b \nabla_c V_a = \left(\partial_b \left(\Gamma^d_{ac}\right) - \partial_c \left(\Gamma^d_{ab}\right) + \Gamma^e_{ac} \Gamma^d_{eb} - \Gamma^e_{ab} \Gamma^d_{ec}\right) V_d.$$

The thing inside the brackets on the rhs is the Riemann tensor, meaning

$$\nabla_c \nabla_b V_a - \nabla_b \nabla_c V_a = R^d_{\ abc} V_d.$$

We can now see that the covariant derivatives will only commute (ie $\nabla_c \nabla_b V_a - \nabla_b \nabla_c V_a$ will only equal zero) when the Riemann tensor itself equals zero, in other words in flat space.

We could have reached the same destination via a slightly different route by starting with a contravariant vector field V^a. The commutator would then be given by

$$\nabla_c \nabla_b V^a - \nabla_b \nabla_c V^a = R^a_{\ dcb} V^d.$$

Symmetries

We can derive these important symmetries of the Riemann tensor by looking at its components in a locally inertial frame (LIF). At the origin of a LIF it's possible to choose coordinates where the first derivatives of the metric are zero, meaning that the connection coefficients are also zero. However, the second derivatives of the metric (and therefore the derivatives of the connection coefficients) are not necessarily zero. We use Greek indices as we're working in spacetime.

Two symmetries

We wish to show

$$R_{\nu\alpha\beta\gamma} = -R_{\alpha\nu\beta\gamma} = -R_{\nu\alpha\gamma\beta}.$$

The Riemann tensor is

$$R^{\mu}{}_{\alpha\beta\gamma} = \partial_\beta \Gamma^{\mu}_{\alpha\gamma} - \partial_\gamma \Gamma^{\mu}_{\alpha\beta} + \Gamma^{\epsilon}_{\alpha\gamma}\Gamma^{\mu}_{\epsilon\beta} - \Gamma^{\epsilon}_{\alpha\beta}\Gamma^{\mu}_{\epsilon\gamma}.$$

At the origin of a LIF

$$R^{\mu}{}_{\alpha\beta\gamma} = \partial_\beta \Gamma^{\mu}_{\alpha\gamma} - \partial_\gamma \Gamma^{\mu}_{\alpha\beta}.$$

The connection coefficients are given by equation (6.2.6)

$$\Gamma^{\mu}_{\alpha\gamma} = \frac{1}{2}g^{\mu\sigma}\left(\frac{\partial g_{\sigma\gamma}}{\partial x^\alpha} + \frac{\partial g_{\alpha\sigma}}{\partial x^\gamma} - \frac{\partial g_{\alpha\gamma}}{\partial x^\sigma}\right).$$

First, rewrite the Riemann tensor as

$$R_{\nu\alpha\beta\gamma} = g_{\mu\nu}R^{\mu}{}_{\alpha\beta\gamma} = g_{\mu\nu}\left(\partial_\beta \Gamma^{\mu}_{\alpha\gamma} - \partial_\gamma \Gamma^{\mu}_{\alpha\beta}\right). \tag{A.3}$$

We need to expand this. Step by step, we start by calculating the derivative

$$\partial_\beta \Gamma^{\mu}_{\alpha\gamma} = \frac{1}{2}\partial_\beta g^{\mu\sigma}\left(\frac{\partial g_{\sigma\gamma}}{\partial x^\alpha} + \frac{\partial g_{\alpha\sigma}}{\partial x^\gamma} - \frac{\partial g_{\alpha\gamma}}{\partial x^\sigma}\right) + \frac{1}{2}g^{\mu\sigma}\left(\partial_\beta \frac{\partial g_{\sigma\gamma}}{\partial x^\alpha} + \partial_\beta \frac{\partial g_{\alpha\sigma}}{\partial x^\gamma} - \partial_\beta \frac{\partial g_{\alpha\gamma}}{\partial x^\sigma}\right).$$

At the origin of a LIF the first term is zero (because all first derivatives of the metric are zero), giving

$$\partial_\beta \Gamma^{\mu}_{\alpha\gamma} = \frac{1}{2}g^{\mu\sigma}\left(\partial_\beta \frac{\partial g_{\sigma\gamma}}{\partial x^\alpha} + \partial_\beta \frac{\partial g_{\alpha\sigma}}{\partial x^\gamma} - \partial_\beta \frac{\partial g_{\alpha\gamma}}{\partial x^\sigma}\right).$$

Multiply this by $g_{\mu\nu}$ to give

$$g_{\mu\nu}\partial_\beta \Gamma^{\mu}_{\alpha\gamma} = \frac{1}{2}g_{\mu\nu}g^{\mu\sigma}\left(\partial_\beta \frac{\partial g_{\sigma\gamma}}{\partial x^\alpha} + \partial_\beta \frac{\partial g_{\alpha\sigma}}{\partial x^\gamma} - \partial_\beta \frac{\partial g_{\alpha\gamma}}{\partial x^\sigma}\right).$$

But $g_{\mu\nu}g^{\mu\sigma} = \delta^{\sigma}_{\nu}$, therefore (with the Kronecker delta acting to relabel the σ index as ν)

$$g_{\mu\nu}\partial_\beta \Gamma^{\mu}_{\alpha\gamma} = \frac{1}{2}\delta^{\sigma}_{\nu}\left(\partial_\beta \frac{\partial g_{\sigma\gamma}}{\partial x^\alpha} + \partial_\beta \frac{\partial g_{\alpha\sigma}}{\partial x^\gamma} - \partial_\beta \frac{\partial g_{\alpha\gamma}}{\partial x^\sigma}\right)$$

$$g_{\mu\nu}\partial_\beta \Gamma^{\mu}_{\alpha\gamma} = \frac{1}{2}\left(\partial_\beta \frac{\partial g_{\nu\gamma}}{\partial x^\alpha} + \partial_\beta \frac{\partial g_{\alpha\nu}}{\partial x^\gamma} - \partial_\beta \frac{\partial g_{\alpha\gamma}}{\partial x^\nu}\right). \tag{A.4}$$

In order to get $\partial_\gamma \Gamma^\mu_{\alpha\beta}$ (the final term in equation (A.3)) we swap the γ and β indices:

$$g_{\mu\nu}\partial_\gamma \Gamma^\mu_{\alpha\beta} = \frac{1}{2}\left(\partial_\gamma \frac{\partial g_{\nu\beta}}{\partial x^\alpha} + \partial_\gamma \frac{\partial g_{\alpha\nu}}{\partial x^\beta} - \partial_\gamma \frac{\partial g_{\alpha\beta}}{\partial x^\nu}\right). \tag{A.5}$$

Next subtract equation (A.5) from equation (A.4):

$$g_{\mu\nu}\partial_\beta \Gamma^\mu_{\alpha\gamma} - g_{\mu\nu}\partial_\gamma \Gamma^\mu_{\alpha\beta} = \frac{1}{2}\left(\partial_\beta \frac{\partial g_{\nu\gamma}}{\partial x^\alpha} + \partial_\beta \frac{\partial g_{\alpha\nu}}{\partial x^\gamma} - \partial_\beta \frac{\partial g_{\alpha\gamma}}{\partial x^\nu} - \partial_\gamma \frac{\partial g_{\nu\beta}}{\partial x^\alpha} - \partial_\gamma \frac{\partial g_{\alpha\nu}}{\partial x^\beta} + \partial_\gamma \frac{\partial g_{\alpha\beta}}{\partial x^\nu}\right)$$

$$R_{\nu\alpha\beta\gamma} = \frac{1}{2}\left(\partial_\beta \frac{\partial g_{\nu\gamma}}{\partial x^\alpha} - \partial_\beta \frac{\partial g_{\alpha\gamma}}{\partial x^\nu} - \partial_\gamma \frac{\partial g_{\nu\beta}}{\partial x^\alpha} + \partial_\gamma \frac{\partial g_{\alpha\beta}}{\partial x^\nu}\right), \tag{A.6}$$

which is only valid at the origin of a LIF.

Now swap the first two indices ν and α :

$$R_{\alpha\nu\beta\gamma} = \frac{1}{2}\left(\partial_\beta \frac{\partial g_{\alpha\gamma}}{\partial x^\nu} - \partial_\beta \frac{\partial g_{\nu\gamma}}{\partial x^\alpha} - \partial_\gamma \frac{\partial g_{\alpha\beta}}{\partial x^\nu} + \partial_\gamma \frac{\partial g_{\nu\beta}}{\partial x^\alpha}\right).$$

The signs in equation (A.6) have reversed, meaning

$$R_{\nu\alpha\beta\gamma} = -R_{\alpha\nu\beta\gamma}.$$

(Although we're considering the Riemann tensor in a locally inertial frame, these, and the other symmetries discussed here, are valid tensor equations and so are true in all frames.)

We can also swap the last two indices β and γ in equation (A.6):

$$R_{\nu\alpha\gamma\beta} = \frac{1}{2}\left(\partial_\gamma \frac{\partial g_{\nu\beta}}{\partial x^\alpha} - \partial_\gamma \frac{\partial g_{\alpha\beta}}{\partial x^\nu} - \partial_\beta \frac{\partial g_{\nu\gamma}}{\partial x^\alpha} + \partial_\beta \frac{\partial g_{\alpha\gamma}}{\partial x^\nu}\right).$$

Again the signs in equation (A.6) have reversed, meaning

$$R_{\nu\alpha\beta\gamma} = -R_{\nu\alpha\gamma\beta}.$$

Another symmetry

We can also swap the first and third (ν and β), and second and fourth (α and γ) indices in equation (A.6):

$$R_{\beta\gamma\nu\alpha} = \frac{1}{2}\left(\partial_\nu \frac{\partial g_{\beta\alpha}}{\partial x^\gamma} - \partial_\nu \frac{\partial g_{\gamma\alpha}}{\partial x^\beta} - \partial_\alpha \frac{\partial g_{\beta\nu}}{\partial x^\gamma} + \partial_\alpha \frac{\partial g_{\gamma\nu}}{\partial x^\beta}\right),$$

meaning

$$R_{\nu\alpha\beta\gamma} = R_{\beta\gamma\nu\alpha}.$$

And one more

We wish to show

$$R_{\nu\alpha\beta\gamma} + R_{\nu\beta\gamma\alpha} + R_{\nu\gamma\alpha\beta} = 0.$$

First take equation (A.6)

$$R_{\nu\alpha\beta\gamma} = \frac{1}{2}\left(\partial_\beta \frac{\partial g_{\nu\gamma}}{\partial x^\alpha} - \partial_\beta \frac{\partial g_{\alpha\gamma}}{\partial x^\nu} - \partial_\gamma \frac{\partial g_{\nu\beta}}{\partial x^\alpha} + \partial_\gamma \frac{\partial g_{\alpha\beta}}{\partial x^\nu}\right).$$

And rename the indices as follows:

$$R_{\nu\alpha\beta\gamma} + R_{\nu\beta\gamma\alpha} + R_{\nu\gamma\alpha\beta} = \frac{1}{2}\left(\partial_\beta \frac{\partial g_{\nu\gamma}}{\partial x^\alpha} - \partial_\beta \frac{\partial g_{\alpha\gamma}}{\partial x^\nu} - \partial_\gamma \frac{\partial g_{\nu\beta}}{\partial x^\alpha} + \partial_\gamma \frac{\partial g_{\alpha\beta}}{\partial x^\nu}\right)$$

$$+ \frac{1}{2}\left(\partial_\gamma \frac{\partial g_{\nu\alpha}}{\partial x^\beta} - \partial_\gamma \frac{\partial g_{\beta\alpha}}{\partial x^\nu} - \partial_\alpha \frac{\partial g_{\nu\gamma}}{\partial x^\beta} + \partial_\alpha \frac{\partial g_{\beta\gamma}}{\partial x^\nu}\right)$$

$$+ \frac{1}{2}\left(\partial_\alpha \frac{\partial g_{\nu\beta}}{\partial x^\gamma} - \partial_\alpha \frac{\partial g_{\gamma\beta}}{\partial x^\nu} - \partial_\beta \frac{\partial g_{\nu\alpha}}{\partial x^\gamma} + \partial_\beta \frac{\partial g_{\gamma\alpha}}{\partial x^\nu}\right).$$

Using $g_{\alpha\beta} = g_{\beta\alpha}$ and the fact that partial derivatives commute means all terms on the rhs disappear and we get

$$R_{\nu\alpha\beta\gamma} + R_{\nu\beta\gamma\alpha} + R_{\nu\gamma\alpha\beta} = 0.$$

And finally ...
If you liked this book, or even (perish the thought) if you didn't, then please consider helping other readers by posting a review on Amazon, Goodreads or other online book review site. All honest reviews are appreciated, whatever the length or rating. Thank you.

Bibliography

[1]   Baez, J. C. & Bunn, E. F. (2006) The Meaning of Einstein's Equation. URL `http://math.ucr.edu/home/baez/einstein/einstein.html`.

[2]   Blau, M. (2011) Lecture Notes on General Relativity. URL `www.blau.itp.unibe.ch/lecturesGR.pdf`.

[3]   Carroll, S. M. (2010) Energy Is Not Conserved. URL `http://blogs.discovermagazine.com/cosmicvariance/2010/02/22/energy-is-not-conserved/`

[4]   Carroll, S. M. (1997) Lecture Notes on General Relativity. URL `http://ned.ipac.caltech.edu/level5/March01/Carroll3/Carroll_contents.html`

[5]   d'Inverno, R. (1992) Introducing Einstein's Relativity.

[6]   Earman, J. & Glymour, C. (1978) Lost in the Tensors: Einstein's Struggles with Covariance Principles 1912-1916. URL `http://www.hss.cmu.edu/philosophy/glymour/earmanglymour1978.pdf`

[7]   Feynman, R. (1994) The Character of Physical Law.

[8]   Feynman, R., Leighton, R. & Sands, M. (2011) Feynman Lectures on Physics, Vol. I.

[9]   Foster, J. & Nightingale, J. D. (2006) A Short Course in General Relativity.

[10]   Harrison, E. (2000) Cosmology: The Science of the Universe.

[11]   GraphFunc Online. URL `http://graph.seriesmathstudy.com/`

[12]   Griest, K. Physics 161: Black Holes: Lectures 1 and 2: 3 and 5 Jan 2011. URL `http://physics.ucsd.edu/students/courses/winter2011/physics161/p161.3-5jan11.pdf`

[13]   Griest, K. Physics 161: Black Holes: Lecture 10: 26 Jan 2011. URL `http://physics.ucsd.edu/students/courses/winter2011/physics161/p161.26jan11.pdf`

[14]   Hawking, S. (2011) A Brief History of Time.

[15]   Hendry, M. (2007) An Introduction to General Relativity, Gravitational Waves and Detection Principles. URL `http://star-www.st-and.ac.uk/~hz4/gr/hendry_GRwaves.pdf`

[16]   Kaku, Michio. (2005) Einstein's Cosmos: How Albert Einstein's Vision Transformed Our Understanding of Space and Time.

[17]   Lambourne, R. J. A. (2010) Relativity, Gravitation and Cosmology.

[18]   Lewis, G. F. & Kwan, J. (2006) No Way Back: Maximizing survival time below the Schwarzschild event horizon. URL `http://arxiv.org/pdf/0705.1029v2`

[19]   Marsh, T. (2009) Notes for PX436, General Relativity. URL `http://www2.warwick.ac.uk/fac/sci/physics/teach/module_home/px436/notes/students.pdf`

[20]   Mathematics Stack Exchange. URL `http://math.stackexchange.com/`

[21]   Math Pages. URL `http://www.mathpages.com/home/index.htm`

[22]   NASA website. URL `http://www.nasa.gov/`

[23]   Misner, C. W., Thorne, K.S., Wheeler, J. A. (1973) Gravitation.

[24]   Paul's Online Math Notes. URL `http://tutorial.math.lamar.edu/`

[25]   Physics Forums. URL `http://www.physicsforums.com/`

[26]   Physics Stack Exchange. URL `http://physics.stackexchange.com/`

[27]   Ryden, B. (2003) Introduction to Cosmology.

[28]   Schutz, B. (2009) A First Course in General Relativity.

[29]   Schutz, B. (2004) Gravity from the Ground Up: An Introductory Guide to Gravity and General Relativity.

[30]   Susskind, L. (2009) Einstein's General Theory of Relativity – Lecture 1 (of 12) video lectures uploaded by Stanford University. URL `http://www.youtube.com/watch?v=hbmf0bB38h0`

[31]   Weiss, M. & Baez, J. Is Energy Conserved in General Relativity? URL `http://math.ucr.edu/home/baez/physics/Relativity/GR/energy_gr.html`

[32]   Winplot for Windows. URL `http://math.exeter.edu/rparris/winplot.html`

[33]   WolframAlpha Calculus and Analysis Calculator. URL `http://www.wolframalpha.com/examples/Calculus.html`

Acknowledgements

Many people, mainly strangers, helped me write this book. First and foremost, I am utterly indebted to the real relativity experts – the authors of the textbooks and online resources listed in the bibliography. Without them, I couldn't even have started, let alone finished this project. Thanks also to the community members at Physics Stack Exchange, Mathematics Stack Exchange and Physics Forums who kindly answered my many questions; James Bowie, Michael Guggenheimer, Walter Vanhimbeeck, Robert Zuidema and the many other eagle-eyed readers who were generous enough to point out errors and confusions in the text; Sam Gralla, Nelson Christensen and David Wittman for their comments; and all those unsung but heroic inspirational teachers I've been fortunate enough to meet along the way. Needless to say, any remaining errors are mine alone. Finally, heartfelt thanks to my partner, Anne, for her inexhaustible – though it was close at times! – patience and support while I shut myself away with my laptop.

My apologies if any acknowledgement or bibliographic citation has been inadvertently omitted. Please contact me and I will be pleased to make the necessary arrangements at the earliest opportunity.

Attribution-Share Alike 3.0 Unported license, URL
`http://en.wikipedia.org/wiki/File:Newton_Cannon.svg`. (5) Earth's gravity as measured by the
GRACE mission, NASA, public domain, URL
`http://earthobservatory.nasa.gov/IOTD/view.php?id=3666`.

Chapter 3: (1) Hermann Minkowski (1864 -1909), public domain, URL
`http://en.wikipedia.org/wiki/File:De_Raum_zeit_Minkowski_Bild.jpg`. (2) A squashed Earth means
you are moving close to the speed of light, NASA, public domain, URL
`http://science.gsfc.nasa.gov/690/Earth.html`.

Chapter 4: (1) Bernhard Riemann (1826 – 1866), public domain, URL
`http://en.wikipedia.org/wiki/File:Georg_Friedrich_Bernhard_Riemann.jpeg`.

Chapter 7: (1) The components of the energy-momentum tensor, author – Bamse, licensed under the
Creative Commons Attribution-Share Alike 3.0 Unported, 2.5 Generic, 2.0 Generic and 1.0 Generic license,
URL `http://en.wikipedia.org/wiki/File:StressEnergyTensor.svg`.

Chapter 9: (1) Karl Schwarzschild (1873–1916), public domain, URL
`http://en.wikipedia.org/wiki/File:Schwarzschild.jpg`. (2) Arthur Stanley Eddington (1882–1944),
public domain, URL `http://en.wikipedia.org/wiki/File:Arthur_Stanley_Eddington.jpg`. (3)
High-precision test of general relativity by the Cassini space probe (artist's impression), NASA, public
domain, URL `http://en.wikipedia.org/wiki/File:Cassini-science-br.jpg`.

Chapter 10: (1) NASA artist's impression of Cygnus X-1 stellar mass black hole, NASA, public domain,
URL `http://www.nasa.gov/mission_pages/chandra/multimedia/cygnusx1.html`. (2) NASA images of
the Active Galaxy NGC 4261, NASA, public domain, URL
`http://imagine.gsfc.nasa.gov/docs/science/know_12/active_galaxies.html`.

Chapter 11: (1) 2dF map of the distribution of galaxies, NASA, public domain, URL
`http://apod.nasa.gov/apod/ap010904.html`. (2) WMAP image of the cosmic microwave background
radiation (CMBR), NASA, public domain, URL `http://apod.nasa.gov/apod/ap050925.html`. (3) Part of
the emission spectrum of hydrogen, author – Merikanto, Adrignola, public domain, URL
`http://commons.wikimedia.org/wiki/File:Emission_spectrum-H_labeled.svg`. (4) Positively curved,
negatively curved and flat two-dimensional surfaces, NASA, public domain, URL
`http://cosmictimes.gsfc.nasa.gov/teachers/guide/2006/guide/universe_seeds.html`. (5) Alexander
Friedmann (1888–1925), this work is not an object of copyright according to Part IV of Civil Code No.
230-FZ of the Russian Federation of December 18, 2006, URL
`http://en.wikipedia.org/wiki/File:Aleksandr_Fridman.png`. (6) NASA representation of the timeline
of the universe, NASA, public domain, URL `http://map.gsfc.nasa.gov/media/060915/index.html`.

Chapter 12: (1) LIGO measurement of gravitational waves, B. P. Abbott et al. (LIGO Scientific
Collaboration and Virgo Collaboration). 'Observation of Gravitational Waves from a Binary Black
Hole Merger'. Phys. Rev. Lett. 116: 061102, URL `https://commons.wikimedia.org/wiki/File:`
`LIGO_measurement_of_gravitational_waves.png`. (2) The gravitational wave spectrum, NASA,
public domain, URL `http://science.gsfc.nasa.gov/663/research/index.html`.

Index

Made in the USA
San Bernardino, CA
07 March 2016